Stomata

Stomata

SECOND EDITION

Colin Willmer
University of Stirling,
UK

and

Mark Fricker
University of Oxford,
UK

SPRINGER-SCIENCE+BUSINESS MEDIA, B.V.

First edition 1983

Second edition 1996

Originally published by Chapman and Hall in 1996
Softcover reprint of the hardcover 2nd edition 1996

Typeset in Garamond 10 5/12pt by Saxon Graphics Ltd, Derby

Printed in Great Britain by St Edmundsbury Press Bury St Edmunds, Suffolk

ISBN 978-94-010-4256-7 ISBN 978-94-011-0579-8 (eBook)
DOI 10.1007/978-94-011-0579-8

A catalogue record for this book is available from the British Library

Printed on permanent acid-free text paper manufactured with ANSI/NISO Z39 48-1992 and ANSI/NISO Z39 48-1984 (Permanence of Paper)

Contents

Preface to the second edition

In 1902 Copeland published a paper (B.E. Copeland, *Ann Bot.*, **16**, 327–364, 1902) in which he stated:

> The movements of stomata have been familiar to botanists for more than a century. From all points of view, anatomical and physiological, the stomata have received more constant and lasting attention probably than any other single vegetative structure in the plant. And yet recent literature on the subject is most contradictory, not only as to the mechanism of their movements, but even as to the conditions which influence their opening and closing.

Essentially these same sentiments can still be applied nearly a century later. Stomata continue to be a source of intense research endeavour and the results of the research often appear contradictory; debate and controversy about various aspects of their function and functioning continues. True, over the last few decades major advances in our understanding of the behaviour of stomata and how stomata function have been made, but much still remains to be discovered. Nevertheless, now seems an appropriate time to update what has been added to our knowledge since the first edition of *Stomata* was published in 1983. Because the subject has become much more complex and encompasses a wide array of topics, including cell anatomy and ultrastructure, physiology, ecology, intermediary metabolism, ion relations and molecular biology, this second edition has expanded significantly and two authors were deemed necessary to do justice to the subject. Thus Mark Fricker co-authors this second edition with Colin Willmer.

Although the basic framework of the first edition exists in the second edition, virtually the whole text has been rewritten. A new chapter has been added on the theory of gas diffusion through stomata; sections on ionic relations of guard cells and on signal perception and transduction by guard cells have been expanded greatly to reflect the recent progress in these areas; and a section on the genetics and molecular biology of guard cells has been inserted in a concluding chapter.

Colin Willmer
University of Stirling
Mark Fricker
University of Oxford
November, 1994

Preface to the first edition

Over the last decade enormous advances in our understanding of the functioning of stomata have been made Today, the overall pattern of events in stomatal functioning is generally understood although some of the finer details of the mechanism remain to be elucidated However, the mechanism whereby changing CO_2 levels within a leaf control stomatal movements remains one of the biggest puzzles

The excellent text *Physiology of Stomata*, by Professors H Meidner and T A Mansfield, is over 12 years old and inevitably sections of it are now outdated Although numerous fine reviews on various aspects of stomata have been written in the meantime, none could be considered as complete, compact appraisals of the subject and thus to have superseded the Meidner and Mansfield book Now is therefore an appropriate time to review the status of our knowledge about stomata

This book encompasses all aspects of the subject except the physical processes involved in the exchange of gases between a leaf and its environment This topic is dealt with very adequately in a variety of other books and reviews, and the subject matter, with minor differences of opinion, remains much the same today as it was perhaps 20 years ago A description of how to measure stomatal apertures and the use of porometers has been restricted to a final chapter which also deals with experiments which can be carried out by students Again, there are numerous texts and papers which deal in greater detail with the theory and use of porometers than that described here

In a text of this type, which is not intended to be a research treatise but a more general, though detailed appraisal of the subject intended for undergraduate and postgraduate students specializing in plant sciences, references must be limited

Choosing the limited number of references was difficult I have tended to quote the more recent references, important and major articles of the past, and review articles Inevitably, however, some statements remain unsupported by a reference Also, in attempts to keep the book as current as possible, I have drawn on information from a number of unpublished works

Where contentious issues arise I have attempted to state each view but space has not always allowed me to discuss the detailed evidence supporting each view The reader, in some cases, is left to read the original reports to decide the most convincing line of argument

I have also tried to be specific about the plant species used in a particular experiment because, although there are certain basic features and processes concerning stomatal functioning and behaviour which are common to all species, there may well be variations on this common theme depending on the species.

Colin M. Willmer
University of Stirling
1982

Acknowledgements

Authors' acknowledgements, second edition

We would like to thank the following people who have reviewed parts of our manuscript or who have advised us and/or supplied us with unpublished information Peter Ayres, University of Lancaster, Michael Blatt, Wye College, University of London, Dudley Bowling, University of Aberdeen, Judy Croxdale, University of Wisconsin, Bill Davies, University of Lancaster, Peter Dittrich, University of Munich, Alistair Hetherington, University of Lancaster, Nick Kruger, University of Oxford, Jane Langdale, University of Oxford, Gerard Lasceve, CEN, Cadarache, Terry Mansfield, FRS, University of Lancaster, Hans Meidner, University of Stirling, Hugh Nimmo, University of Glasgow, Bill Outlaw, University of Florida State, Tallahassee, Robin Phillips, University of Stirling, Klaus Raschke, University of Gottingen, Robert Scotland, University of Oxford, Ken-ichiro Shimazaki, Kyushu University, Fukuoka, Japan, Andrew Smith, University of Oxford, Jonathan Weyers, University of Dundee, Alain Vavasseur, CEN, Cadarache, Eduardo Zeiger, University of California, Los Angeles

We would like to emphasize that any errors, misinformation or faulty reasoning which remain in the book are solely our responsibility

Author's acknowledgements, first edition

I am particularly indebted to Professor D A Baker, who edited my manuscript and gave valuable advice A special thanks goes to Professors J E Varner and H Meidner, both of whom read much of my manuscript and a number of improvements resulted from their criticisms

Thanks are also due to the following Dr W H Outlaw Jr of the Department of Biological Sciences, Florida State University, Tallahassee, Drs P Brown, J Maniotis, Barbara Pickard and others within the Department of Biology, Washington University, St Louis, all of whom helped me in their individual ways, colleagues within the Department of Biology, University of Stirling, for their comments and opinions about various points, T Forrest, W Struthers and L Taylor of the Department of Biology, University of Stirling for EM and photographic work, numerous people, whose names are mentioned in the figure or table legends, who gave me permission to use their published and unpublished data I also thank the following bodies who gave me permission to reproduce some of their published material the American Society of Plant Physiology, Springer-Verlag, Cambridge University Press, Blackwell Scientific Publications Ltd, the Institute of Biology, London, and the National Research Council of Canada

However, I am completely responsible for the views expressed in this book, for the presentation of its contents and any errors that may be found therein

I would also like to acknowledge that this book would not have been written without the influence of Professor TA Mansfield, who, in the very beginning, gave me the opportunity to study and research into stomata and who initiated my interest in the subject

Publisher's acknowledgements, second edition

We are indebted to the following for permission to reproduce copyright material Academic Press Ltd and the authors for our Figs 6 2 and 6 11 from Figs 11 and 4 (Willmer, 1988) and our Table 5 1 and Fig 8 5 from Table 8 1 and Fig 8 5 (Nobel, 1991), The American Society of Plant Physiologists for our Fig 8 10 from Fig 3 (Allan *et al* , 1994), our Figs 5 6 and 5 7 from Fig 9 (Ting and Loomis, 1965), our Figs 6 5 and 6 9 from Figs 4 and 5 (Hsiao *et al* , 1973), our Fig 6 13 from Fig 1 (Cockburn *et al* , 1979), our Fig 9 2 (Gotow *et al* , 1985), our Fig 6 18a from Fig 2 (Radin, 1984), our Fig 8B 2 from Figs 2 and 3 (Hedrich and Schroeder, 1989) and our Fig 6 18b from Figs 1 and 2 (Radin *et al* , 1982) all © American Society of Plant Physiologists, The *Australian Journal of Plant Physiology* for our Fig 6 20 from Fig 3 (Constable and Rawson, 1980), The *Botanical Journal of Scotland* for our Fig 3 9 from Fig 4 (Willmer, 1993), The *American Journal of Botany* for our Fig 5 5 from Fig 2 (Ting and Loomis, 1963) and our Fig 3 12 from Figs 2, 3, 5 and 6 (Sack and Paolillo, 1983), Cambridge University Press for our Fig 6 15 from Fig 2 (Davies *et al* , 1981) and our Figs 6 3 and 6 6 from Figs 5 4 and 5 2 (Jones, 1992), The Company of Biologists Ltd for our Fig 3 13 from Fig 31 (Galatis *et al* , 1986) and for our Fig 3B 1 from Fig 2 (Sholton, 1989), both in the *Journal of Cell Science*, The Institute of Biology for our Fig 6 17 from Fig 10 (Mansfield, 1971), The Japanese Society of Plant Physiologists for our Tables 9 4 and 9 5 from Tables 3 and 4 (Sato, 1985), Kluwer Academic Publishers for our Fig 6 7 from Fig 11 2 (Leshem and Willmer, 1992) and for our Fig 5 9 from Fig 1 (Wigley and Clark, 1974), Longman Group Ltd for our Figs 4 4 and 5 8 and Table 4 1 from Figs 2 4b, 3 5 and Table 2 2 (*Methods in Stomatal Research*, Weyers and Meidner, 1990), Macmillan Magazines Ltd and the authors for our Figures 2 4a and b from Figs 1 and 2 (Woodward, 1982), our Fig 8 12 from Fig 2 (Gilroy *et al* , 1990) and our Fig 8 9 from Fig 3 (McAinsh *et al* , 1990), *all from Nature*, The National Academy of Science, USA for our Fig 8 17 from Fig 5 (Thiel *et al* , 1993), The National Research Council of Canada for our Fig 3 15 from Fig 2 (Allaway and Setterfield, 1972), our Fig 3 6d from Fig 30 (Dayanadan and Kaufman, 1973) and our Fig 7 7 from Fig 3 (Allaway and Mansfield, 1970) (all from the *Canadian Journal of Botany*), *New Phytologist* for our Fig 2 3 from Fig 2 (Miranda *et al* , 1981), our Fig 6 19 from Figs 1b and 2b (Martin and Meidner, 1971), our Fig 7 9 from Fig 2 (Plumbe and Willmer, 1986) and our Fig 8 8 from Fig 4 (De Silva *et al* , 1985), Oxford University Press for our Fig 6 1 from Fig 2 (Heath and Russell, 1954), our Fig 6 4 from Fig 2 (Travis and Mansfield, 1984), our Fig 6 6 from Figure 1A (Pemadasa, 1982), our Fig 6 21 from Fig 3B (Willmer *et al* , 1988), our Fig 8 2 from Figs 2 and 3 (Bowling, 1987) and our Fig 4 1 from Fig 2 (Edwards and Meidner, 1979), all from the *Journal of Experimental Biology, Physiologia Plantarum* for our Fig 6 10 from Fig 2 (Fitzsimons and Weyers, 1986), Blackwell Science Ltd for our Fig 2 6 (Smith *et al* , 1989), our Figs 3 16 and 3 6(c) from Figs 5C and 1C (Rutter and Willmer, 1979) and our Fig 6 12 from Fig 1a (Morison and Jarvis, 1983), all from *Plant Cell and Environment*, Springer-Verlag and the authors for our Fig 4 6 from Raschke (1979) Movements using Turgor Mechanisms, *Physiology of Movements* (eds W Haupt and M E Feileib), **4**, Fig 6a–c, our Fig 6 16 from Tenhunen *et al* (1980), *Oecologia*, **47**, 365–367, Fig 2, our Fig 3 10 from Marc *et al* (1989), *Planta*, **179**, 516–529, Fig 19a–e, our Fig 7 1 from Wright (1977), *Planta*, **134**, 183–189, Fig 5, our Fig 8 2 from Penny and Bowling (1974), *Planta*, **119**, 17–25, Figs 2–5, Penny and Bowling (1975), *Planta*, **122**, 209–212, Figs 1 and 2 and Penny *et al* (1976), *Planta*, **130**, 291, Figs 1 and 2, our Fig 7 3 from Willmer *et al* (1978), *Planta*, **139**, 281–287, Fig 4, our Fig 3 11 from Cleary and Hardman (1989), *Protoplasma*, **149**, 76–81, Fig 40, our Fig 3 19a–e from Willmer and Sexton (1979), *Protoplasma*, **100**, 113–124, Figs 1, 3, 5, 6 and 7 and our Fig 4 9 from Weyers and Fitzsimons (1985), *The Physiological Properties of Plant Protoplasts* (ed Pilet), p 158, Fig 4 In addition, we are also indebted to Longman Group Ltd for permission to reproduce material from the first edition of this book

Special acknowledgement

Advances in research are often due to an international exchange of ideas, and each generation of scientists pushes the frontiers of knowledge further forward. It is upon this foundation that the success of future generations of scientists depends. Perhaps nowhere is this more evident than in the field of stomatal physiology.

Additionally, throughout the ages, it is often the case that certain scientists working in a particular field stand out and their names remain indelibly associated with the subject of their endeavours. Three such names associated with research into stomatal behaviour and functioning, all currently Professors Emeriti, are 'Peter' Heath, Hans Meidner and Klaus Raschke. We would like to acknowledge their life-long fascination with the workings of stomata, their major contribution to our knowledge of the subject and the inspiration they have given to ourselves and others to continue the quest for a fuller understanding of stomata and their impact on the growth of plants.

1 Introduction

The story about stomata starts almost 400 million years ago in the late Silurian–early Devonian Period when plants left their aquatic environment and invaded land In order to survive, the plants had to develop features which would prevent excessive water loss whilst allowing CO_2 to enter the plant for photosynthesis Thus, it is considered that a cuticle, stomata and vascular tissue developed almost simultaneously, and these three features were the key to the emergence and development of large terrestrial plants (e g Chaloner, 1970)

1.1 General considerations about stomata

Stomata are pores formed by a pair of specialized cells, the guard cells, which are found in the surface of aerial parts of most higher plants and which can open and close to control gas exchange between a plant and its environment Thus, they are the portals for entry of CO_2 into the leaf for photosynthesis and an exit for water vapour from the transpiration stream Their major function is to allow sufficient CO_2 to enter the leaf to optimize photosynthesis under the prevailing conditions, while conserving as much water as possible (see Chapter 6) In addition, under some conditions, evaporative cooling of the leaf by water loss via the transpiration stream may be important in lowering leaf temperature (see Nobel, 1991) The flux of water may also aid uptake and transport of salts necessary for the nutrition of the plant (e g Epstein, 1972), although there is evidence that transpiration *per se* is not required for long-distance transport of ions (e g Kramer, 1983, Tanner and Beevers, 1990) However, the transpiration stream is potentially an important upward route for many signals from the root to the shoot, such as abscisic acid (see Chapter 7), and transpiration may have an important role for full integration of the systems within the plant body

In contrast to CO_2, oxygen exchange between the plant and its environment is not greatly affected by stomata The atmospheric oxygen concentration (about 21%) is very high relative to the CO_2 concentration (about 0 036%) and inward diffusion of the O_2 will readily occur, furthermore, in the light oxygen is produced during photosynthesis within the leaf

Some other interesting roles of stomata have been proposed recently. For example, it has been suggested that phototropism of coleoptiles and possibly other plant organs is mediated by a light-induced increase in stomatal transpiration. Thus, on the illuminated side of a plant organ possessing stomata, transpirational water loss will be increased, which will reduce cell turgor and slow expansion and growth of these cells, resulting in a curvature of organ growth to the light source (McIntyre, 1994).

Another study (Wilkins, 1993) found that stomata of the crassulacean acid metabolism (CAM) plant, *Bryophyllum fedtschenkoi*, influenced the circadian rhythm of CO_2-fixation in the mesophyll tissue. In other CAM plants studied, the rhythm is apparently localized to the mesophyll tissue and independent of the epidermal tissue (Chapter 6).

1.2 Historical aspects

An excellent appraisal of the history of studies on stomata and plant water relations has been written by Meidner (1987). Here, some of the landmark findings and developments will be highlighted.

The word stoma (plural stomata) is Greek for mouth and de Candolle adopted this as the name initially proposed by Heinrich Link to describe the pores in leaves (de Candolle, 1827). Present day convention, however, normally uses the term *stoma* to include the pore and the surrounding pair of guard cells. Likewise, a stomatal complex refers to the guard cells and neighbouring subsidiary cells.

The first recorded investigations concerning stomata were made in 1660 by Edme Marriotte, a physicist, (reported by de Candolle, 1832) who measured evaporation rates from leaves. This was followed by the discovery of pore-like structures in the surface of leaves which coincided with the introduction of the simple microscope and the observation of cells by Robert Hooke (1635–1703). The first published observations of stomata were made by Malpighi (1628–94) in his manuscript, *Anatome Plantarum* (1675), however, he could make nothing of their function. Malpighi was better known for his studies of the skin and circulation system in animals, although some of his major scientific contributions are on plant anatomy.

At about the same time Nehemiah Grew (1641–1712) made similar observations and drawings of stomata to those of Malpighi and his series of lectures given to the Royal Society in 1680, when Sir Isaac Newton was president of the Society, were published in 1682. Figures 1.1 and 1.2 are reproductions of some of his drawings. Grew called stomata 'orifices' or 'pass-ports' and believed that they opened up into xylem vessels which he called 'aer vessels'. He considered that the whole system of stomata and xylem vessels acted as an aeration pathway much like the spiracles and trachea of insects.

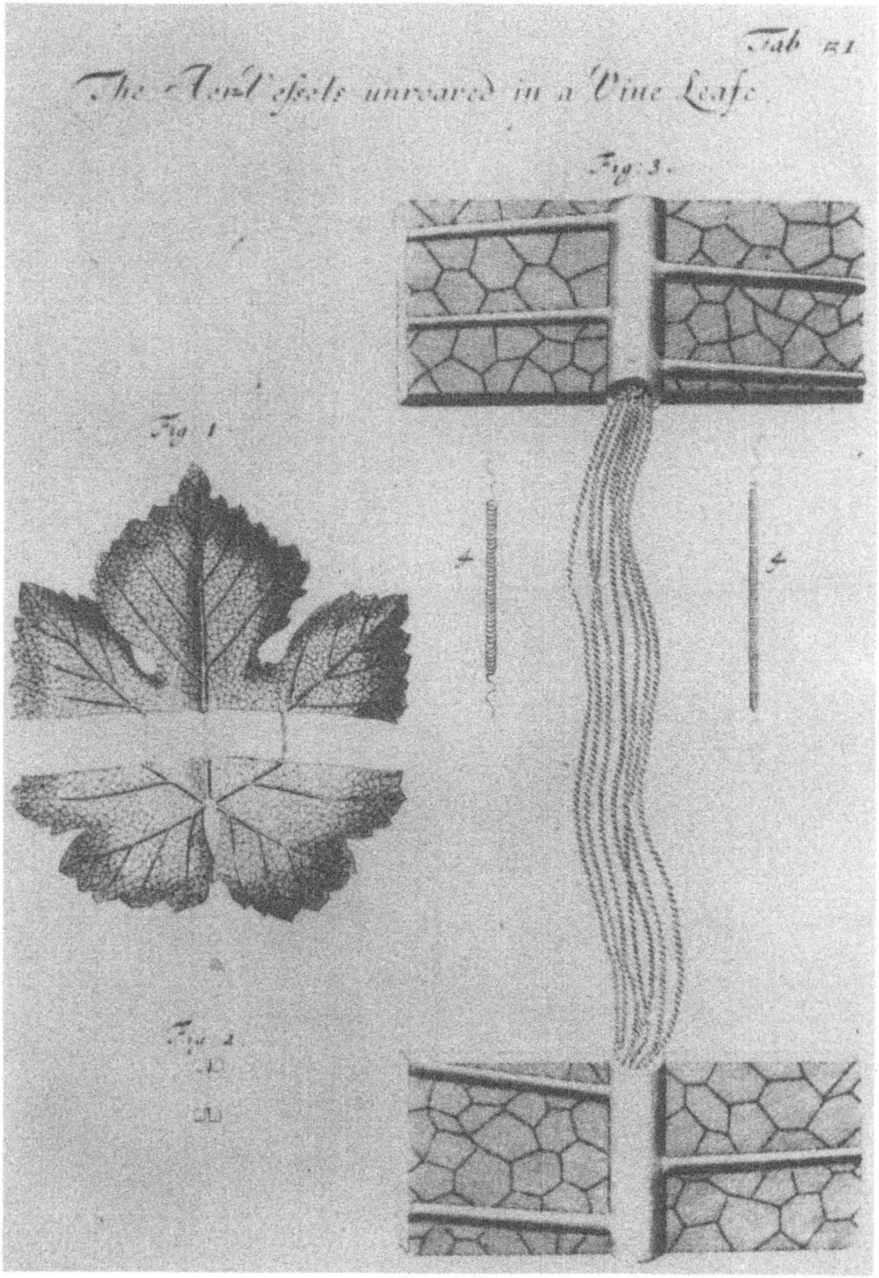

Figure 1.1 'Aer vessels' in a vine leaf (from Grew, 1682).

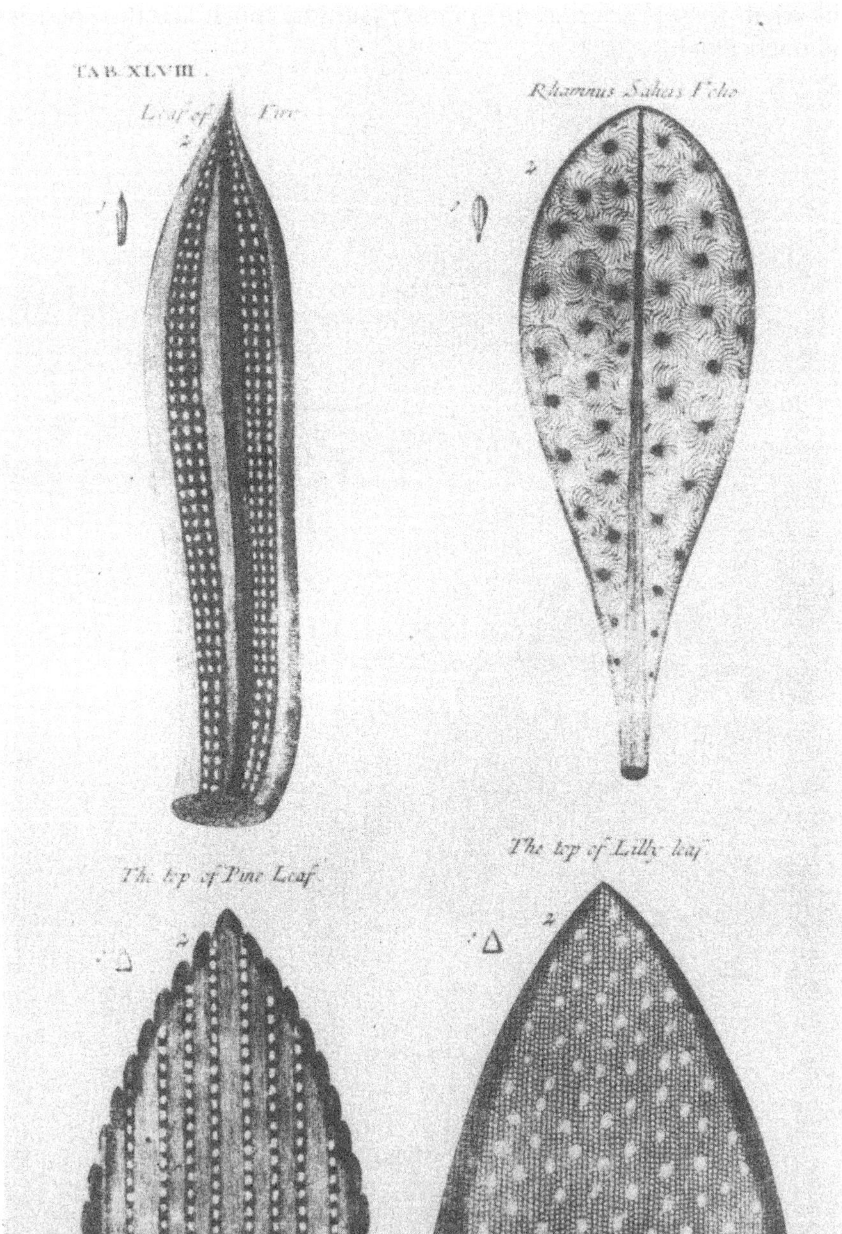

Figure 1 2 Low power drawings of leaf surfaces showing the patterns and locations of stomata (from Grew 1682)

Stomata were still commonly called 'breathing holes' well into the 19th century and it was not until 1841 that the experiments of Williams, a lecturer in forensic medicine, finally dispelled a general view that a membrane existed over the pore across which gas diffused

A number of eminent scientists studied gas exchange and specifically transpirational water loss by plants including Stephen Hales (1677–1761), de Saussure (1767–1845) and Dutrochet (1776–1847), but it was de Candolle (1827) who confirmed an earlier suggestion by Hedwig (1793) that stomatal apertures were variable

Outstanding studies into the structure and function of stomata were made by von Mohl (1805–1872) Not only did he appreciate that stomatal movements were due to turgor changes in guard cells which were osmotically regulated, but he made many anatomical studies of stomata (von Mohl, 1856) Later, Muller (1872), Schwendener (1881) and Haberlandt (1887) complemented this work with further anatomical studies, and they attempted to classify stomatal types and explain the mechanics of stomatal movements on an anatomical basis

As knowledge about the anatomy, morphology and function of stomata increased so there was a desire to understand more about their behaviour Initially stomatal apertures were measured under a microscope, but automated systems (porometers) were developed to measure leaf porosity Darwin and Pertz (1911) were the first to publish a method of measuring stomatal resistances (more correctly, leaf resistance) using a mass flow porometer, and since then other instruments have been developed and refined (e g Gregory and Pearse, 1934, Alvim, 1965) Such porometers measure the rate at which air or other gas can be pushed or sucked through a leaf when a pressure difference is applied across the leaf, i e they measure resistance to viscous flow Another type of porometer, known as a diffusion porometer, has also been developed A prototype was made by Muller (1870) and much more sophisticated ones with humidity sensor elements were developed more recently (e g see Wallihan, 1964, van Bavel *et al* , 1965, Kannemasu *et al* , 1969) Diffusion porometers measure the rate of diffusion of a gas across a leaf through the upper and lower surfaces, as in the nitrous oxide diffusion porometer (e g Slatyer and Jarvis, 1966), or of water vapour out of a leaf (e g Kanemasu *et al* , 1969)

With the availability of various devices for monitoring stomatal behaviour, so a variety of reponses were observed Thus, Loftfield (1921) noted midday closure, Darwin (1916) and Knight (1916) observed transient stomatal movements, while Darwin (1898) and Lloyd (1908) pioneered work on endogenous diurnal (circadian) rhythms Freudenberger (1940) and Heath (1948) established the importance of the CO_2 concentration in the leaf intercellular airspace in regulating stomata, Wilson (1947) worked on the effects of light, temperature and humidity on stomatal responses, while Heath and Russell (1954) clarified much of the earlier, often contradictory, work on the effects of the interaction of light and CO_2 on stomatal responses

While all these studies were being conducted work into how stomata functioned was under way Many ingenious hypotheses have been proposed over the last two centuries to explain how stomata open and close [see Heath and Mansfield (1969) and Meidner and Mansfield (1968) for more comprehensive reviews] and, although the early ones are mainly of historic interest only, they established a framework for our current understanding and should not be forgotten

One of the earliest was the 'photosynthetic' hypothesis proposed by von Mohl (1856) which suggested that the products of guard cell photosynthesis were responsible for the observed osmotic potential changes in guard cells This was extended and elaborated to form the starch–sugar hypothesis first conceived by Kohl (1895) and subsequently developed by Lloyd (1908), which was generally accepted to explain stomatal functioning up to the mid-1960s (see Heath and Mansfield, 1969) Basically the hypothesis considered that guard cell starch was converted to soluble sugars in the light, which acted as the osmoticum to increase cell turgor and drive stomatal opening The reverse occurred during stomatal closure in darkness Although the detail of the starch–sugar hypothesis has subsequently been found inadequate, the concept of the starch–sugar interconversion is correct, since we now know that carbon flows between starch, sugar and malate during stomatal movements (see Chapter 9) Additionally, there is still support for sucrose, produced in guard cells via photosynthesis, acting as an osmoticum in guard cell functioning (see Chapter 9) Nevertheless, the importance of starch–sugar interconversions as the only mechanism generating changes in osmotic pressure of guard cells has diminished with increased knowledge about their ionic relations and the involvement of K^+ as a major osmoticum

In the early 1900s another hypothesis to explain stomatal movements evolved which was attributed to Linsbauer (1916) and may best be described as the permeability hypothesis Guard cell 'membrane permeability' was gauged by the rate of dye uptake by cells in the light or dark and, although there was some conflict of view as to which condition resulted in increased permeability, it was generally conceded that uptake was greatest in the light Thus, it was concluded that solutes entered guard cells more readily in the light, decreased the osmotic potential and brought about stomatal opening The hypothesis was always lacking in detail, in part because knowledge of transport phenomena were not available at that time, but is loosely linked to current views of solute transport phenomena For example, today we know that a variety of opening stimuli, including light, can ultimately promote uptake of potassium through ion channels that effectively increase the membrane conductance (permeability) (Chapter 8).

Another hypothesis was proposed by Scarth (1929), which may be called the amphoteric colloid hypothesis. Scarth proposed that in the light the increase in 'guard cell pH' during opening resulted in the swelling of amphoteric colloids within the cytoplasm and increased guard cell turgor. In the dark the reverse situation was considered to occur with a contraction of the colloids, a decrease in guard cell turgor and stomata closed. Again, however, our increasing knowledge has not supported this hypothesis.

A scheme to explain stomatal movements which arose in the 1960s was the so-called glycollate hypothesis proposed by Zelitch (1963). It was suggested that a glycollate–glyoxyllate system in the guard cells was linked to non-cyclic photophosphorylation. ATP generated in the system could then be used to drive active ion 'pumps' integral to the stomatal mechanism. Alternatively, glycollate could be used as a source of carbon for production of carbohydrates available to decrease guard cell osmotic potentials. This scheme also subsequently lacked experimental support.

All of the above hypotheses have proved inadequate to greater or lesser extents and a mechanism to explain stomatal movements based on the fluxes of K^+ and H^+ across guard cells and upon organic acid synthesis in guard cells is now the centre of attraction (see Chapter 8). There is still considerable controversy about the finer details of this mechanism and it is important to realize that the vast majority of detailed information on guard cell physiology has been obtained from remarkably few species, principally *Commelina communis*, *Vicia faba* and *Zea mays*. There are considerable differences even between these three species, and it is likely that the diversity of form discussed in Chapters 2 and 3 is paralleled by an equal variation in physiological adaptations based upon this common theme.

1.3 The importance of research into stomatal functioning and behaviour

Interest in stomata has expanded over the years and now a vast amount of information on their structure, function and ecology is available. Stomata have a fundamental role in control of two of the most important plant processes, i.e. photosynthesis and transpiration (see Chapter 6). They are major factors in determining rates of dry matter accumulation through regulation of CO_2 uptake and are one of a number of key factors in modelling of crop productivity. Thus a knowledge of stomatal behaviour is of increasing relevance in studies of potential effects of CO_2 enrichment of the biosphere, alteration of spectral quality from ozone depletion and increases in atmospheric pollutants. The extent of water vapour exchange between the leaf and the atmosphere also has far-reaching meteorological consequences, and can significantly affect local and global weather patterns.

On a different scale, stomatal guard cells have provided one of the best experimental systems to study fundamental aspects of plant physiology, and cell development and differentiation. The distribution, timing and orientation of cell divisions during differentiation of a stomatal complex present an excellent system to study the organization of cell polarity and control of asymmetric cell divisions that are inherent in the differentiation of most plant tissue (Chapter 3). The mature guard cell has to respond to and continuously integrate a diverse range of electromagnetic, chemical and mechanical stimuli to give a co-ordinated response (Chapters 6 and 7). Their phenomenal sensitivity, and dramatic and readily measurable responses have established them as one of the most studied sensory systems in plants. Indeed, no other higher plant cell has been as intensively investigated in terms of its membrane transport events, and guard cells are currently the best characterized model system for ion transport and signal transduction (Chapter 8). This has arisen in part from the massive ion fluxes involved in stomatal movements and also from their almost unique status amongst higher plant cells in that they lack functional plasmodesmata and are therefore physiologically isolated from the rest of the plant body.

From a biochemical viewpoint guard cells are also proving to be a fascinating subject, as they appear to have modified the control of a number of conventional biochemical pathways to suit their specific metabolic needs. For example, the control of carbon flux is altered to allow starch breakdown in the light and synthesis to occur in the dark, the reverse of what happens in mesophyll cells of C3 plants (see Chapter 9).

Studies on the molecular biology of guard cells have also begun recently; for example a search for the genes which are involved in the control of stomatal frequency and sensitivity is under way (see Chapter 10). With the convergence of all of the approaches mentioned above, genetic engineering may provide both a useful tool to dissect the guard cell system further and also a means to improve crop productivity and stress tolerance via genetic manipulation of stomata in the future.

References

Alvim, P. de T. (1965) A new type of porometer for measuring stomatal opening and its use in irrigation studies. *UNESCO Arid Zone Res.*, **25**, 325–329.

van Bavel, C H M , Nakayama, F S and Ehrler, WL (1965) Measuring transpiration resistance of leaves *Plant Physiol* , **40**, 535–540

de Candolle, A P (1827) Chapter 6, in *Organographie Vegetale, ou description raisonee des organes des plantes*, Vol 1, Deterville, Paris, pp 78–88

de Candolle, A P (1832) De l emanation ou exhalation aqueuse des vegetaux vasculaires, in *Physiologie vegetale*, Bechet, Paris, pp 107–178

Chaloner, WG (1970) The rise of the first land plants *Biol Rev Cambridge Phil Soc* , **45**, 353–377

Darwin, F (1898) Observations on stomata *Phil Trans Roy Soc Lond* , *Ser B*, **190**, 531–621

Darwin, F (1916) On the relation between transpiration and stomatal aperture *Phil Trans Roy Soc Lond* , *Ser B*, **207**, 413–437

Darwin, F and Pertz, D F M (1911) On a new method of estimating the aperture of stomata *Proc Roy Soc Lond* , *Ser B*, **84**, 136–154

Epstein, E (1972) *Mineral Nutrition of Plants Principles and Perspectives*, John Wiley, New York

Gregory, F G and Pearse, H L (1934) The resistance porometer and its application to the study of stomatal movement *Proc Roy Soc Lond* , *Ser B*, **114**, 477–493

Freudenberger, H (1940) Die Reaktion der Schliesszellen auf Kohlensaure und Sauerstoffentzug *Protoplasma*, **35**, 15–54

Grew, N (1682) *Anatomy of Plants*, London

Haberlandt, G (1887) Zur Kenntnis des Spaltoffnungsapparates *Flora*, **45**, 97–109

Hales, S (1727) *Vegetable Staticks*, London

Heath, O V S (1948) Control of stomatal movement by a reduction in the normal [CO_2] of the air *Nature*, **161**, 179–181

Heath, O V S and Mansfield, T A (1969) The movement of stomata, in *Physiology of Plant Growth and Development*, (ed M Wilkins), McGraw-Hill, London

Heath, O V S and Russell, J (1954) An investigation of the light responses of wheat stomata with attempted elimination of control by the mesophyll *J Exp Bot* , **5**, 1–15, 269–292

Hedwig, J (1793) *Sammlung seiner zerstreuten*, Abhandlungen 1, Crusius, Leipzig, p 126

Kanemasu, E T , Thurtell, G W and Tanner, C B (1969) Design, calibration and field use of a stomatal diffusion porometer *Plant Physiol* , **44**, 881–885

Knight, R C (1916) On the use of the porometer in stomatal investigations *Ann Bot* , **30**, 57–76

Kohl, F G (1895) Uber Assimilationsenergie und Spaltoffnungsmechanik *Botanisches Centralblatt*, **64**, 109–110

Kramer, P (1983) *Water Relations of Plants*, Academic Press, Orlando, FL

Linsbauer, K (1916) Beitrage zur Kenntnis der Spaltoffnungsbewegung *Flora*, **9**, 100–143

Lloyd, F E (1908) The physiology of stomata *Publ Carnegie Inst Wash 82*

Loftfield, J VG (1921) The behaviour of stomata *Publ Carnegie Inst Wash 314*

Malpighi, M (1675) *Anatome Plantarum*, Royal Society, London

McIntyre, G (1994) The role of transpiration in phototropism of the *Avena* coleoptile *Aust J Plant Physiol* , **21**, 359–375

Meidner, H (1987) Three hundred years of research into stomata, in *Stomatal Function*, (eds E Zeiger, G D Farquhar and I R Cowan), Stanford University Press, Stanford, CA, pp 7–27

Meidner, H and Mansfield, T A (1968) *Physiology of Stomata*, McGraw-Hill, London

von Mohl, H (1856) Welche Ursachen bewirken die Erweiterung und Verengung der Spaltoffnungen? *Bot Ztg* , **14**, 697–704, 713–720

Muller, N J C (1870) Uber den Durchgang von Wasserdampf durch die geschlossene Epidermiszelle *Jahrb Wissenschaft Botanik*, **7**, 193–199

Muller, N J C (1872) Die Anatomie und die Mechanik der Spaltoffnungen *Jahrb Wissenschaft Botanik*, **8**, 75–116

Nobel, PS (1991) *Physicochemical and Environmental Plant Physiology*, Academic Press, San Diego, CA

Scarth, G W (1929) The influence of H-ion concentration on the turgor and movement of plant cells with special reference to stomatal behaviour *Proc Int Conf Plant Sci* , **2**, 1151–1162

Schwendener, S (1881) *Uber Bau und Mechanik der Spaltoffnungen* Monatsberichte der Koniglichen-Preusssischen Akademie der Wissenschaften zu Berlin, Physikalisch-mathematische Klasse, pp 833–867

Slatyer, R O and Jarvis, PG (1966) Gaseous diffusion porometer for continuous measurement of diffusive resistance of leaves *Science*, **151**, 574–576

Tanner, W and Beevers, H (1990) Does transpiration have an essential function in long-distance ion transport in plants? *Plant Cell Environ* , **13**, 745–750

Wallihan, E F (1964) Modification and use of an electrical hygrometer for estimating relative stomatal aperture *Plant Physiol* , **39**, 86–90

Wilkins, M B (1993) The role of stomata in the generation of circadian rhythms in plant tissues *J Exp Bot Suppl* , **44**, 2

Williams, T (1841) On the structure and uses of the stomata *Microscopic J Struct Record*, **1**, 118–121

Wilson, C.C. (1947) The effects of some environmental factors on the movements of guard cells. *Plant Physiol.*, **23**, 5–37.

Zelitch, I. (1963) The control and mechanism of stomatal movements, in *Stomata and Water Relations in Plants (Connecticut Experimental Station Bull. 664)*, Newhaven, CT, pp. 18–36.

2 The distribution of stomata

2.1 Location and occurrence of stomata

One of the earliest recorded vascular plants is *Cooksonia pertoni* and its fossil remains show the presence of stomata (Edwards *et al*, 1992) (Fig 2 1) Such fossil records suggest that stomata were relatively large in early plants In *Zosterphyllum myretonianum*, stomata up to 120 μm long were recorded (Lele and Walton, 1960–61) These are the largest stomata that have been measured in living or extinct plants

In extant plants, stomata are found at the evolutionary level of the Division Bryophyta (containing the mosses, liverworts and hornworts) and all other vascular plants Within the mosses stomata are present in the capsules (diploid sporophyte generation) of most species and also in two genera of the class Anthocerotae (hornworts, also known as horned liverworts) In at least some species of hornworts stomata are found in the lower surface of thalli and, although they may not be functional, this is the only instance of stomata being located in the gametophyte (haploid) generation of plants Stomata are not found in liverworts but, instead, pores, which remain permanently open, exist in the upper surface of the thalli

Stomata are found in all groups of the Division Pteridophyta (containing the ferns, the clubmosses and quillworts, the horsetails and the Psilotatae) In the Psilotatae (e g *Psilotum*) large stomata are mainly located in grooves between the stem ridges The Lycopodiatae (clubmosses and quillworts) also have large stomata in their leaves and stems, but a notable exception occurs in the quillwort, *Stylites andicola*, which does not possess stomata, but accumulates CO_2 via its roots (Keeley *et al*, 1984) In the Equisetatae (horsetails and scouring rushes) the stomata are usually confined to the longitudinal grooves of the aerial stems, whilst in the Filicatae (ferns), stomata are large and abundant in the leaves, usually only on the lower surface

Stomata have the most prolific distribution in the Division Spermatophyta (higher plants) and may be found in the epidermal layers of most aerial parts (Fig 2 2A–L) They are present in floral parts, including petals, sepals, stamens and gynoecia, and in awns, paleas and lemmas of

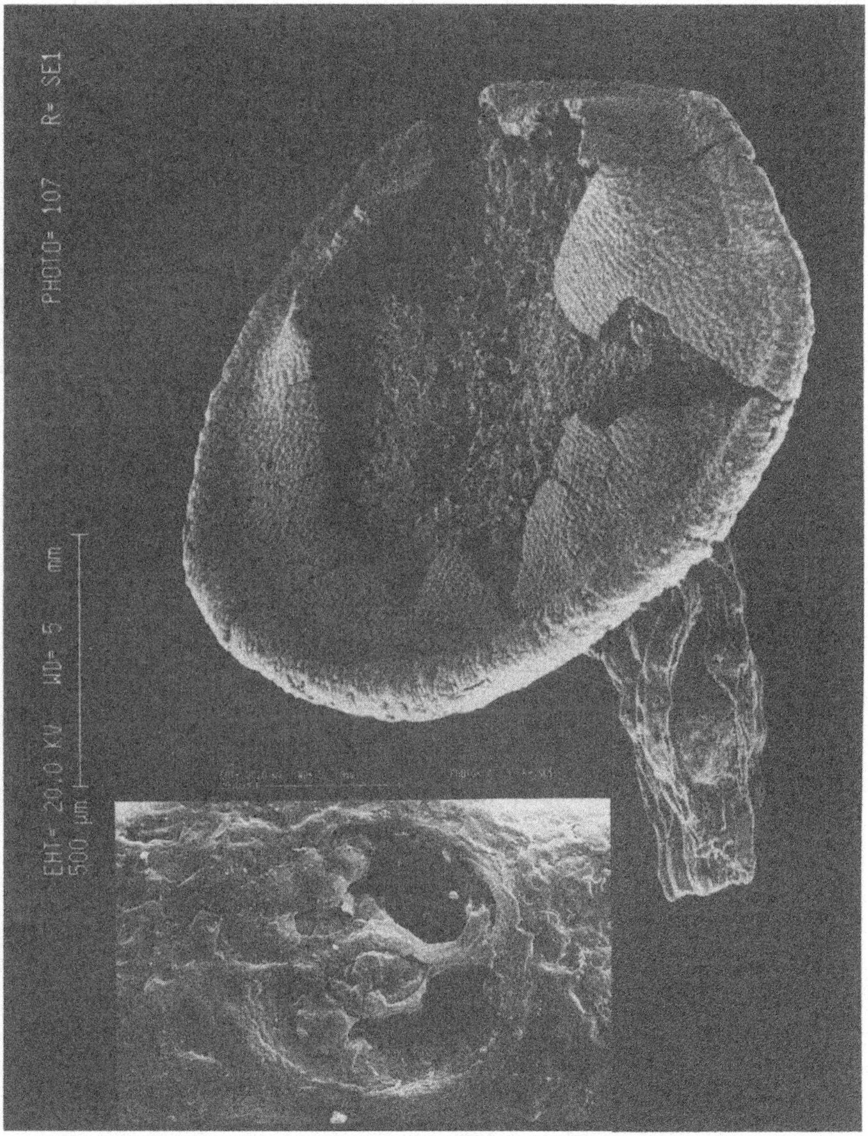

Figure 2.1 Scanning electron micrographs of a fossilized specimen of *Cooksonia pertoni* spp *apiculispora* Gedinnian Shropshire Inset shows a stoma located on the short stalk. Courtesy of Edwards *et al.*, 1992.

cereal inflorescences. Stomata are also found in the surfaces of developing fruits such as quince, apple, grape, banana and tomato, and in the inner and outer surfaces of pods, and have also been recorded in the surface of seeds inside pods. In some cases, as the fruits mature, the stomata on their surfaces may develop into lenticels. According to Blanke and Bonn

(1985), in the early stages of fruit development in 'Golden Delicious' apples, the stomata function in a similar manner to those in leaves They observed that the stomata were 20–30 µm long and 10–25 µm wide with a frequency of about 25 mm , decreasing to less than 1 mm when the fruit reached full size Additionally, stomata have been recorded in rhizomes and the ligules of some grass species (e g Chaffey, 1982), and even on the primary roots of *Pisum sativum* (Lefebvre, 1985) and in potato tubers In growing potato tubers the stomata also develop into lenticels (Adams, 1975) More typically stomata are found in the surface of leaves, green stems and modified leaves such as bracts and tendrils Stomata are not found in the chlorophyll-free plants, *Monotropa* or *Neottia*, but have been found in the stems of *Orobanche* Mutant albino plants apparently possess non-functional stomata and stomatal functioning may also be limited in some of the locations mentioned above, such as petals

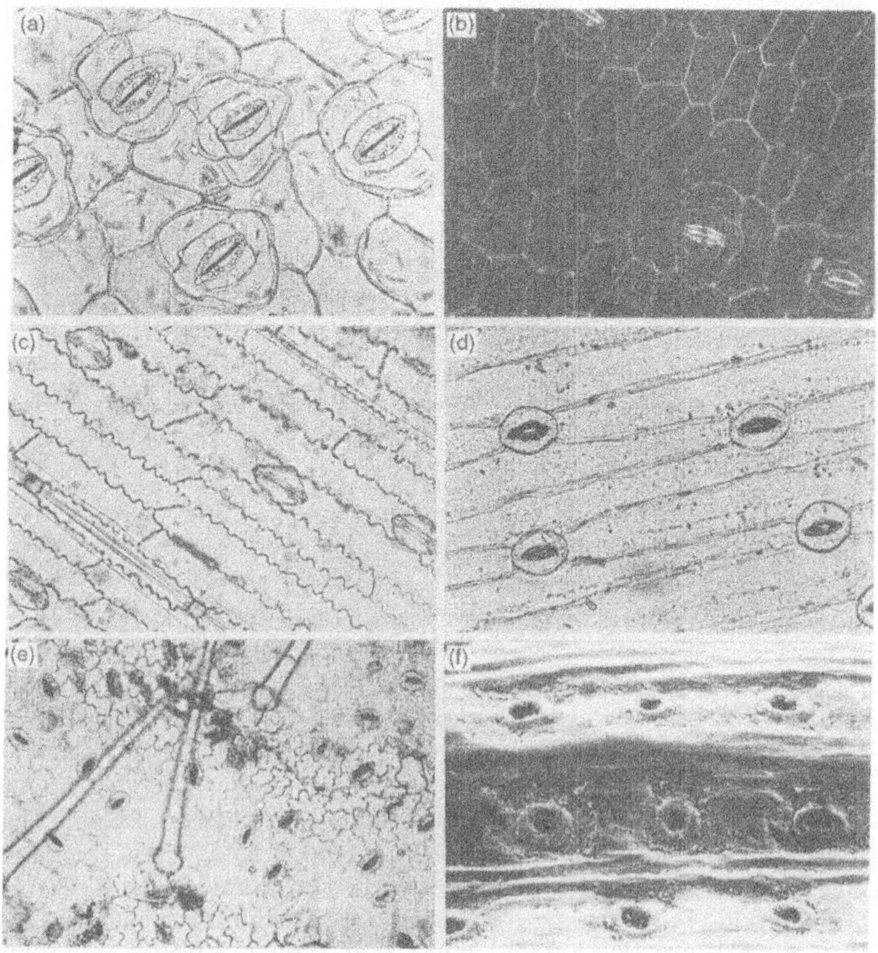

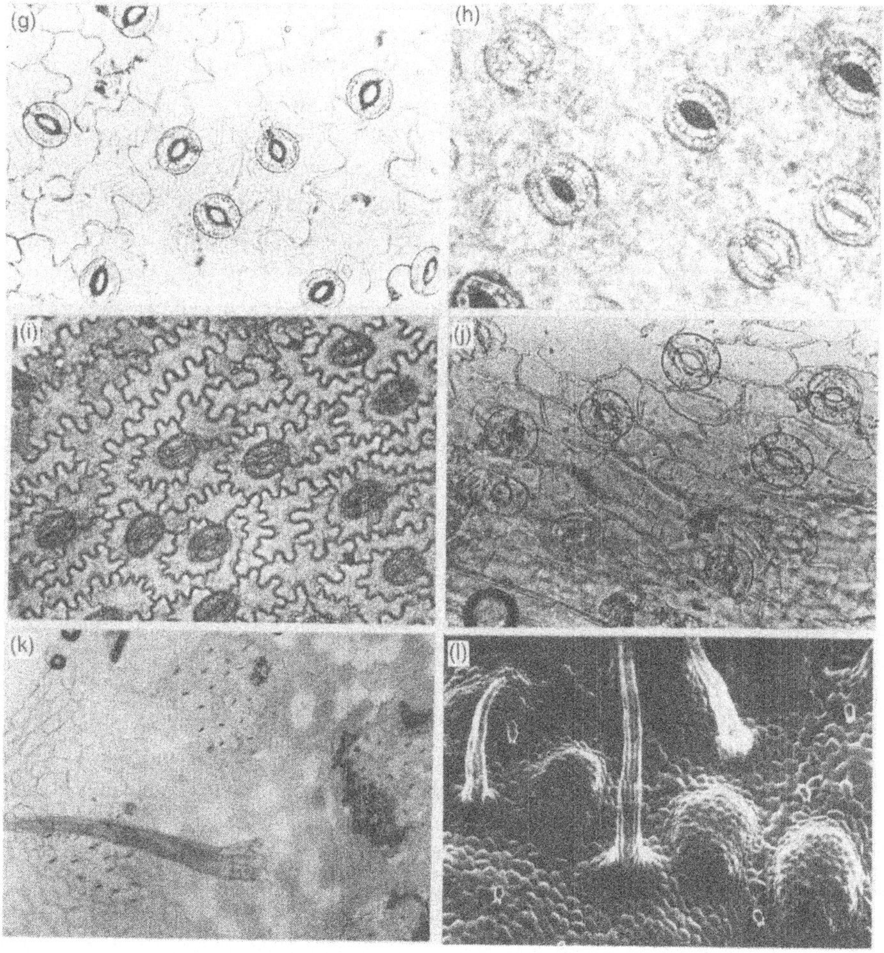

Figure 2 2 Stomatal patterns in the surface of leaves and of a petal showing a wide variety of types (a) *Commelina communis* lower epidermis (b) *C communis* upper epidermis (c) maize (*Zea mays*) epidermis (d) onion (*Allium cepa*) epidermis (e) geranium (*Pelargonium zonale*) epidermis (f) scanning electron micrograph of the surface of a pine (*Pinus sylvestris*) needle (g) broad bean (*Vicia faba*) lower epidermis (h) *Paphiopedilum harrisseanum* lower epidermis (i) Hart s tongue fern (*Phyllitis scolopendrium*) lower epidermis (k) and (l) lower epidermis and scanning electron micrograph of the leaf surface of *Saxifraga stolonifera*

More details about the evolution, morphology and anatomy of stomata can be found in Ziegler (1987) and Willmer (1993)

In the leaves of most herbaceous plants stomata are found in both the upper (adaxial) and the lower (abaxial) surfaces, but usually there are fewer stomata on the upper surface Such leaves are termed amphistomatous Leaves with stomata in the lower leaf surface only, as are found in virtually all tree species, are called hypostomatous Aquatic plants with floating leaves, such as water-lilies, have stomata on the upper surface only and such leaves are called epistomatous (or hyperstomatous) Also, in some aquatic plants stomata may remain permanently open In the water fern, *Marsilea* (an aquatic Pteridophyte which can also grow on land), stomata are restricted to the upper surface of floating leaves, while aerial leaves have stomata on both surfaces Most submerged aquatic plants do not possess stomata and diffusion of gases occurs between the water and the plant tissues across a thin cuticle

The adaptive significance of stomata occurring on one or both leaf surfaces is unclear, although hypostomaty is considered to be an evolutionary primitive character and, as plants adapted to growth in more open sunlit habitats, amphistomaty may have evolved almost simultaneously (see Mott *et al*, 1982) An unsubstantiated belief is that hypostomatous leaves are better adapted to dry conditions than amphistomatous leaves Certainly, the side of a leaf facing the sun may be slightly warmer than the opposite side in shade and this would lead to more evaporative water loss from the side exposed to the sun However, amphistomaty may increase CO_2 uptake by reducing the length of the CO_2 diffusion pathway to the mesophyll (Parkhurst, 1978) and via its influence on stomatal resistances and boundary layer resistances (Mott *et al*, 1982) It has been hypothesized that amphistomaty benefits most those species with high maximum photosynthetic rates growing in sunny habitats, particularly if they are thick-leaved species (Mott *et al*, 1982) The observation that most pioneer species are amphistomatous and have higher photosynthetic rates than species characteristic of later successional stages does not counter this view

There is evidence that stomata in the two leaf surfaces can be regulated relatively independently of each other (e g Turner, 1970, Smith, 1981) Thus, Smith (1981) observed that abaxial stomata opened later in the morning and closed later in the evening than adaxial stomata in several broad-leaved, understorey herbs, while Foster and Smith (1986) recorded full closure of adaxial stomata of *Populus angustifolia* and several *Salix* species when abaxial stomata remained partially open under conditions of large leaf to air water vapour pressure deficits Foster and Smith considered therefore that leaves may exhibit 'functional hypostomaty and regulate gas exchange in the same manner as 'morphological hypostomaty'

2.2 Stomatal frequency

Stomatal frequency or density (the number of stomata per unit area of one leaf surface) can vary significantly within leaves, plants or individuals of a single species within a community and can be modified by environmental factors, leaf morphology and genetic composition. Pioneer work on many of these aspects was conducted by Salisbury (1927) and an excellent study of the subject was made by Schoch (1978) in more recent times. The wide range of stomatal frequencies and guard cell dimensions which occur among species are presented in Table 2.1. Tree species generally have high stomatal frequencies, while xerophytes generally have low stomatal frequencies.

Stomatal frequencies often vary according to cell size and smaller guard cells are usually associated with higher stomatal frequencies. Thus, Salisbury (1927) introduced the term 'stomatal index' which relates the number of stomata per unit leaf area to the number of epidermal cells plus guard cells per unit leaf area:

$$\text{stomatal index} = \frac{\text{no. of stomata per unit leaf area}}{\text{no. of stomata per unit leaf area} + \text{no. of epidermal cells per unit leaf area}} \times 100 \quad (2.1)$$

The stomatal index was considered to be fairly constant within the leaves of a single plant. However, this has not been confirmed in all species investigated. In some species the light intensity and quality received by mature leaves of a plant can affect the stomatal index of still developing leaves. For example, in *Vigna sinensis* grown at high irradiance, a day of shade causes a decrease in the stomatal index of those leaves that are at a 'critical' period of development (about 6 days preceding the major phase of unfolding of the leaf lamina) (Schoch *et al.*, 1980).

2.2.1 Variations of stomatal frequency within a leaf

Stomatal frequency has been found to vary within a leaf, often in a specific way. Salisbury (1927) found that in some broad-leaved plants frequencies were greater at the leaf margins than near the mid-rib, while in some monocotyledonous leaves which have an intercalary leaf meristem, frequencies increased from the base, reaching a maximum value approximately in the middle of the leaf lamina, before decreasing in frequency towards the tip. More recent studies on the monocot, maize, however, show a different pattern of stomatal frequency along the leaf. Thus, in young developing (Miranda *et al.*, 1981) and relatively mature leaves (Heichel *et al.*, 1971), frequencies initially decreased from the

Table 2.1 Stomatal frequencies, guard cell dimensions and pore area as a percentage of leaf area when stomatal apertures are 6 μm

Species	Stomatal frequency (mm^{-2})* lower epidermis	Stomatal frequency (mm^{-2})* upper epidermis	Pore area (% of leaf area with apertures of 6 μm)	Guard cell dimensions for lower epidermis* (μm) length	Guard cell dimensions for lower epidermis* (μm) width	Reference
Ferns						
Phyllitis scolopendrium (hart s tongue-fern)	59	0	0.55	77	21	Meidner and Mansfield (1968)
Osmunda regalis (Royal fern)	67	0	0.50	56	19	
Herbaceous plants						
Monocots						
Commelina communis	67	19	–[†]	48	12	Willmer unpublished
Zea mays (maize)	108	98	0.70	43	12	
Triticum vulgare (wheat)	40	50	0.63	53	14	
Hordeum vulgare (barley)	85	70	0.65	38	11	Meidner and Mansfield (1968)
Avena sativa (oat)	45	50	0.50	56	13	
Allium cepa (onion)	175	175	0.63	42	19	
Paphiopedilum venustum (Lady slipper orchid)	26	0	–	67	53	Rutter and Willmer (1979)
Tradescantia virginiana	23	7	0.35	70	21	Meidner and Mansfield (1968)
Dicots						
Helianthus annuus (sunflower)	175	120	1.10	32	15	
Vicia faba (broad bean)	75	65	1.00	46	13	
Sedum spectablis	35	28	0.32	21	10	Meidner and Mansfield (1968)
Nicotiana tabacum (tobacco)	190	50	0.80	31	13	
Xanthium pennsylvanicum	173	177	0	39	12	
Pisum sativum (pea)	216	101	–	–	–	
Phaseolus vulgaris (French bean)	281	40	–	–	–	Spector (1956)
Nymphaea alba (water lily)	0	460	–	–	–	
Lycopersicum esculentum (tomato)	130	120	–	–	–	
Arabidopsis thaliana (bitter cress)	194	103	–	20	6	Lasceve, unpublished
Veronica cookiana	2200	200	–	–	–	Salisbury (1927)
Small trees and shrubs						
Corylus americana (hazelnut)	347	0	–	37	–	
Canius florida (flowering dogwood)	83	0	–	32	–	Carpenter and Smith (1975)
Rhus copallina (shining sumac)	731	0	–	6	–	

Species	Stomatal frequency (mm $^{-2}$)*		Pore area (% of leaf area with apertures of 6 µm)	Guard cell dimensions for lower epidermis* (µm)		Reference
	lower epidermis	upper epidermis		length	width	
Large trees						
Angiosperms						
Quercus palustris (pin oak)	909	0	–	10	–	Carpenter and Smith (1975)
Quercus valutina (black oak)	405	0	–	50	–	
Quercus triloba (Spanish oak)	1192	0	–	–	–	Spector (1956)
Tilia americana (American basswood)	891	0	–	26	–	Carpenter and Smith (1975)
Tilia europa (European lime)	370	0	0 90	25	9	Meidner and Mansfield (1968)
Gymnosperms						
Pinus sylvestris (Scots pine)	120	120	1 20	28	14	
Larix decidua (larch)	16	14	0 15	42	13	
Other						
Welwitschia mirabilis (a gymnosperm with CAM)	222	222	–	–	–	Bornman (1972)

* Stomatal frequencies and dimensions depend on a variety of factors including their location on the leaf the leaf insertion and growth conditions of the plants and the values given here should not be accepted as definitive ones
–† no figures available

base along the sheath up to the ligule and then slowly increased, reaching highest values towards the leaf tip (Fig. 2.3). Miranda *et al.* (1981) assumed the initial decrease in stomatal frequency was due to cell and tissue expansion; presumably, also, the increasing frequency which occurred along the lamina from the ligule was due in part, at least, to changes in the rate of formation of stomatal initials. In an extensive study with the monocotyledonous plant, *Commelina communis*, Smith *et al.* (1989) have mapped stomatal frequencies using iso-frequency contours (see Fig. 2.6a) and discovered that highest frequencies were near the leaf margins and lowest frequencies next to the central main vein.

2.2.2 Variation of stomatal frequency within a plant and within genotypes

Stomatal frequencies may also vary among mature leaves of a plant, highest frequencies often being found at higher insertion levels (e.g. Cole and Dobrenz, 1970; Miskin and Rasmusson, 1970; Turner and Begg, 1973) (Table 2.2). However, as Salisbury (1927) pointed out, leaves are usually smaller at the top of a plant with smaller cells resulting in a higher stomatal frequency, whilst the stomatal index may remain fairly constant. Moreover, environmental factors can influence stomatal

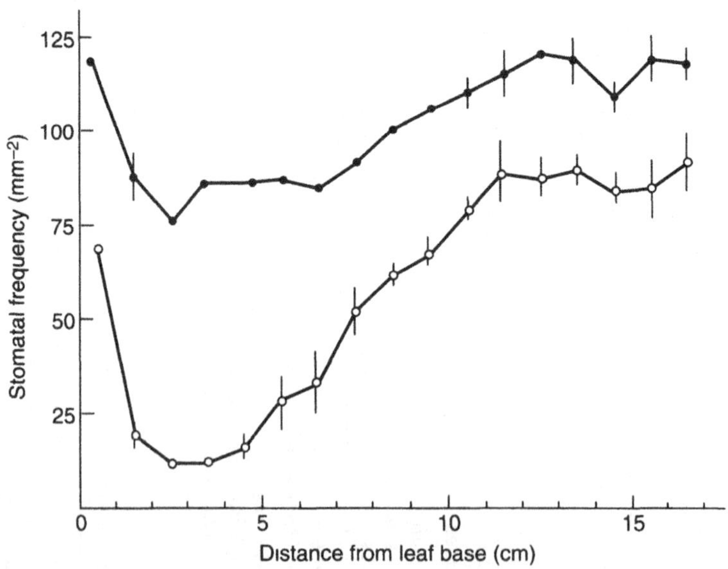

Figure 2.3 Changes in stomatal frequency along the upper surface (○) and the lower surface (●) of a maize leaf Vertical bars indicate twice the standard error of the mean, n=5 From Miranda *et al* (1981)

frequency gradients. Thus, Salisbury (1927) found that in *Mercurialis perennis* the stomatal frequency gradient, with highest values at the top of the plant, tends to disappear when plants are grown in a moist environment.

Table 2.2 Changes in stomatal frequency with height of leaf insertion on a plant (from Meidner and Mansfield, 1968)

Species	Height of insertion (cm)	Stomatal frequency (mm^{-2}) upper epidermis	lower epidermis
Polygonatum multiflorum (after Salisbury, 1927)	21 5	–	54
	27 0	–	59
	30 0	–	76
	50 0	–	73
	56 5	–	91
Tilia europa	500	–	370
	1500	–	420
Triticum vulgare	third node below flag leaf	39	30
	first node below flag leaf	50	40

In crop plants stomatal frequency can vary greatly among different genotypes of the same species growing under identical conditions (Table 2.3).

Table 2.3 Mean stomatal frequency and guard cell length of field grown soybeans (after Ciha and Brun, 1975)

Genotype	Mean stomatal frequency (mm^{-2})		Guard cell length (μm)	
	adaxial	abaxial	adaxial	abaxial
M-62-93	170	385	22 3	20 8
Corsoy	149	369	22 0	20 0
M-61-93	157	363	21 5	19 5
A K Harrow	143	344	21 9	20 9
Blackhawk	130	327	22 5	19 9
Manchu	121	327	22 5	20 2
M-62-56	120	300	21 5	19 6
M-62-263	104	287	21 9	19 4
O Mandarin	135	263	22 2	20 6
Flambeau	81	242	21 8	20 6

The genotypes have been placed in order of decreasing stomatal frequencies on the abaxial leaf surface The same trend, though less marked, is observed in the adaxial surface Stomatal size (in this case represented by guard cell length) usually decreases with increasing frequency but this feature is not observed in the soybean genotypes

2.2.3 The influence of environmental factors on stomatal frequency

Environmental factors are also involved in the morphogenesis of stomata Water availability, light intensity, temperature and CO_2 concentration have all been observed to affect stomatal frequency Stomatal frequencies are usually higher in plants grown in full sunlight or high photon flux density (PFD) than in plants grown in shade (low PFD) (e g Salisbury, 1927, Cooper and Qualls 1967, Miskin and Rasmusson, 1970, Friend and Pomeroy 1970, Wild and Wolf 1980) However, there is often a change in leaf area associated with a change in leaf irradiance, so that overall stomatal numbers per leaf may not be significantly different (Kubinova, 1991, see also below) Thus, leaf area has been reported to increase in low irradiance (Friend and Pomeroy, 1970), remain unchanged (Dale *et al* , 1972), or even increase in high PFD (Lichtenthaler, 1985) The reasons for these apparently conflicting results are not clear, but could be due to species differences and/or that the values for high and low PFD were not standardized between groups, for example, Miskin and Rasmussen (1970) compared effects of sunlight (full sunlight is about 2000 μmol m s^{-1}) and deep shade while Kubinova (1991) used a PFD of 200 and 25 μmol m s^{-1} Also, plants growing in a dry soil and low humidity generally have higher stomatal frequencies than plants growing in wet soil and high humidities (e g Penfound, 1931)

Changing atmospheric CO_2 concentration may also affect stomatal frequencies Some studies have shown that as CO_2 concentrations increase so stomatal frequencies decrease (Bristow and Looi, 1968, Woodward, 1987, Woodward and Bazzaz, 1988) In some pioneering studies Bristow and Looi (1968) found that in a terrestrial form of *Marsilea* stomatal frequencies reached a maximum at about 0 03% CO_2 and declined as CO_2 concentrations decreased or increased on either side of this value Also, using herbarium and present day specimens, Woodward (1987) found that stomatal frequencies have declined by about 40% over the last 200 years while the CO_2 concentration increased by about 60 ppm over the same period (see Fig 2 4) Woodward also found that stomatal frequencies of plants experimentally grown in low CO_2 concentration increased

Figure 2 4 (a) Abaxial stomatal densities of herbarium stored leaves of *Acer pseudoplatanus Carpinus betulus Fagus sylvatica Populus nigra Quercus petraea Q robur Rhamnus catharticus* and *Tilia cordata* Leaves had been stored in the herbarium in the Department of Botany University of Cambridge Only leaves on reproductive shoots were sampled with the assumption that these leaves had developed in full irradiance Five leaves of each species were collected from different dates back to AD 1750 and from collections made in the midlands of England Stomatal densities varied between species by a factor of about 2 however the changes in stomatal densities relative to the recent collections (1970 1981) were similar for all species Reconstructed changes in atmospheric CO_2 based on ice core studies are also included The linear regression line with 95% confidence limits shows a 40% reduction in the ratio of stomatal densities over a period of 200 years *r* 0 828 (b) Comparison of the experimental effects of a change

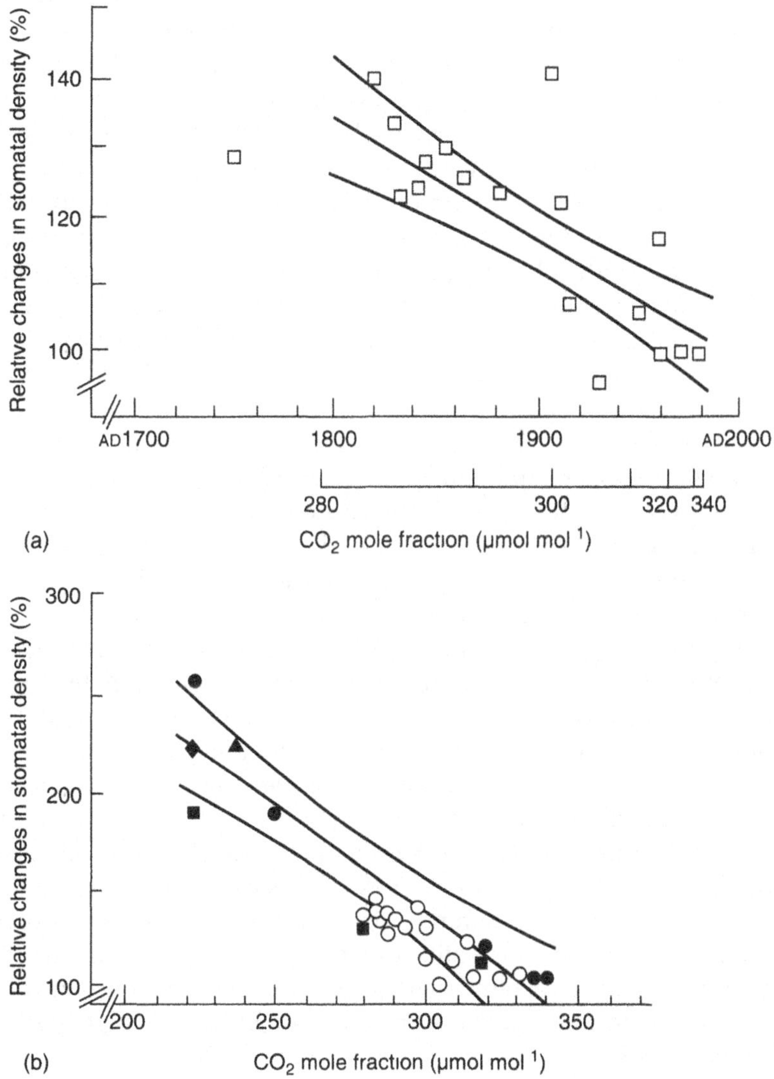

in the CO_2 mole fraction on stomatal density with the putative effects of CO_2 shown in herbarium material (○) Experimental observations were made on *A pseudoplatanus* (●) *Q robur* (◆) *R catharticus* (▲) and *Rumex crispus* (■) Plants were grown in small air tight enclosures for a period of three weeks in a 16 h photoperiod at 18°C a mean leaf water vapour pressure deficit of 1 2 kPa and an irradiance of 373 μmol m ^{2} s ^{1} The night temperature was 7 C The CO_2 mole fractions supplied to the enclosures was controlled with a gas diluter and measured with an IR gas analyser The linear regression for the experimental observations on leaves which had been initiated and developed in the treatments had a slope of 1 12% ± 1 5% μmol ^{1} mol CO_2 (95% confidence limit) (*r* = –0 940) The linear regression coefficient for the herbarium material (○) was –0 61% ± 0 8% μmol ^{1} mol CO (*r* – 0 858) From Woodward (1987)

However, the relationship between stomatal frequency and CO_2 concentration may not be a direct or simple one. Using *Arabidopsis thaliana*, Woodward (1988) found that as CO_2 concentrations increased there was a marked decrease in adaxial stomatal frequency whilst the decrease in abaxial stomatal frequency was smaller. Woodward (1987) also considered that for trees and shrubs the trend of decreasing stomatal frequency with increasing CO_2 concentration does not continue above current CO_2 concentrations (about 365 ppm). Stomatal frequency is not necessarily the best indicator of the changing morphogenesis of stomata. For example, the frequency may decrease as a result of greater expansion or number of epidermal cells. However, stomatal index takes such factors into consideration (see Section 2.2) and, according to Woodward (1987), both stomatal index and frequency decreased as CO_2 concentration increased. Not all studies, however, demonstrate that changing CO_2 concentrations affect stomatal frequencies (e.g. Oberbauer *et al.*, 1985).

If high CO_2 concentrations do decrease stomatal frequency, then this could have important consequences for plant growth in the future since the CO_2 concentration of the Earth's atmosphere is rapidly increasing. Moreover, many experiments show that the water-use efficiency (the ratio of CO_2 uptake against water loss) is increased by increasing CO_2.

As indicated above, although stomatal frequency per unit leaf area may vary, since environmental factors may also change leaf areas the total number of stomata per leaf may remain relatively unchanged. For example, Ciha and Brun (1975) found in soybean leaves that increasing temperature caused decreased stomatal frequency per unit leaf area in the abaxial leaf surface, while the frequency increased in both leaf surfaces with increased light intensity. However, because of changing leaf areas brought about by the different light intensities or temperatures, the stomatal population per leaflet did not differ significantly. In the same study, water stress resulted in a greater stomatal frequency, smaller leaf area and significantly lower stomatal numbers per leaflet than in non-stressed leaves.

2.3 Stomatal patterning

The spacing of stomata in the epidermis is characteristic of the species and can vary greatly. Stomata may be fairly evenly spaced throughout a leaf (e.g. Fig. 2.2A, E, G–J), located in regular rows along the length of a leaf as in most monocotyledons and Gymnosperms (Fig. 2.2C and F), or they may be clustered in patches. For example, in leaves of *Saxifraga stolonifera* (formerly *sarmentosa)* raised patches of stomata sharing a common substomatal chamber are separated by stomatal-free regions

(Fig. 2.2K and L). Raised patches also occurs in the surface of young woody stems of some species and the stem internodes of some other species such as *Coleus blumei* (Turner and Lersten, 1983). In some instances a single, raised stoma is present. Some of the variations in stomatal arrangement between species are illustrated in Fig. 2.2(A–L).

The mechanism determining that stomata are roughly equidistant apart either in rows or generally scattered about the leaf surface is not completely understood. Bünning and Sagromsky (1948) proposed that in dicots stomatal meristemoids inhibited a zone of surrounding neighbouring cells from becoming guard mother cells, possibly via a substance diffusing from them. Others have qualified this idea and assumed that the inhibition zone only affects cells contiguous to the guard mother cells where cell contact and interchange of diffusable substances would most readily control cell initiation events (e.g. Korn, 1981, 1993).

In the monocots, Bünning and Sagromsky (1948) suggested that the placement of stomata is determined by an ordered series of divisions to produce a series of stomatal initials. In this model, cell lineage plays a dominant role in defining, first, which file of cells will differentiate guard cells and, second, the spacing of the guard cells within the file. The simplest pattern arises from polarized and asymmetric division of each cell to give an alternating series of guard cells and epidermal cells, with the guard cells usually sited distal to the leaf base. The guard cells may induce additional divisions in neighbouring cells (Stebbins and Jain, 1960; Stebbins and Shah, 1960). However, Croxdale *et al.* (1992) found that only some aspects of stomatal patterning in *Tradescantia* could be explained by such a cell lineage model. Although formation of stomatal initials was highly irregular, particular sequences of stomata separated by one epidermal cell (termed strings) tended to occur more often than expected. Croxdale *et al.* (1992) suggested such strings might arise if cells in a particular lineage were partially synchronized in their response to an inductive stimulus, triggering formation of stomatal initials. In addition, studies in which cells were killed with a laser beam revealed that events which determine whether a stoma will develop occured very early, near the meristem (Croxdale *et al.*, 1992). Thus, neighbouring cells did not take over the role of a killed cell (Croxdale *et al.*, 1992), unlike in most other differentiating tissues (Sachs, 1991). The lineage model could also not account for the two-dimensional pattern that determines which files of cells would contain stomata.

A finding that may have far reaching consequences has arisen from studies on chimeras in maize, which indicate that guard cells can differentiate from periclinal divisions of underlying mesophyll tissues, followed by 'invasion' of these cells into the epidermis (Sinha and Hake, 1990). Since these cells are derived from a different lineage to the epidermal tissue, the cues for differentiation in this instance must be position dependent rather than based on lineage (Hake and Sinha, 1991).

There is currently little information on the molecular control of stomatal initiation and spacing patterns. However, Zeiger and Stebbins (1972) report that the pleiotropic, wax deficient mutant (*eceriferum-g*) in barley has abnormal stomatal development with a high percentage of double and triple guard cells in rows and abnormal subsidiary cells. Also, in *Arabidopsis* there is a mutant termed 'too many mouths – *tmm*' with clusters of adjoining stomata, two to eight in number, that maintain the normal spacing pattern of a single stoma found in the wild type plants (Yang and Sack, 1993).

Light regulation of stomatal initiation and development may be mediated by phytochrome at least in some species. For example, in mustard (*Sinapis alba* L.) cotyledons, stomata do not fully develop in the dark while exposure to far-red light results in rapid maturation of stomata and expansion of epidermal cells (Kleiber and Mohr, 1963). Also, in *Vigna sinensis* L. leaves, the stomatal index depends on the level of active phytochrome in the plant at the beginning of stomatal differentiation (Schoch *et al.*, 1977).

The control of stomatal spacing often breaks down in tissue culture. In callus tissue and in plantlets cultured in media supplemented with various salts and hormones, stomata often develop in an uncontrolled fashion. For example, when potato plantlets are formed from callus tissue in continuous light and in the presence of an exogenous carbon source, such as sucrose, exceptionally large stomata form (Fig. 2.5). Furthermore, these stomata appear to be locked open and neither darkness nor abscisic acid (ABA) close them. Interestingly, if ABA is added to the culture medium the stomata appear to develop normally (Willmer, Costa and Li, unpublished).

The inability of stomata to close in leaves on plantlets grown in culture has also been observed in roses (Short *et al.*, 1981; Sallanon *et al.*, 1990), apple (Brainerd and Fuchigami, 1982), carnations (Ziv *et al.*, 1987) and *Delphinium* (Santamaria *et al.*, 1993). In roses the lack of closure was attributed to the high levels of K which the guard cells accumulated from the salt-rich culture medium (Short *et al.*, 1981). With micropropagated *Delphinium* plants, Santamaria *et al.* (1993) reported that, although the stomata responded to ABA, CO_2, light and water potential, they failed to close fully.

2.4 Heterogeneity in stomatal characteristics and responses

The causes of a general variability in stomatal size and aperture under identical environmental conditions are expected to be due to a mixture of endogenous and exogenous factors that generate natural heterogeneity in biological systems. Such aspects, including 'patchy' stomatal behaviour are discussed in this section.

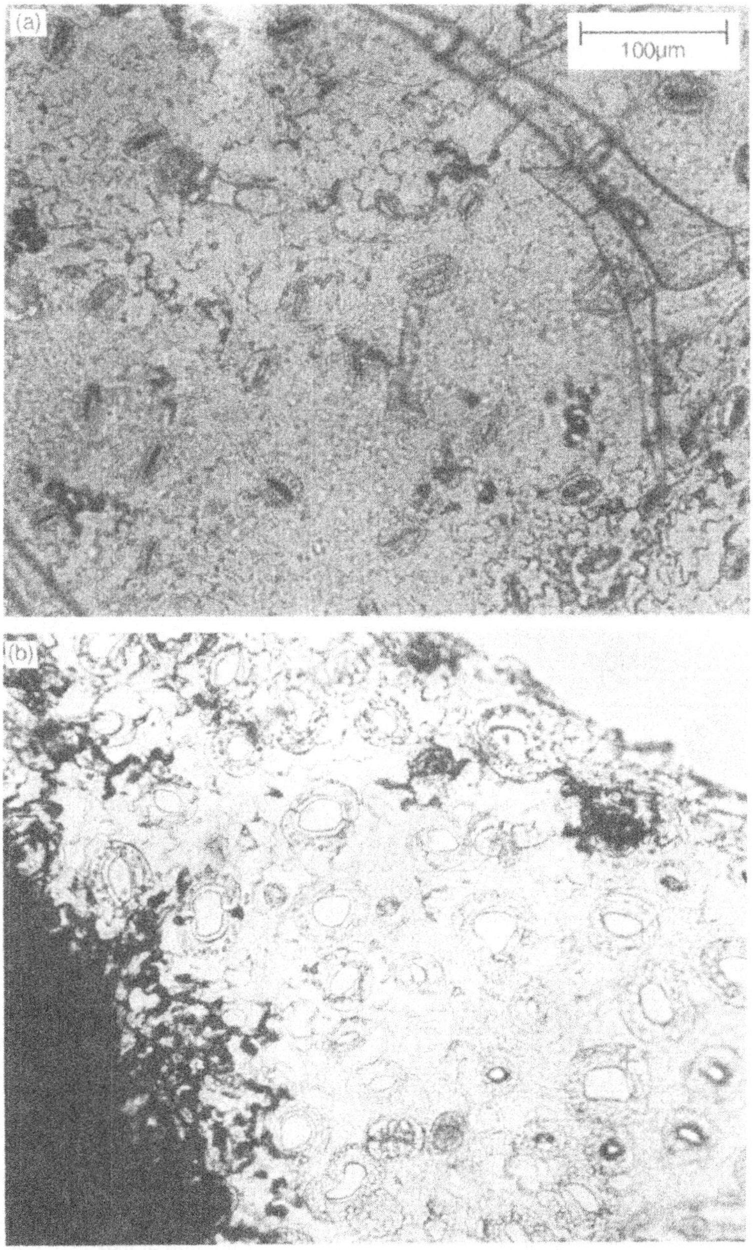

Figure 2.5 (a) Abaxial stomata in leaflets of potato (*Solanum tuberosum*) plants grown in the soil in a greenhouse. (b) Abaxial stomata in leaflets of potato plantlets developed from callus tissue grown in agar containing sucrose as a carbon source and in continuous light. Note the high frequency of very large stomata, many of which have developed abnormally with pores 'locked' open in (b).

2.4.1 Variation in stomatal size

As Table 2.1 indicates, there can be a considerable variation in stomatal dimension even within mature areas of the same leaf. Indeed, in mature leaves of *Arabidopsis thaliana* there are still some developing stomata, and guard mother cells can even be observed (Lascève, unpublished). Figure 2.6(a and b) shows the wide variation in guard cell width and stomatal length in mature leaves of *C. communis*. Generally, however, smaller stomata occur at higher frequency. This results in the total possible pore area for a leaf being similar for most species (see Table 2.1). The pore area of a leaf can reach as much as 5% when the stomata are very wide open, but normally the value remains below 2%. Why there are such large variations in stomatal frequency and guard cell dimensions, and what the functional significance of these variations is, are not fully understood.

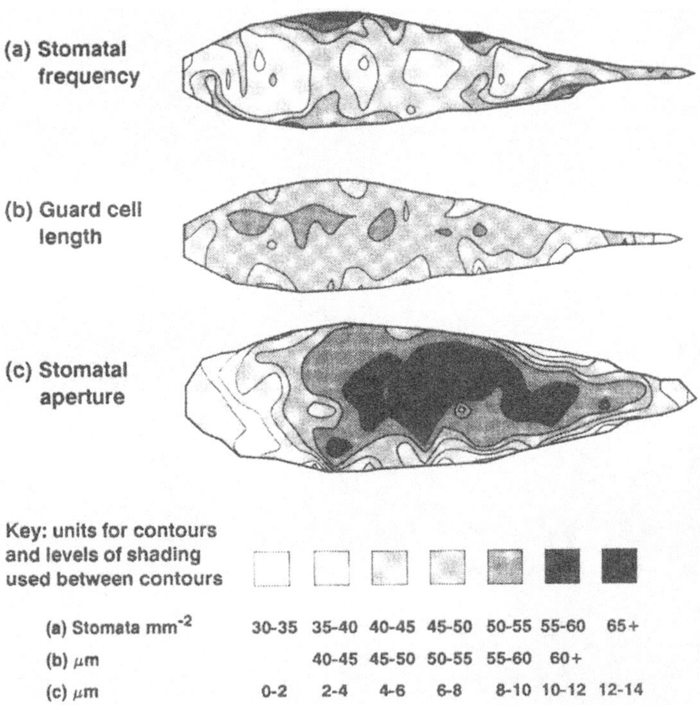

Figure 2.6 Variation in stomatal characteristics over the surface of leaves of *Commelina communis*. (a) Leaf map showing iso-frequency contours for fields of view between veins; (b) leaf map showing iso-size contours for the length of guard cells of closed stomata; and (c) leaf map showing iso-aperture contours for a leaf sampled from a glasshouse at 1300 h. In each case, the contours were produced by a computer program from mean values of samples from over 200 areas of 9 mm^2 marked out on silicone rubber impressions of the entire leaf surface. Maps taken from Smith *et al.* (1989), where full details of methods are given Scale in (a) and (b) the original leaf length was 115 mm and in (c) it was 87 mm

The ploidy level of plants also influences stomatal characteristics: as ploidy increases stomatal frequency generally decreases while guard cell size increases (e.g. Laptev *et al.*, 1976). Indeed, guard cell size has been used to predict the haploid level of primitive angiosperms and the extent of polyploidy in present-day angiosperms. Since cell size has been correlated with DNA content and thus chromosome number, Masterton (1994) estimated the ploidy level of extinct angiosperms by measuring dimensions of guard cells in fossilized specimens. It was concluded that the primitive haploid number of chromosomes was between seven and nine per cell and that most angiosperms (about 70%) have polyploidy in their history.

2.4.2 Variations of stomatal aperture within a leaf

Stomatal pores are remarkably symmetrical, suggesting that there is a high degree of co-ordination between the daughter guard cells. For graminaceous stomata this is inevitable because large pores exist in the bulbous heads of the guard cells joining the protoplasts of a guard-cell pair. In contrast, there is considerable variation between pore aperture and other stomatal characteristics over a single leaf (Weyers *et al.*, 1995) and even sometimes between adjacent stomata.

There may be three types of patterns in the variation of stomatal aperture which are observed within a single leaf. First there is the natural variation of stomatal aperture across a leaf which may give a large scatter of pore widths depending on the species (e.g. Laisk *et al.*, 1980; Spence, 1987; van Gardingen *et al.*, 1989). Secondly, superimposed on this 'noise', there may be a patchy pattern of aperture in which areas of a leaf may have much wider open stomata than other areas of guard cells with similar apertures. Thirdly, pore widths may change gradually in trends covering larger regions of the leaf such that stomatal aperture at leaf margins or along the length of the leaf lamina may be considerably different from apertures in other parts of a leaf (e.g. Smith *et al.*, 1989) (Fig. 2.6C).

Some of the variation in apertures can be linked to variation in stomatal size and frequency. Thus, higher stomatal frequencies (often occurring towards the leaf tip in monocots, but see Section 2.2.1) are associated with smaller stomata and smaller apertures. Nevertheless it is often observed that there is a wide variation in stomatal aperture in parts of a leaf with the same stomatal frequency and dimensions. This can be observed in intact leaves (e.g. Slavik, 1961; Laisk *et al.*, 1980; Kappen *et al.*, 1987; van Gardingen *et al.*, 1989) and in epidermal strips (Spence, 1987; Smith *et al.*, 1989), as all who have worked with such material will be aware. Indeed, in intact leaves and more occasionally in epidermal strips, a closed stoma can be situated next to a wide open one. Kappen *et al.* (1987), for example, found a wide spectrum of stomatal apertures in a leaf of *V. faba* which were measured using video microscopy although all cells responded in the same direction and with similar magnitudes to changes in humidity and CO_2.

In general, stomata tend to have lower apertures and close more readily at the leaf margins compared to those in the leaf centre, although stomata near main veins also tend not to open widely (e.g. Smith *et al.*, 1989). The causes for this are probably related to the different rates of water availability and loss in these different tissue areas (e.g. Hashimoto *et al.*, 1984). Also, stomata near veins often appear restricted in movement due to the rigidity the veins confer on the epidermis.

'Patchy' stomatal opening is often more pronounced as a result of sudden changes in environmental conditions and stomata within an area respond in concert and apparently independent from neighbouring areas (Mott *et al.*, 1993). It has been suggested that a specific leaf anatomy predisposes this pattern of behaviour. Terashima *et al.* (1988), for example, observed 'patchy' opening in heterobaric leaves. Heterobaric leaves are those which have bundle sheath extensions which extend to the epidermis so that the mesophyll is separated laterally into patches, as opposed to homobaric leaves which lack a bundle sheath extension (see Larcher, 1980). A general view is that heterobaric leaves also have stomata approximately opposite each other on both leaf surfaces between which run air-space channels (see Meidner and Mansfield, 1968). Another suggestion is that the 'patches' of stomata are in a transitory stage 'hunting' some new condition in response to sudden environmental changes and eventually an optimized aperture will be reached by all stomata (see Cardon *et al.*, 1994). Not all species exhibit 'patchy' responses, however, and the extent to which they occur may depend critically on the severity of the imposed stress and the speed of application (Gunasekera and Berkowitz, 1992).

Nevertheless, the realization that stomata exhibit non-uniform and heterogeneic behaviour has led to important modifications of predictive models of transpiration and photosynthesis. A most notable example is the demonstration that 'patchy' stomatal behaviour can account for the previously erroneous conclusion that ABA affects photosynthesis directly rather than acting via stomatal closure (e.g. Downton *et al.*, 1988; Terashima *et al.*, 1988) (also see Chapter 7).

References

Adams, M.J. (1975) The development of lenticels on potato tubers. *Ann. Appl. Biol.*, **79**, 264–273.

Blanke, M. and Bonn, F.L. (1985) Spaltöffnungen, Fruchtoberflache und Transpiration wachsender Apelfrüchte der Sorte 'Golden Delicious'. *Erwebsobstbau*, **27**, 139–143.

Brainerd, K.E. and Fuchigami, H. (1982) Stomatal functioning of *in vitro* and greenhouse apple leaves in darkness, mannitol and CO_2. *J. Exp. Bot.*, **33**, 388–392.

Bristow, J.M. and Looi, A.-S. (1968) Effects of carbon dioxide on the growth and morphogenesis of Marsilea. *Am. J. Bot.*, **55**, 884–889.

Bornman, C.H. (1972) *Welwitschia mirabilis:* paradox of the Namib desert. *Endeavour*, **31**, 95–99.

Bünning, E. and Sagromsky, H. (1948) Die Bildung des Spaltöffnunsmusters in der Blattepidermis. *Z. Naturforsch.*, **38**, 203–216.

Cardon, Z.G., Mott, K.A. and Berry, J.A. (1994) Dynamics of patchy stomatal movements, and their contribution to steady-state and oscillating stomatal conductance calculated with gas-exchange techniques. *Plant Cell Environ.*, **17**, 995–1007.

Carpenter, S.B. and Smith, N.D. (1975) Stomatal distribution and size in southern Appalachian hardwoods. *Can. J. Bot.*, **53**, 1153–1156.

Chaffey, N.J. (1982) Presence of stomata-like structures in the ligule of *Agrostis gigantea* Roth. *Ann. Bot.*, **50**, 717–720.

Charlton, W.A. (1990) Differentiation in leaf epidermis of *Chlorophytum comosum.* Baker. *Ann. Bot.*, **66**, 567–578.

Ciha, A.J. and Brun, W.A. (1975) Stomatal size and frequency in soybeans. *Crop Sci.*, **15**, 309–313.

Cole, D.F. and Dobrenz, A.K. (1970) Stomatal density of alfalfa (*Medicago sativa* L.). *Crop Sci.*, **10**, 20–24.

Cooper, C.S. and Qualls, M. (1967) Morphology and chlorophyll content of shade and sun leaves of two legumes. *Crop Sci.*, **7**, 672–673.

Croxdale, J., Smith, J., Yandell, B. and Johnson, B. (1992) Stomatal patterning in *Tradescantia*: an evaluation of the cell lineage theory. *Dev. Biol.*, **149**, 158–167.

Dale, J.E., Felippe, G.M. and Fletcher, G.M. (1972) Effect of shading the first leaf on growth of barley plants. *Ann. Bot.*, **36**, 385–395.

Downton, W.J.S., Loveys, B.R. and Grant, W.J.R. (1988) Stomatal closure fully accounts for the inhibition of photosynthesis by abscisic acid. *New Phytol.*, **108**, 263–266.

Edwards. D., Davies, K.L. and Axe, L. (1992) A vascular conducting strand in the early land plant *Cooksonia*. *Nature*, **357**, 683–685.

Foster, J.R. and Smith, W.K. (1986) Influence of stomatal distribution on transpiration in low wind environments. *Plant Cell Environ.*, **9**, 751–759.

Friend, D.J.C. and Pomeroy, M.E. (1970) Changes in cell size and number associated with effects of light intensity and temperature on the leaf morphology of wheat. *Can. J. Bot.*, **48**, 85–90.

van Gardingen, PR , Jeffree, C E and Grace, J (1989) Variation in stomatal aperture in leaves of *Avena fatua* L observed by low-temperature scanning electron microscopy *Plant Cell Environ* , **12**, 887–898

Gunasekera, D and Berkowitz, G A (1992) Heterogenous stomatal closure in response to leaf water deficits is not a universal phenomenon *Plant Physiol* , **98**, 660–665

Hake, S and Sinha, N (1991) Genetic analysis of leaf development *Oxford Surv Plant Mol Cell Biol* , **7**, 187–254

Hashimoto, Y, Ino, T , Kramer, PJ *et al* (1984) Dynamic analysis of water stress of sunflower leaves by means of a thermal image processing system *Plant Physiol* , **76**, 266–269

Heichel, G (1971) Stomatal movements, frequencies, and resistances in two maize varieties differing in photosynthetic capacity *J Exp Bot* , **22**, 644–649

Kappen, L , Andresen, G and Losch, R (1987) *In situ* observations of stomatal movements *J Exp Bot* , **38**, 126–141

Keeley, J E , Osmond, C B and Raven, J A (1984) *Stylites*, a vascular land plant without stomata absorbs CO_2 via its roots *Nature*, **310**, 694–695

Kleiber, H and Mohr, H (1963) Der Einfluss sichtsaber Strahlung auf die Stomata-Bildung in der Epidermis der Kotyledonen von *Sinapis alba* L *Z Bot* , **52**, 78–85

Korn, R W (1981) A neighboring-inhibition model for stomate patterning *Dev Biol* , **88**, 115–120

Korn, R W (1993) Evidence in dicots for stomatal patterning by inhibition *Int J Plant Sci* , **154**, 367–377

Kubinova, L (1991) Stomata and mesophyll characteristics of barley leaf as affected by light stereological analysis *J Exp Bot* , **42**, 995–1001

Laisk, A (1983) Calculation of leaf photosynthetic parameters considering the statistical distribution of stomatal apertures *J Exp Bot* , **34**, 1627–1635

Laisk, A , Oja, V and Kull, K (1980) Statistical distribution of stomatal apertures of *Vicia faba* and *Hordeum vulgare* and the Spannungsphase of stomatal opening *J Exp Bot* , **31**, 49–58

Laptev, YP, Makarov, PP, Glazova, M V *et al* (1976) Stomata and pollen as indicators of the ploidy of plants *Genetika* (translated from Russian), **12**, 47–55

Larcher, W (1980) *Physiological Plant Ecology*, 2nd edn, Springer, Berlin, p 252

Lefebvre, D D (1985) Stomata on the primary root of *Pisum sativum* L *Ann Bot* , **55**, 337–341

Lele, K M and Walton, J (1960–61) Contributions to the knowledge of *Zosterophyllum myretonianum* Penhallow from the lower old red sandstone of Angus *Trans Roy Soc Ed* , **64**, 469–477

Lichtenthaler, H K (1985) Differences in morphology and chemical composition of leaves grown at different light intensities and qualities, in *Control of Leaf Growth*, (eds WJ Baker, WJ Davies and C K Ong), Cambridge University Press, Cambridge, pp 201–221

Marx, A and Sachs, T (1977) The determination of stomatal pattern and frequency in *Anagallis Bot Gaz* , **138**, 385–392

Masterson, J (1994) Stomatal size in fossil plants evidence for ploy-ploidy in majority of angiosperms *Science*, **264**, 421–424

Meidner, H and Mansfield, TA (1968) *Physiology of Stomata*, McGraw-Hill, London

Miranda, V, Baker, N R and Long, S P (1981) Anatomical variation along the length of the *Zea mays* leaf in relation to photosynthesis *New Phytol* , **88**, 595–605

Miskin, K E and Rasmusson, D C (1970) Frequency and distribution of stomata in barley *Crop Sci* , **10**, 575–578

Mott, K A , Cardon, Z G and Berry, J A (1993) Asymmetric patchy stomatal closure for the two surfaces of *Xanthium strumarium* L leaves at low humidity *Plant Cell Environ* , **16**, 25–34

Mott, K A , Gibson, A C and O'Leary, J W (1982) The adaptive significance of amphistomatous leaves *Plant Cell Environ* , **5**, 455–460

Oberbauer, S F , Strain, B R and Fetcher, N (1985) Effect of CO_2 enrichment on seedling physiology and growth of two tropical species *Physiol Plant* , **65**, 352–356

Parkhurst, D F (1978) The adaptive significance of stomatal occurrence on one or both surfaces of leaves *J Ecol* , **66**, 367–383

Penfound, WT (1931) Plant anatomy as conditioned by light intensity and soil moisture *Am J Bot* **18**, 197–209

Rutter, J C and Willmer, C M (1979) A light and electron microscopic study of the epidermis of *Paphiopedilum* spp with emphasis on stomatal ultrastructure *Plant Cell Environ* , **2**, 211–219

Sachs, T (1991) *Pattern Formation in Plant Tissues*, Cambridge University Press, Cambridge

Salisbury, E J (1927) On the causes and ecological significance of stomatal frequency with special reference to woodland flora *Phil Trans Roy Soc Lond* , *Ser B*, **216**, 1–65

Sallanon, H , Laffray, D and Coudret, A (1991) Ultrastructure and functioning of guard cells of *in vitro* cultured rose plants *Plant Physiol Biochem* , **29**, 333–339

Santamaria, J M , Davies, WJ and Atkinson, C J (1993) Stomata on micropropagated delphinium plants respond to ABA, CO_2, light and water potential but fail to close fully *J Exp Bot* , **44**, 99–107

Schoch, P-G (1978) Differentiation numerique des stomates du *Vigna sinensis* L et de quelques autres especes Doctoral thesis, INRA, Stations de Bioclimatologie de Guadeloupe et d'Avignon-Montfavet

Schoch, P-G, Lecharny, A, Jaques, R and Zinsou, C (1977) Phytochrome et indice stomatique des feuilles du *Vigna sinensis* L *C R Acad Sci Paris*, **285**, 877–879

Schoch, P-G, Zinsou, C and Sibi, M (1980) Dependence of the stomatal index on environmental factors during stomatal differentiation in leaves of *Vigna sinensis* L *J Exp Bot*, **31**, 1211–1216

Short, K C, Price, L and Roberts, A V (1981) Micropropagation of roses, in *The Rose Annual* (ed J Harkness), The Royal National Rose Society, London

Sinha, N and Hake, S (1990) Mutant characters of *Knotted* maize leaves are determined in the innermost tissue layers *Dev Biol*, **141**, 203–210

Slavik, B (1963) The distribution pattern of transpiration, water saturation deficit, stomata number and size, photosynthetic and respiration rate in the area of the tobacco leaf blade *Biol Plant*, **5**, 143–153

Smith, WK (1981) Temperature and water relations patterns in subalpine understorey plants *Oecologia*, **48**, 353–359

Smith, S, Weyers, J D B and Berry, WG (1989) Variation in stomatal characteristics over the lower surface of *Commelina communis* leaves *Plant Cell Environ*, **12**, 653–659

Spector, WS (1956) In *Handbook of Biological Data XXXVI*, WB Sanders Co, Philadelphia, PA, p 146

Spence, R D (1987) The problem of variability in stomatal responses, particularly aperture variance, to environmental and experimental conditions *New Phytol*, **107**, 303–315

Stebbins, G L and Jain, S K (1960) Developmental studies of cell differentiation in the epidermis of monocotyledons *Dev Biol*, **2**, 409–426

Stebbins, G L and Shah, S S (1960) Developmental studies of cell differentiation in the epidermis of monocotyledons *Dev Biol*, **2**, 477–500

Terashima, I, Wong, S-C, Osmond, C B and Farquhar, G D (1988) Characterisation of non-uniform photosynthesis induced by abscisic acid in leaves having different mesophyll anatomies *Plant Cell Physiol*, **29**, 385–394

Turner, N C (1970) Responses of adaxial and abaxial stomata to light *New Phytol*, **69**, 647–653

Turner, N C and Begg, J E (1973) Stomatal behaviour and water status of maize, sorghum and tobacco under field conditions I At high soil water potential *Plant Physiol*, **51**, 31–36

Turner, G W and Lersten N R (1983) Raised stomatal clusters on *Coleus* (Lamiaceae) stems *Am J Bot*, **70**, 975–977

Weyers, J D B, Lawson, T and Peng, Z Y (1995) Variation in stomatal characteristics at the whole-leaf level, in *Scaling Up*, (eds P Van Gardingen, G Foody and P Curran), Cambridge University Press, Cambridge, pp 65–8

Wild, A and Wolf, G (1980) The effect of different light intensities on the frequency and size of stomata, the size of cells, the number, size and chlorophyll content of chloroplasts in the mesophyll and the guard cells during the ontogeny of primary leaves of *Sinapis alba* *Z Pflanz*, **97**, 325–342

Willmer, C M (1993) The evolution, structure and functioning of stomata *Bot J Scot*, **46**, 33–445

Woodward F I (1987) Stomatal numbers are sensitive to increases in CO_2 from preindustrial levels *Nature*, **327**, 617–618

Woodward, F I (1988) The responses of stomata to changes in atmospheric levels of CO_2 *Plants Today*, **1**, 132–135

Woodward, F I and Bazzaz, F (1988) The responses of stomatal density to CO_2 partial pressure *J Exp Bot*, **39**, 1771–1781

Yang, M and Sack, F (1993) An Arabidopsis mutant with multiple stomata *Plant Physiol Suppl*, **102**, 122

Zeiger, E and Stebbins, G L (1972) Developmental genetics in barley a mutant for stomatal development *Am J Bot*, **59**, 143–148

Ziegler, H (1987) The evolution of stomata, in *Stomatal Function*, (eds E Zeiger, G D Farquhar and I R Cowan), Stanford University Press Stanford, CA, pp 29–58

Ziv, M, Schwartz, A and Fleminger, D (1987) Malfunctioning stomata in vitreous leaves of carnation (*Dianthus caryophyllus*) plants propagated *in vitro*, implications for hardening *Plant Sci*, **52**, 127–134

3 The structure and development of stomata

3.1 Anatomy and morphology of the leaf epidermis

The epidermis is the outermost cell layer or layers of the leaf lamina and serves to protect against excessive and uncontrolled water loss from the leaf It also acts as a physical barrier, reducing infection by fungal pathogens and bacteria, and minimizes mechanical damage to the mesophyll tissue The epidermal tissue is not normally photosynthetic, but can significantly affect the radiation received by the underlying tissues Wax secretion (e g Mulroy, 1979) and epidermal structures, such as trichomes and salt glands (e g Mooney *et al* , 1977), can result in large changes in leaf spectral characteristics, such as increased leaf reflectance, particularly of UV wavelengths The epidermis is also the major site of absorption of UV radiation due mainly to a range of different flavenoid and phenolic pigments which are contained in the cells (Robberecht and Caldwell, 1978, Robberecht *et al* , 1980) and waxes on the surface of the epidermis Figure 3 1 shows the absorption spectra of ethanol extracts of epidermal and mesophyll tissues from *Commelina* and illustrates the high UV absorption of the former tissue relative to the latter In many cases the upper and lower epidermes do not have the same spectral characteristics, with higher UV absorbance from the adaxial surface (Donkin and Martin, 1981, Weissenbock *et al* , 1986, Shimazaki *et al* , 1988) The epidermis attenuates transmission of photosynthetically active radiation (PAR) to a small degree, although scattering and reflection from the abaxial epidermis also prevents the light from escaping once it has entered the leaf (Lin and Ehleringer, 1983) This may increase the overall efficiency of light harvesting and green leaves typically absorb about 85% of the sunlight in the photosynthetically active waveband, 400–700 nm (e g Ehleringer, 1981) In contrast, the epidermis absorbs remarkably little IR radiation below a wavelength of 2 μm, but is remarkably efficient in absorbing or emitting longer wavelengths than this The wavelength distribution for solar radiation tails off at wavelengths greater than 1 μm and is very low above 2 μm

Thus, in practice, the leaf absorbs little of the incident thermal energy below 2 μm, but can radiate a significant amount of thermal energy at longer wavelengths. This can be a major factor in the energy balance of the leaf (Nobel, 1991). In certain cases, the epidermis may be an important site of light perception entraining circadian rhythms and photoperiodic responses (e.g. Schwabe, 1968; Mayer *et al.*, 1973).

The morphology and anatomy of the epidermal layer varies greatly between species and the epidermis may contain a variety of cell types, such as trichomes, epidermal cells (including cork cells, silica cells, bulliform cells and long cells in grass species), subsidiary cells and guard cells. Cells in the epidermal layer comprise a relatively small percentage of the total in a leaf, although in several species of cereals they can be as much as 10% of the total cells of the flag leaf (the first, uppermost leaf) (Jellings and Leech, 1982). Usually, guard cells comprise probably less than 2% of the volume of the epidermal layer. The anatomy and morphology of a variety of leaf types is shown in Fig. 3.2(A–F), and enables a comparison of the sizes and shapes of different cell types and of stomatal locations to be made. Further description of the form and function of the various cell types can be found in Esau (1977) or Fahn (1982).

3.1.1 Cuticle

The epidermal layer is covered by a layer of cuticle which varies in thickness both between different species and within an individual species

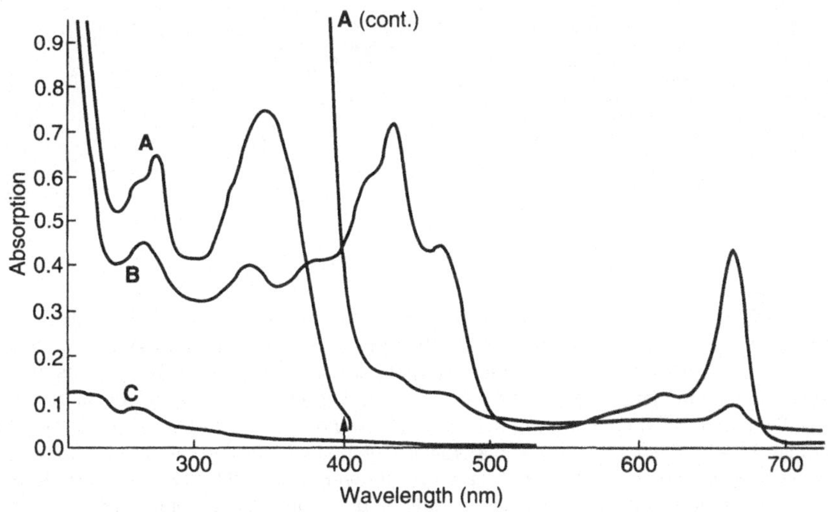

Figure 3.1 Absorption spectra of ethanol extracts of epidermal tissue (A) and mesophyll tissue (B) of *Commelina communis* and an ethanol blank (C). To the left of the arrow the epidermal tissue extract was diluted × 5.

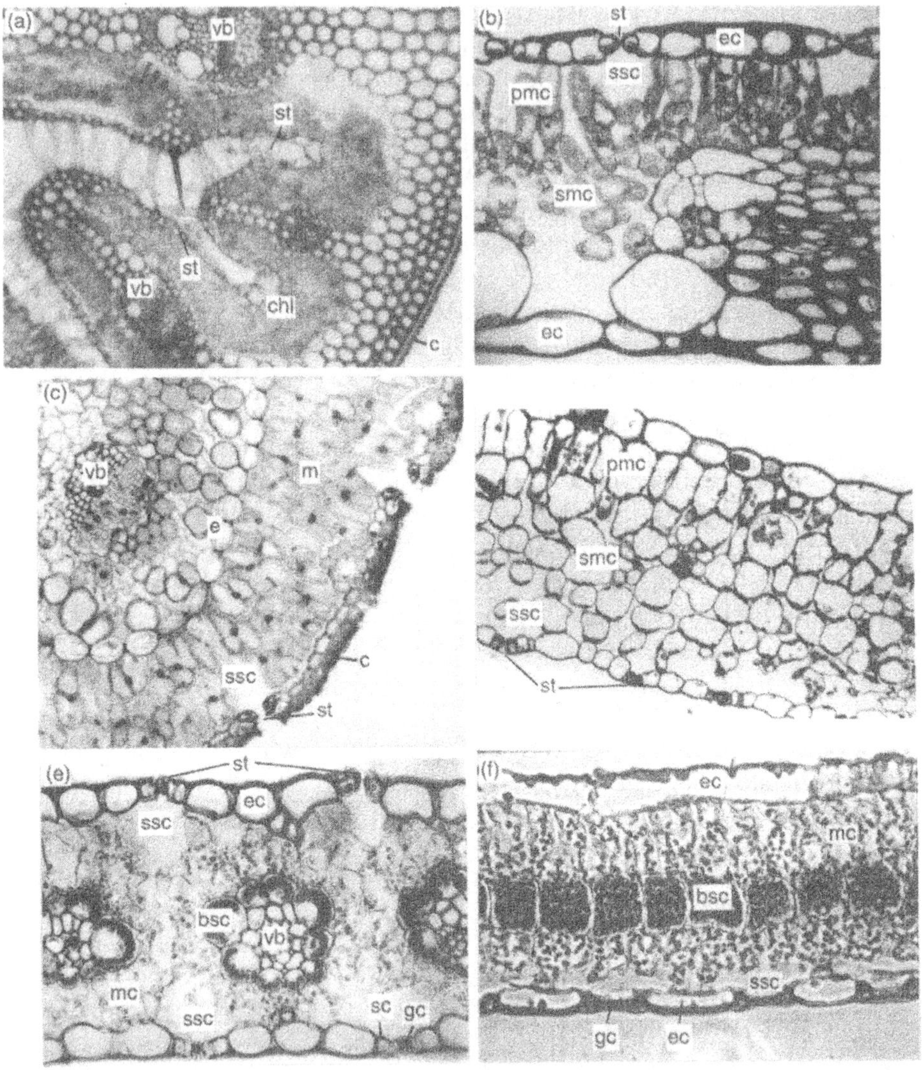

Figure 3.2 Sections of leaves from (a) marram grass (*Ammophila arenaria*) with stomata located at the bottom of deep grooves of the upper surface; (b) peanut (*Arachis hypogea*) showing stomata on the upper surface (stomata are present at lower frequencies on the lower surface) and large water-filled cells below the spongy mesophyll layer; (c) Scots pine (*Pinus sylvestris*) showing the thick cuticularized epidermis and sunken stomata; (d) geranium (*Pelargonium zonale*) with a typical C3-type leaf structure; (e and f) transverse and longitudinal sections of maize (*Zea mays*) showing a typical C4-type leaf structure and graminaceous-type stomata. Key: c, cuticle; gc, guard cell; ec, epidermal cell; mc, mesophyll cell; bsc, bundle sheath cell; vb, vascular bundle; sc, subsidiary cell; st, stoma; ssc, substomatal cavity; pmc, palisade mesophyll cell; smc, spongy mesophyll cell; chl, chlorenchyma; m, mesophyll; e, endodermis.

depending on the growth conditions of the plants (Martin and Juniper, 1970, Holloway, 1982) The cuticle is composed of large polymers of hydroxylated C16 and C18 fatty acids forming a cutin matrix, interspersed amongst cellulose microfibrils, oligosaccharides and wax (Kolattukudy, 1981) The precise structure, chemistry and synthesis of such a complex mixture has been difficult to elucidate Morphologically, the cuticle has been split up into zones, each zone being slightly different from the other in terms of their staining pattern and appearance in electron micrographs (e g Holloway, 1982) The lower zone impregnates the outer cellulose cell wall layers, while wax of varying thickness is deposited on the outer surface of the cuticle (Martin and Juniper, 1970) Guard cells tend to have less wax deposits than neighbouring epidermal cells (Juniper, personal communication), but a thicker cuticle (Palevitz, 1981a) Equally important, however, are observations of regions of localized guard cell wall covered by a very thin cuticle in some species, such as *Quercus* and *Pinus* (e g Appleby and Davies, 1983) These may act as sites for peristomatal transpiration and provide a means to sense humidity changes (see Chapter 6)

It is considered by many that a cuticular layer completely surrounds cell wall surfaces in direct contact with air Thus, the cuticle extends from the leaf surface through stomatal pores and then, as a very thin layer, covers walls exposed to air in the substomatal cavity and other leaf air spaces (Martin and Juniper, 1970) However, this is a controversial topic and Nonami *et al* (1990), for example, observed that in *Tradescantia* the cuticle extended from the outside of the leaf into the stomatal cavity to cover the guard, subsidiary and some epidermal walls but not mesophyll cell walls

Humid conditions may cause the cuticle to swell and become more permeable to water and water may even be excreted through microscopic pores, particularly above hydathodes, by a process known as guttation

The rate of diffusion of water vapour through the cuticle is normally very low and contributes 1–5% of the flux through fully open stomata (see Jones, 1993)

3.1.2 Trichomes

Trichomes show great variation in form and structure, from unicellular to multicellular and either secretory or non-secretory (see Esau, 1977) Their purpose may be to excrete unwanted materials from leaves, such as NaCl which may accumulate in halophytes, or to secrete aromatic compounds either to attract pollinators or distract herbivores Trichomes may also decrease water loss from leaves by increasing the boundary layer resistance Some authorities, however, consider that surface appendages and protruberances contribute little to the boundary layer resistance because they are simply not large enough Thus values for the thickness of the boundary layer vary from about 50 µm for a small leaf in high winds to 10 mm for a large leaf in relatively still air (Nobel, 1991) Hence trichomes will only add a significant contribution to the boundary layer at higher wind speeds (see Chapter 5 for further details)

In certain metabolic aspects trichomes may resemble guard cells (Willmer *et al* , 1987), they are often highly active cells, having a dense cytoplasmic content with many mitochondria and often accumulate high concentrations of K

3.1.3 Epidermal cells

Epidermal cells are the most numerous and usually the largest cell type within the epidermal layer, although they show considerable variation in size and shape They are highly vacuolate and the peripheral cytoplasm contains a low frequency of organelles In the grasses, however, there are more specialized epidermal cells called cork cells and silica cells, which are relatively small and have dense cytoplasmic contents Chloroplasts are normally absent or occur at a very low frequency, with the notable exception of ferns, where they are very abundant and presumably contribute to net photosynthesis The vacuoles of epidermal cells may often contain large numbers of inclusions, such as the calcium oxalate crystals found in *Commelina communis*, or anthocyanins, flavonoids and alkaloids

The functions of the epidermal cells are many and varied They may act as depositories for unwanted material such as excess calcium or as a storage site for water The large watery vacuoles also contribute to the spectral properties discussed in Section 3 1 The thicker, cuticularized outer walls of the epidermal cells also resist attack by insects and other pests and pathogens as well as preventing excessive water loss The bulliform epidermal cells prevalent in grasses are large, thin-walled and highly vacuolated Their exact function is not certain but they may be involved in unrolling developing leaves, in the opening and closing movements of mature leaves, or in water storage

3.1.4 Subsidiary cells

Subsidiary cells (also known as accessory cells) are specialized epidermal cells which are different in size and shape from other epidermal cells, and lie adjacent to the guard cells or in close proximity where more than one subsidiary cell exists Although subsidiary cells have usually been defined on a morphological basis, over the years there have been debates about whether they should be defined on an ontogenetical basis. This is because some subsidiary cells originate from the same protodermal cell that the guard cells develop from, while in other cases they develop completely independently from an adjacent protodermal cell (see Section 3.2).

In some species subsidiary cells lie above the guard cells, as in pine (see Fig. 3.6A) and *Equisetum* (see Fig. 3.6D). Subsidiary cells are more complex in terms of structure and function than epidermal cells. They are usually smaller and contain a more dense cytoplasmic content with a greater frequency of cell organelles. They do not normally contain chloroplasts, anthocyanins or crystalline inclusions. In the grasses and some other species subsidiary cells are intimately involved in the stomatal mechanism both mechanically (see Chapter 4) and as repositories for ions (see Chapter 8).

3.1.5 Guard cells

Guard cells are not only the most complicated cells biochemically and anatomatically within the epidermal layer, but are also unique amongst the cells of the plant body. They possess a specialized set of metabolic pathways designed to bring about rapid changes in osmotic potential within the cells when stimulated by relatively minor changes in the external environment. Guard cells have a high metabolic activity relative to most other cell types and they contain an abundance of mitochondria (see Section 3.3) and protein synthesizing machinery (Fig. 3.3).

Guard cells are relatively small cells, but there is considerable variation in their size between species. The dimensions of the outer limits of the walls may vary from under 10 to almost 80 μm in length and from a few micrometres to about 50 μm in width (see Chapter 2, Table 2.1), though the dimensions of width may vary according to the stomatal aperture (see Chapter 4). Table 3.1 gives some other dimensions and characteristics of guard cells and compares them with values from other cell types of the same species.

The unique morphology and the specialized function of guard cells are characteristic of 'terminally differentiated' cells that are no longer totipotent and able to de-differentiate into any other cell type or structure. Thus, guard cell protoplasts (GCP) from *Vicia faba* remained viable under conditions that minimized swelling and bursting for up to 40 days with some wall formation, but no cell division (Herscovich *et al.*, 1992). Recently, however, GCP of *Nicotiana glauca* (Graham) have been shown to divide in culture, synthesize new cell walls, form callus tissue after about 4–5 weeks incubation in an appropriate complex medium (Cupples *et al.*, 1991) and, ultimately, generate whole plants (Sahgal *et al.*, 1993).

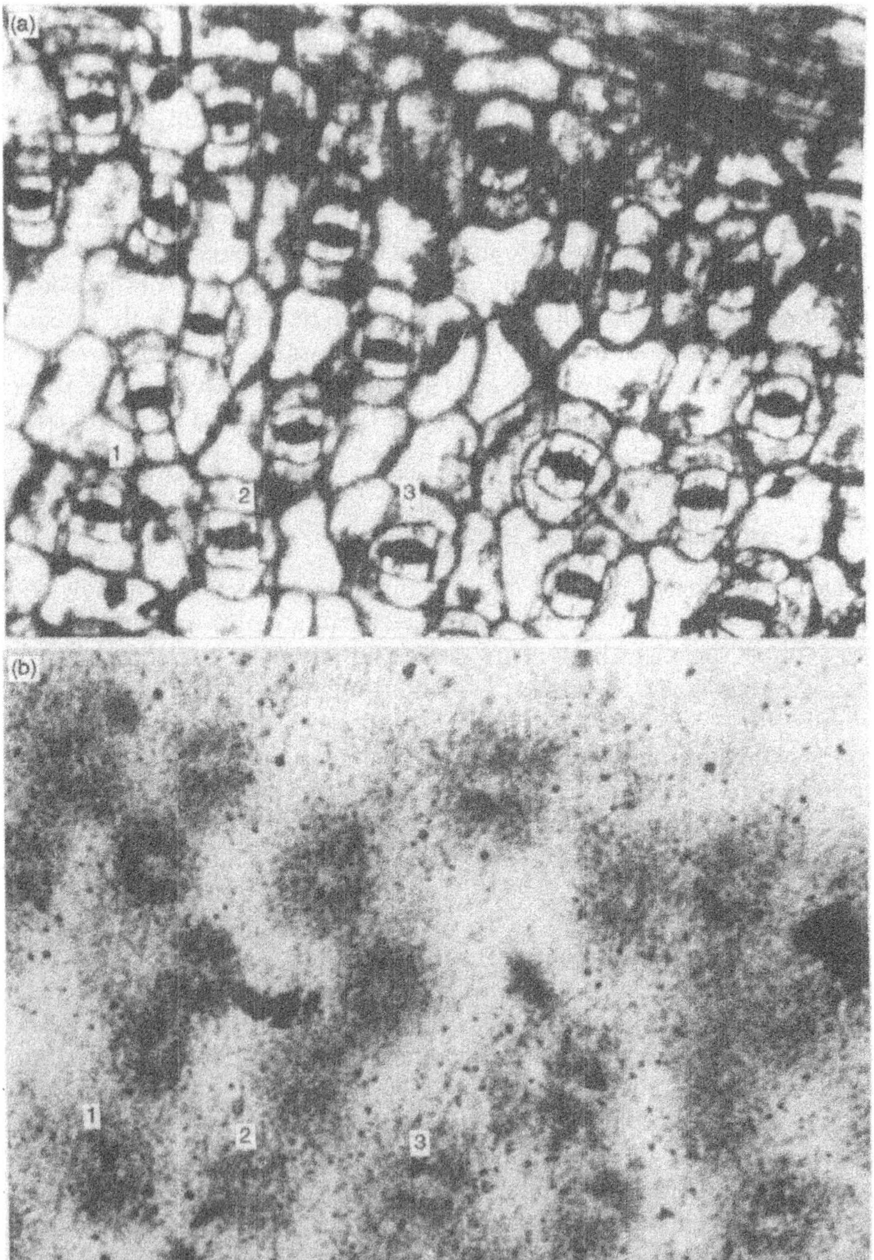

Figure 3.3 (a) Freeze dried epidermis of *Commelina communis* which has been exposed to [^{3}H] leucine. (b) Autoradiograph of the tissue after extraction of all ethanol soluble material. The blackened areas which appear over the stomata (stomata 1, 2, 3 correspond to blackened areas 1, 2, 3) represent location of high protein synthesis. From Willmer (1983).

Table 3.1 Some characteristics of guard cells, palisade cells and spongy mesophyll cells of a variety of species

Cell type	Species			
	Commelina cyanea	*Vicia faba*	*Commelina communis*	*Pisum sativum*
Guard cell				
Protoplast volume (pl) at				
varying stomatal apertures*		2 5–5 0[e], 5 0[i]	4–9[m]	1 1–2 2[k]
stoma open to 10 μm	7 2[a]	2 8[a]		
stoma closed (0 μm)	3 2[a]	1 5[a]		
wall volume (pl)		5 0[i]		
Dry weight (ng)		3 0[c]		
stoma open to 10 μm		3 4[b]		
stoma closed		2 9[b]		
Chloroplasts/cell	10[a]	8–10[a,c]	10[g]	
Chloroplast volume (fl)	17[a]	12[a]		
Chlorophyll (pg)/chloroplast		0 3–0 5[c]	0 47[g]	
Chlorophyll (pg)/cell		2[o], 2 5[j], 1[h], 2 4–5 0[c], 10 7[n]	3 0[l], 3 3[f], 4 8[g]	0 9[k], 0 75[h]
Chlorophyll *a/b* ratio		2 76[e]	1 78[g], 2 6[f]	
Soluble protein (pg)/cell		166[e]	266[g], 152[f]	
Soluble protein chlorophyll		80[o], 35 1[c]	46 2[f], 54 8[g]	
DNA (pg)/cell			1 69[g]	
Palisade cell				
Protoplast volume (pl)	57[a]	46[a]		
Dry weight (ng)		11 6[d]		
Chloroplasts/cell	41[a]	59[a]	64[g†]	
Chloroplast volume (fl)	64[a]	33[a]		
Chlorophyll (pg)/chloroplast		2 1[c]	3 59[g†]	
Chlorophyll (pg)/cell		158[o†], 71[j†], 184[n]	120[l†], 230[g†], 127[f†]	76[k†]
Chlorophyll *a/b* ratio			2 83[g†], 2 4[f†]	
Soluble protein (pg)/cell			2622[g†], 1499[f†]	
Soluble protein chlorophyll		20[e,j]	11 4[g†], 11 8[f†]	
DNA (pg)/cell			14 5[g†]	
Spongy mesophyll cell				
Protoplast volume (pl)	48[a]	42[a]		
Dry weight (ng)		13 8[d]		
Chloroplasts/cell	28[a]	24[a]		
Chlorophyll (pg)/cell				
Chlorophyll *a/b* ratio		2 3[e†]	2 4[f†]	
Soluble protein (pg)/cell				
Soluble protein chlorophyll		40 1[f], 39[e,j]	11 8[f†]	
Chlorophyll (pg)/chloroplast				
Epidermal cell				
Protoplast volume (pl)			56[m]	
Subsidiary cell				
Protoplast volume (pl)			8 1–19[m]	

*1pl = 1000 μm^3

† Palisade and spongy mesophyll cells not differentiated

[a] Pearson and Milthorpe 1974 [b] Outlaw and Lowry 1977 [c] Outlaw *et al* 1980 [d] Jones *et al* 1977 [e] Outlaw *et al* 1981 [f] Fitzsimons and Weyers 1983 [g] Birkenhead and Willmer 1986 [h] Hedrich *et al* 1985 [i] Raschke 1975 [j] Outlaw *et al* 1976 [k] Reckmann *et al* 1990 [l] Gautier *et al* 1991 [m] MacRobbie and Lettau 1980 [n] Sato 1985 [o] Shimazaki and Zeiger 1985

From an anatomical viewpoint there are two basic types of stomata, although intermediates may exist in gymnosperms and sedges (Cyperaceae). One type has kidney-shaped guard cells and elliptical stomata (Fig. 3.4A and B), while the other type, known as the grass- or graminaceous-type, has dumb-bell-shaped guard cells (Fig. 3.4C and D). The dumb-bell-shaped guard cells are restricted to the monocotyledons and, although found in all Gramineae, are also reported to be in the Cyperaceae, Lepidocaryoid and Arecoid Palmae, Flagellariaceae, Rapateaceae, Marantaceae, Anarthriaceae and Loweriaceae.

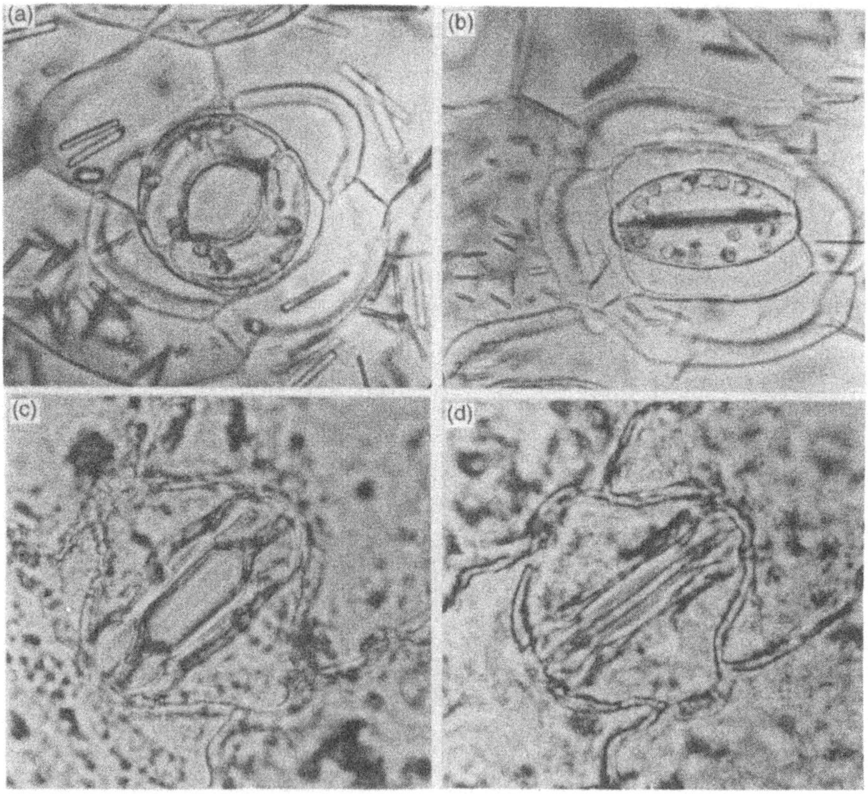

Figure 3.4 (a) Open and (b) closed stoma of *Commelina communis* which has kidney- or sausage-shaped guard cells and elliptical stomata, (c) open and (d) closed stoma of *Zea mays* which has dumb-bell-shaped guard cells and 'grass-type' stomata Note also that *C communis* has two lateral pairs of subsidiary cells and a terminal pair When stomata open the inner lateral pair become squashed by the guard cells The long, rectangular crystals in all but the guard cells and inner lateral subsidiary cells are of calcium oxalate Notable features in *Z mays* are the pair of triangular-shaped subsidiary cells and the ventral walls of each guard cell which remain parallel to each other as the stoma opens From Willmer (1983)

Figure 3.5 shows a cross-section through a typical elliptical stoma with kidney-shaped guard cells. Conventionally, plant anatomists have called walls parallel to the surface, periclinal or paradermal, and walls at 90° to the surface, anticlinal. The anticlinal wall facing the pore of the guard cell is termed the ventral wall, while the opposite wall juxtaposed to the neighbouring cell is termed the dorsal wall. The walls parallel to the surface of the leaf facing the atmosphere and the substomatal cavity are known as the inner and outer lateral walls, respectively. Some authors have used the term lateral anticlinal wall to refer to the dorsal wall (Cleary and Hardham, 1989).

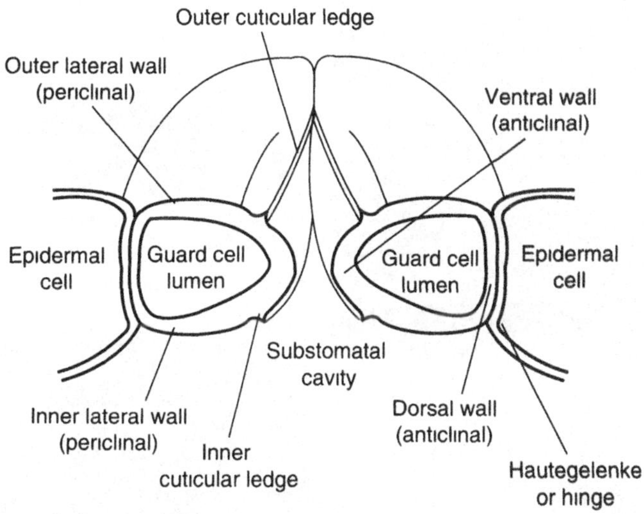

Figure 3.5 Cross section through a typical elliptical stoma showing the guard cell with respect to epidermal cells or subsidiary cells and the variable wall thickness on different surfaces of the guard cell.

Guard cell wall thickening varies greatly according to species and location around the cell. In kidney-shaped guard cells the dorsal wall is usually thin, while the ventral wall is usually heavily thickened and the surface facing the pore may be sculptured in a variety of ways. The pore is usually guarded by a pair of lips or ledges (usually heavily cuticularized) protruding from the upper edge of the ventral wall. The aperture formed between such lips has been called the eisoidal aperture and may actually be the limiting one at small pore (throat) apertures. In some species a lower pair of lips may also exist, while, in some other species, lips may be entirely absent. The term stomatal aperture normally refers to the width of the pore at the throat, rather than the width between the lips or the width of an antechamber which is found in some species and precedes the stomatal pore (Fig. 3.6A–D).

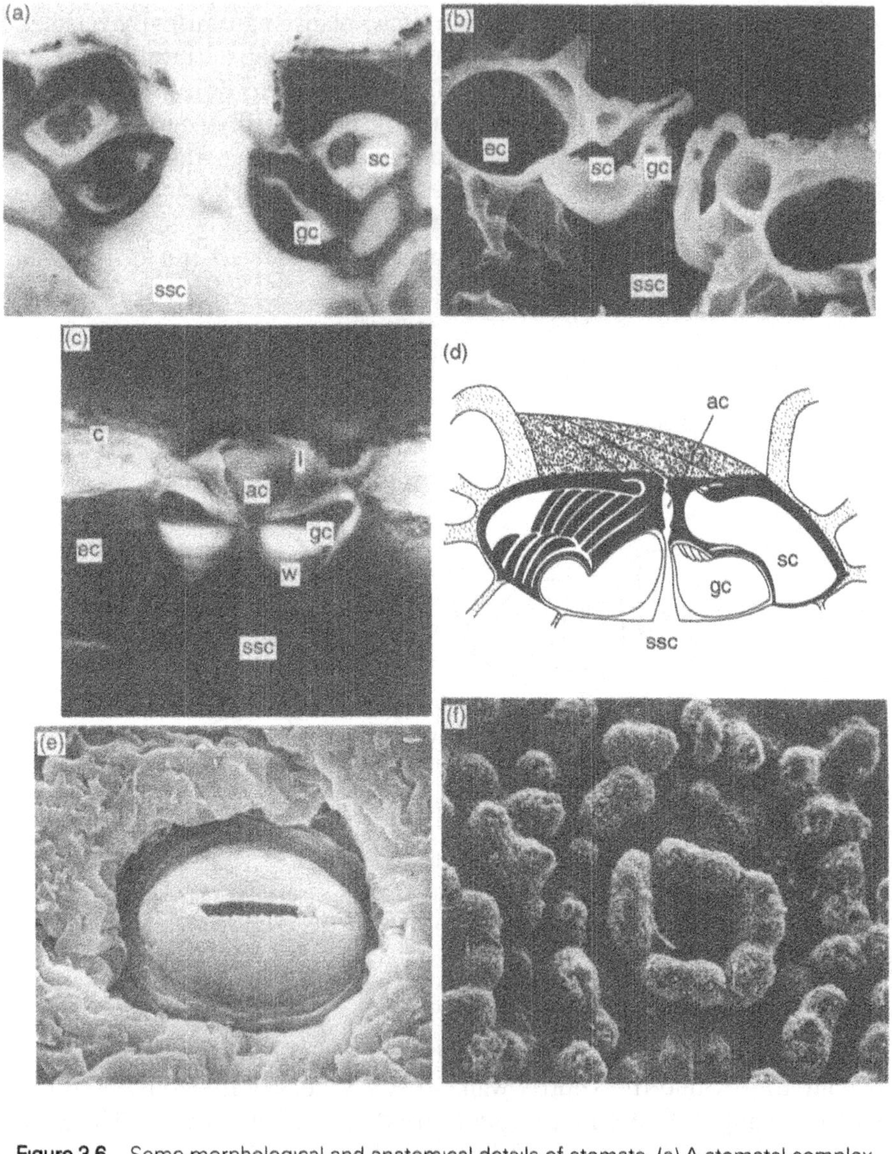

Figure 3 6 Some morphological and anatomical details of stomata (a) A stomatal complex of *Pinus sylvestris* (Scots pine) showing a sunken stoma and the lignified (darkened regions) and non lignified areas in the guard and subsidiary cell walls (b) a scanning electron micrograph of a freeze sectioned stomatal complex of *Oryza sativa* (rice) emphasizing the smallness of guard cells relative to other cell types (courtesy of Dizes Bois Lasceve and Couchat (c) a stoma of *Paphiopedilum insigne* (hybrid) under Normarski interference contrast illumination Note the prominent cuticularized lips extending over a large antechamber the thick cuticle and the greatly thickened guard cell walls particularly the inner lateral ones (from Rutter and Willmer 1979) (d) a diagrammatic representation of a stomatal complex of *Equisetum* spp in the subgenus *Hippochaete* Note the ridges on the common wall between the subsidiary and guard cells and the interlocking walls at

In grasses with dumb-bell-shaped guard cells the bulbous ends are relatively thin-walled and are connected to each other by a thick walled middle section In *Zea mays* and probably other grasses the middle portion has thin ventral and dorsal walls, but very thick inner and outer lateral walls (Fig 3 7)

Some stomata and stomatal complexes are very elaborately structured In *Equisetum* conspicuous ridges exist on the joint subsidiary cell/guard cell wall and numerous infoldings and inter-locking ridges occur in the subsidiary cell wall (Fig 3 6D) In *Pinus* the wall thickenings of the stomatal complex are also unusual, in that there are regions of lignified material in the upper and lower lateral guard cell walls with a very thin wall area facing the pore and heavily lignified subsidiary cell walls (Fig 3 6A) Normally, guard cells are attached to other cells by their dorsal wall only, but in some genera such as *Pinus* and *Equisetum* a number of cells, or just one subsidiary cell, envelop the guard cell, making a relatively rigid structure (Figs 3 6A and D) In such cases, as the stomata develop and lignification increases, they may become non-functional

Additional xerophytic adaptations to increase boundary layer resistances include location of stomata in pits as in marram grass (Fig 3 2A) or sunken below the surface of the leaf as in conifers (Figs 3 2C and 3 6A) Also, in the conifers the pits leading to the stomatal pores gradually fill up with intermeshed wax tubules Often sunken stomata have subsidiary and/or epidermal cells growing above the guard cells as in *Equisetum* and gymnosperms (Fig 3 6A and D) and the opening to the leaf surface may be surrounded by a complex surface sculpturing of wax crystals or wall protruberances (e g Fig 3 6B, E and F) All these features may assist in decreasing water loss from plants by increasing the boundary layer resistance (but see Section 3 1 2) In some species, e g *Saxifraga stolonifera* (formerly *sarmentosa)*, stomata project above the level of the leaf surface (see Chapter 2, Fig 2 2K and L) which would be expected to increase transpirational water loss Stomata may also be able to move in the leaf surface as though in sockets 'Hautgelenke' or hinges located at the top and bottom of the dorsal wall of the guard cell have been reported (Schwendener, 1881) which are purported to facilitate such guard cell movements

the entrance to the antechamber preceding the stomatal pore (after Dayanadan and Kaufman 1973) (e) a scanning electron micrograph of a stoma in a leaf of *Hebe* spp situated in a slight depression within the epidermal surface covered in a thick wax layer (courtesy of J Croxdale) (f) a scanning electron micrograph of the surface of a needle of *Taxus baccata* (yew) showing a pit entry at the bottom of which will be a stoma Note the large protuberances from the epidermal cells Key ac antechamber c cuticle gc guard cell ec epidermal cell l cuticularized lips sc subsidiary cell ssc substomatal cavity w wall

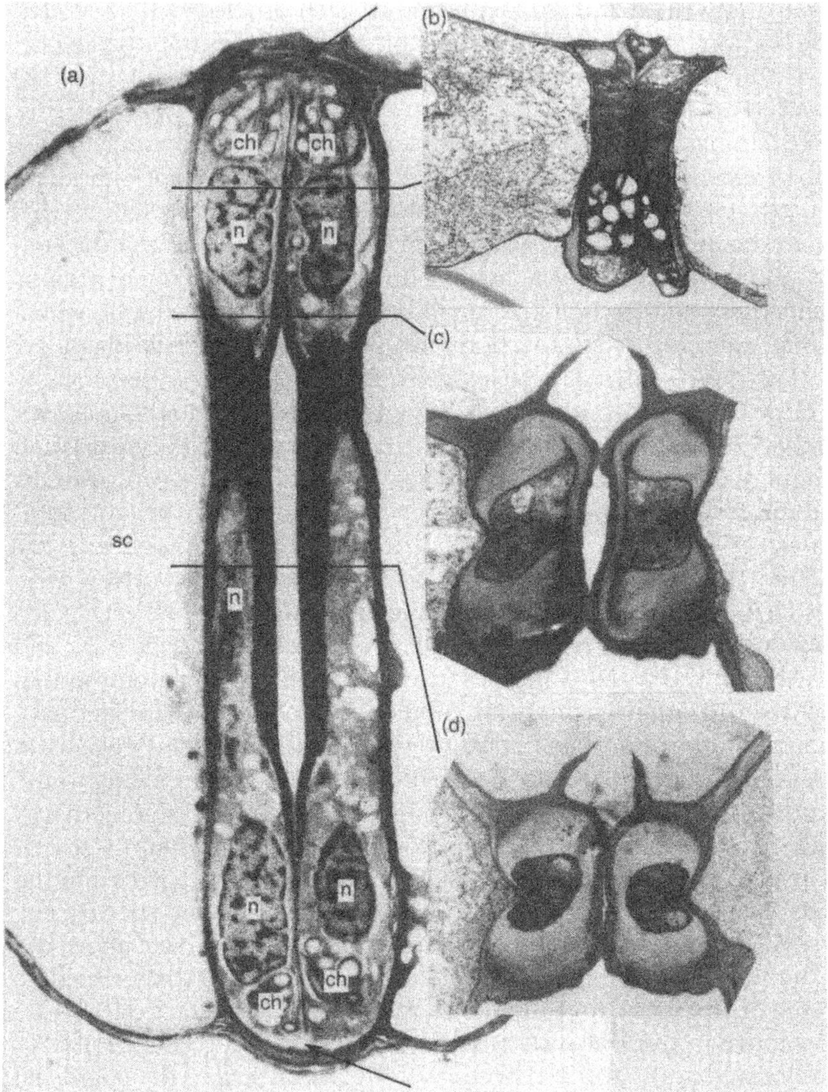

Figure 3 7 Electron micrographs of stomata of *Zea mays* (a) A nearly closed stoma sectioned in the plane of the epidermis Note the incomplete wall between the bulbous ends of the guard cells (arrowed) (b c and d) Transverse sections normal to the plane of the epidermis at the locations indicated by the lines (b) A section through the bulbous ends showing that the two guard cells are not joined over all their depth allowing a partial separation when they inflate (c and d) Sections through the middle portions of the guard cells showing the heavy wall thickening making them rigid Note the horns of cytoplasm which permit an expansion of the protoplasts in the corners of the guard cells as the pore opens (After Raschke 1979 part a courtesy of Jappe Lasceve and Vavasseur) Key n nucleoplasm ch chloroplast sc subsidiary cell

3.2 Development and differentiation of stomata

The development and differentiation of guard cells and stomatal complexes has proved invaluable to our understanding of several fundamental cellular processes in plants. Stomatal complexes form from a series of predictable divisions with a well-defined geometry in a single sheet of cells that can be readily observed using light or electron microscopy (but see also Section 2.3). Key elements associated with differentiation in most organisms, such as nuclear migration, cytoskeletal rearrangements and asymmetric divisions, can be readily investigated and manipulated (Palevitz, 1981a, 1982, 1993, Sack, 1987). The regular spacing of stomata is also a manifestation of underlying mechanisms that control spatial patterning in plants and potentially provides an accessible marker for molecular analysis (e.g. Zeiger and Stebbins, 1972, Yang and Sack, 1993).

3.2.1 Classification of stomatal complexes

Leaves from monocotyledons, particularly the grasses, have been very useful in developmental studies as the cells arise in files from a basal meristem and provide a continuous sequence of events from the youngest tissue at the base to fully differentiated tissue towards the tip. Fortuitously, examples can be found within the monocots that illustrate the two major developmental pathways leading to formation of elliptical guard cells or dumb-bell-shaped guard cells. These systems have provided most of the detailed information on guard cell development. There is, however, considerable variation in the number of divisions associated with development of a functional stomatal complex depending on the number and position of the subsidiary cells. The precise ontogeny, in conjunction with other morphological features, can be used to classify stomata into particular types (see, e.g. Vesque, 1889, Florin, 1931, Metcalfe and Chalk, 1950, van Cotthem, 1970, Rasmussen, 1981, Baranova, 1992). Subsidiary cells (or cells neighbouring the guard cells) may arise from divisions of the guard mother cell (GMC) or from a neighbouring protodermal cell. Three categories have been devised based on the ontogenetic relationship of the guard cells and neighbouring cells, although such classifications require developmental studies because the mature stomatal pattern does not necessarily reveal the ontogenetic relationship of the cells (see Fig. 3.8).

1. Agenous. The GMC divides symmetrically once to form a pair of guard cells and there are no associated subsidiary cells as in *Allium cepa* (a monocot) and *V. faba* (a dicot).
2. Mesogenous. Guard cells and neighbouring cells have a common origin from a GMC. Thus in the *Sedum* genus an asymmetric division of

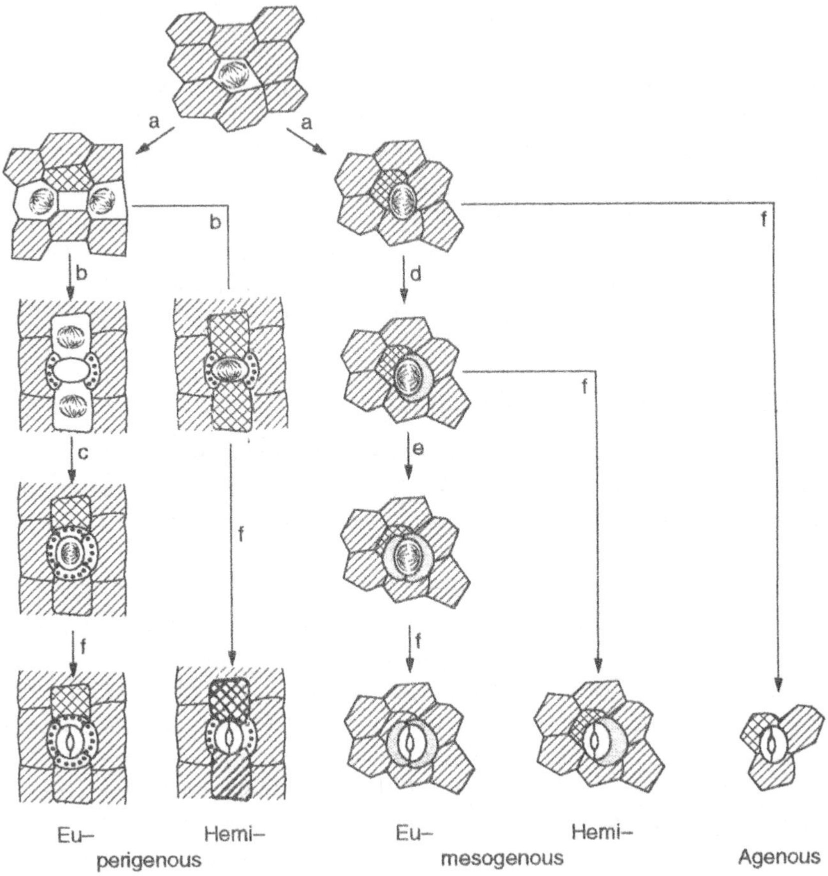

Figure 3.8 Developmental pathways of stomatal complexes Protodermal → epidermal/neighbouring cells are cross-hatched, stomatal meristemoids → guard cells are unshaded, perigene subsidiary cells are stippled and mesogene subsidiary cells are shaded Illustrating development of (a) stomatal meristemoid, (b) lateral perigene subsidiary cells, (c) polar perigene subsidiary cells, (d) mesogene subsidiary cell, (e) second mesogene subsidiary cell and (f) guard cell pairs If the guard cells are completely surrounded and contacted by subsidiary cells, their developmental pathway is prefixed *eu-*, if they are surrounded and contacted by both subsidiary and neighbouring cells, they are prefixed *hemi-* Very occasionally guard cells may be found which are associated with both perigenous and mesogenous subsidiary cells and their developmental pathway is referred to as *mesoperigenous* After Martin *et al* (1983)

a protodermal cell first produces the subsidiary cells before one more division leads to formation of the two guard cells

3 Perigenous Guard cells and neighbouring cells have no common origin, i e the guard cells develop from a GMC while the subsidiary cells develop from another meristemoid, a subsidiary mother cell (SMC) This is the situation in the grasses where an unequal division of a protodermal cell produces the GMC but, before the GMC divides to produce the guard cells, the contents of the protodermal cells (subsidiary mother cells) on either side of the GMC polarize An asymmetric division of the subsidiary mother cell produces a smaller cell with dense cytoplasmic contents which develops into a subsidiary cell and a larger more vacuolated cell destined to become a non-specialized epidermal cell In the development of stomatal complexes in the Commelinaceae the lateral subsidiary cells are always formed first followed by additional divisions of proximal and distal epidermal cells to give the terminal subsidiary cells before division of the GMC finally occurs The lateral subsidiary cells then divide again to produce inner and outer lateral subsidiary cells

Occasionally *mesoperigenous* development occurs in which at least one neighbouring cell has a common origin with the guard cell Thus, the stomatal complex is both perigenous and mesogenous

Stomata have also been classified according to the arrangement of neighbouring cells around the guard cells Such classifications have been primarily used to assist in taxonomic groupings For example, Metcalfe and Chalk (1950) identify four main types within the dicotyledons

1 Anomocytic (also known as Ranunculaceous type or agenous type) in which the guard cells are surrounded by cells indistinguishable from other epidermal cells, i e no subsidiary cells are present
2 Anisocytic (also known as Cruciferous type or hemimesogenous type) in which guard cells are surrounded by three unequally-sized subsidiary cells and a stoma is derived from a sequence of usually two unequal divisions
3 Paracytic (also known as Rubeaceous type) in which each guard cell is accompanied by one or more subsidiary cells parallel to its long axis
4 Diacytic (also known as Caryophyllaceous type), in which each stoma is surrounded by two subsidiary cells, their common walls being at right angles to the long axis of the stoma

In monocotyledons Stebbins and Khush (1961) distinguish the following types (see also Fig 3 8)

1 A stoma surrounded by four to six subsidiary cells as in the Commelinaceae

2 A stoma surrounded by four to six cells of which two are roundish and smaller than the rest and are situated at the ends of the stoma Examples are found in the Palmae
3 A stoma which possesses one lateral subsidiary cell per guard cell as in the grasses
4 A stoma without subsidiary cells as in the Filiaceae This type is considered more primitive than the rest

In angiosperms, gymnosperms and pteridophytes, variations on the types described above have been reported and the classifications have been modified and extended according to different authorities (e g Metcalfe and Chalke, 1950, Paliwell, 1969, van Cotthem, 1970, Stevens and Martin, 1978b)

3.2.2 Development of elliptical stomata: division of the GMC

The basic features of the division of the GMC giving rise to elliptical stomata are probably common to all species (Palevitz, 1981a, 1982, Sack, 1987) The cellular events leading to formation of an elliptical stoma have been extensively detailed for *Allium*, where the process is not complicated by the formation of subsidiary cells Similar, but less complete reports, are available for development of elliptical stomata in species such as *Pisum* (Singh and Srivastava, 1973)

The GMC is formed from an asymmetric division of a protodermal cell The apical (distal) end of the cell becomes polarized and accumulates most of the cytoplasm and organelles, including the nucleus A broad band of microtubules forms transversely near the apical end of the cell and then narrows to define the future site where the cell plate will fuse (Mineyuki and Palevitz, 1990) This band of microtubules is known as the pre-prophase band (PPB) and was discovered by Pickett-Heaps and Northcote in developing stomata of wheat (Pickett-Heaps and Northcote, 1966) It encircles the cortex of most higher plant cells before mitosis and marks the zone at which the future cell wall will be formed (e g Wick, 1991)

A broad band of actin microfilaments is also associated with the PPB Treatment with Cytochalasin D, a drug that disrupts microfilaments, interferes with the apical positioning and prevents narrowing of the PPB, but does not alter its orientation (Mineyuki and Palevitz, 1990) Thus microfibrils and microtubules act together to establish the position and orientation of the division plane Division takes place at 90° to the long axis of the cell to give two cells of unequal size (Bunning and Biegert, 1953) The smaller distal cell with dense cytoplasmic contents becomes the GMC (sometimes called the guard cell mother cell or stomatal meristemoid), while the larger proximal cell becomes a less specialized epidermal cell (Palevitz and Hepler, 1974a)

At interphase the GMCs are usually rectangular in shape and about 9 × 14 µm in size They contain a normal complement of cell organelles including lipid bodies, microbodies and proplastids which lack starch in *Allium* (Palevitz and Hepler, 1974a), although in most other species starch grains are present even at this early stage of development (e g Singh and Srivastava, 1973) The mature guard cell vacuole is also unusual in *Allium* as it displays a green intrinsic fluorescence when illuminated with blue light, through the accumulation of flavin or flavoprotein pigments (Zeiger and Hepler, 1979) This allows direct visualization of vacuolar dynamics in living GMCs during maturation (Palevitz *et al* , 1981) The vacuole develops from discrete globular cisternae in young GMCs to a reticulum of interlinked tubules and small chambers in the periphery of the cell and sheathing the nucleus, before forming large vacuolar chambers at maturity Accumulation of K^+ in the GMC starts at this stage (Palevitz and Hepler, 1976) Dictyosomes and polysomes are also abundant and recent evidence has revealed high expression of a small GTP-binding protein with homology to the Ypt/Rab family (involved in vesicle mediated transport and secretion) in developing GMCs of *Arabidopsis* and *Nicotiana* (Terryn *et al* , 1993)

At interphase the microtubules are randomly distributed in the cortical cytoplasm adjacent to the cell walls, with few microtubules associated with the nuclear envelope As the GMCs continue to develop there is a transient increase in the number of microtubules around the nucleus before formation of a broad longitudinal PPB band aligned at 90° to the previous division plane and parallel to the leaf axis (Mineyuki *et al* , 1989) In some species of leguminosae, there is a limited amount of wall deposition associated with this PPB, causing local wall thickening (Galatis *et al* , 1982) The chromosomes then start to condense and the nucleolus breaks down (Palevitz and Hepler, 1974a) The PPB microtubules disassemble during prophase and reform as a transverse prophase spindle (Mineyuki *et al* , 1988) Gamma-tubulin is associated with the spindle poles (McDonald *et al* , 1993), where it may function as part of the microtubule organizing centre Nuclear envelope breakdown occurs and the condensed chromosomes congress to the metaphase plate between the spindle poles At this stage the reticulate vacuole network is divided in two and compressed into the corners of the cell away from the spindle poles (Palevitz *et al* , 1981) During metaphase, space restrictions in the small cell force the spindle and metaphase plate into an oblique orientation (Palevitz 1986, Mineyuki *et al* , 1988)

At anaphase the daughter chromosomes separate to opposite corners of the GMC The separation may also involve one nuclear mass moving to the outer periclinal surface of the cell while the other nuclear mass moves towards the inner periclinal surface Thus, the nuclei come to be located diagonally opposite from the top corner to the bottom corner of the cell

At late anaphase/early telophase large numbers of vesicles accumulate near the centre of the interzone region and fuse to form the developing cell plate (also known as the phragmoplast) The developing plate may appear as a single structure or as two shorter separate pieces which fuse before growing centrifugally towards the corners The phragmoplast microtubules appear to align themselves in the direction of the daughter nuclei and even contact them Initially the phragmoplast follows the oblique orientation of the preceding metaphase plate, but starts to rotate back to a longitudinal position In part this is driven by the movement of the telophase nuclei down the anticlinal walls, but the phragmoplast also interacts with the cortex or plasma membrane at the previous site occupied by the PPB to ensure alignment (Palevitz, 1986) The nuclei are then opposite each other in the same plane

The degree of flexibility observed for the guard cell mitotic apparatus also occurs in a diverse range of other plant cell types Palevitz (1993) has argued that, just as morphological plasticity of the mitotic apparatus is required to achieve separation of chromosomes within the restricted confines of a walled cell, the mechanisms associated with prior determination of the division plane provide a means to compensate for the ensuing distortion and deformation of the cell plate and ensure its correct alignment

At telophase, nuclear envelopes reform from elements of the endoplasmic reticulum around each nuclear mass and the cell plate continues growth until the GMC is completely divided longitudinally The young cell plate often has a wrinkled appearance and it is often wider or more bulbous at its junction with the original GMC wall The vacuoles in each daughter cell begin to coalesce from the reticulate structures that persist through cytokinesis, eventually reforming larger globules and mature vacuoles (Palevitz *et al* , 1981) Figure 3 9 summarizes diagrammatically the events which occur as a GMC divides to form a pair of guard cells

3.2.3 Development of elliptical stomata: pore formation

The phragmoplast microtubules disappear rapidly after contact of the cell plate with the parental wall but re-polymerize from a new planar microtubular organizing centre on the ventral wall whose location corresponds to the future pore site (Marc *et al* , 1989a) Gamma-tubulin is also located in this region, but no other sites in the cortex, and has been implicated in establishment of a microtubular organizing centre (McDonald *et al* , 1993) Microtubules radiate away from this zone in an anticlinal direction and fan out on the periclinal walls (Fig 3 10a–e) Similar arrays are reported in other species (e g Singh and Srivastava,

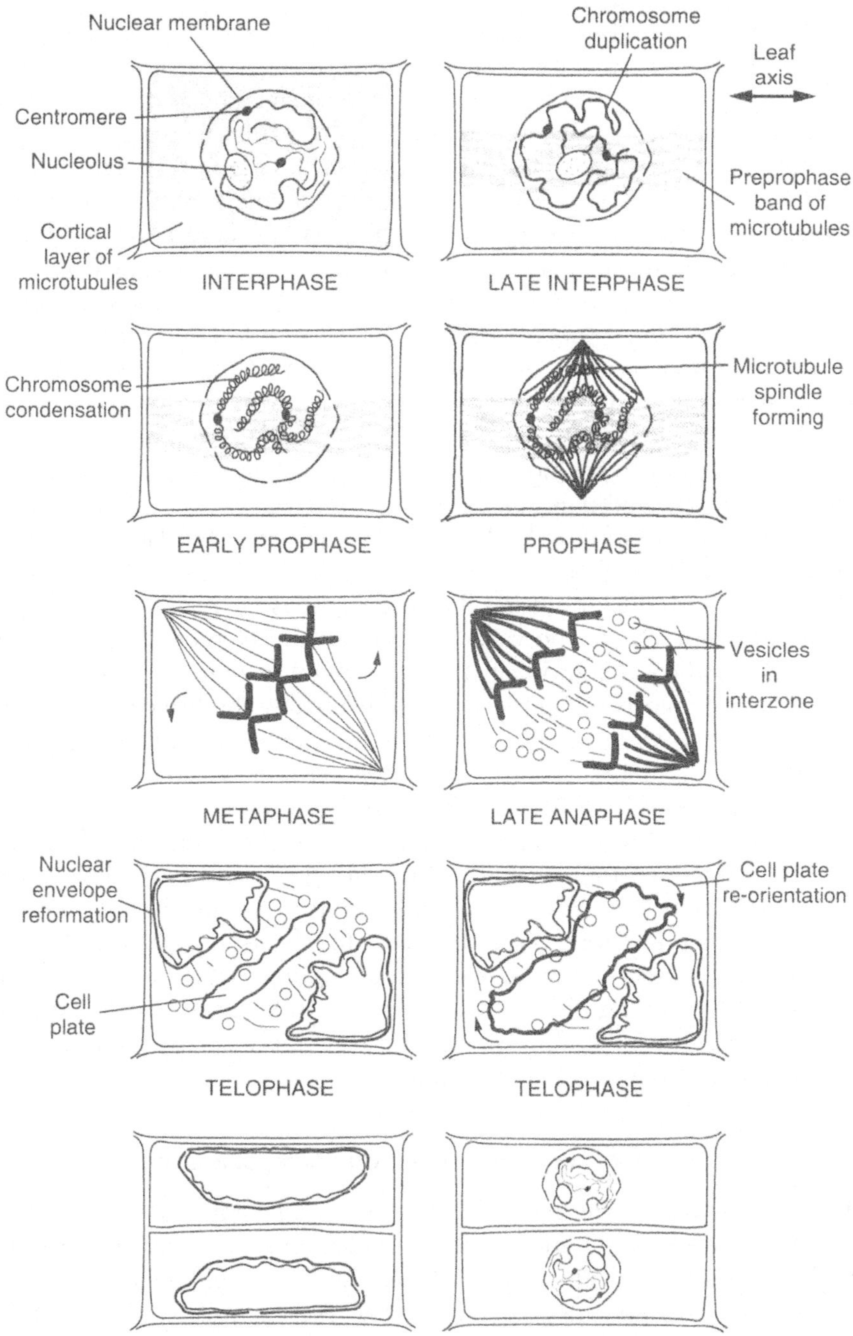

Figure 3.9 Mitosis of guard mother cells to form guard cells From Willmer (1993)

1973), including the ferns *Adiantum capillus veneris* (Galatis *et al* , 1983) Marc *et al* (1989b) found that depolymerization and repolymerization treatments of cells resulted in microtubules reassembling at the same site in the region of the microtubule organizing centre and in the same direction Thus they concluded that the planar microtubular organizing centre can both initiate microtubule polymerization and control the local orientation of the arrays (Marc *et al* , 1989b) The planar microtubular organizing centre is specifically associated with the cortex or plasma membrane near the ventral wall, rather than the nuclear envelope, as centrifugation of the nucleus away from its mid-position does not affect the microtubule array (Marc *et al* , 1989b) Furthermore, although Cytochalasin D disrupts the position of the ventral wall, the radial array still forms correctly in accordance with the misaligned wall (Marc and Palevitz, 1990)

The pattern of deposition of cellulose microfibrils follows the radiating fan of microtubules and a causal link has been inferred from colchicine disruption of both patterns (Palevitz and Hepler, 1986) Localized thickening occurs adjacent to the future pore The driving force for cell expansion during the shaping phase may come from rapid accumulation of K in the vacuole (Palevitz and Hepler, 1976, Palevitz, 1981a), but the separation of the two daughter guard cells to form a stoma is not completely understood Presumably the breakdown of the pectinaceous middle lamella occurs due to secretion of pectinases from each of the daughter guard cells Stevens and Martin (1978a) also believe that, at least in *Polypodium*, increased pressure within the guard cells causes the final mechanical separation of the walls which are still attached in the region of what will be the upper and lower lips of the guard cells Why the common wall between the two guard cells does not split along its entire length is not known In developing guard cells from *Pisum*, a layer of electrontranslucent wall material is observed along this region of wall during development (Singh and Srivastava, 1973) and, as in the formation of leaf air spaces, the separation of the wall may be restricted, for unknown reasons, to this region

3.2.4 Development of subsidiary cells around elliptical stomata

In *Allium*, the protodermal cell giving rise to the GMC and the cells bordering the developing guard cells are indistinguishable from other epidermal cells In a wide variety of species the neighbouring cells are morphologically and functionally distinct For example, Galatis and coworkers have characterized mesogenous and mesoperigenous differentiation of stomatal complexes in the dicot, *Vigna sinensis*

In mesogenous complexes, a protodermal cell divides asymmetrically and cuts off a triangular meristemoid The meristemoid sequentially

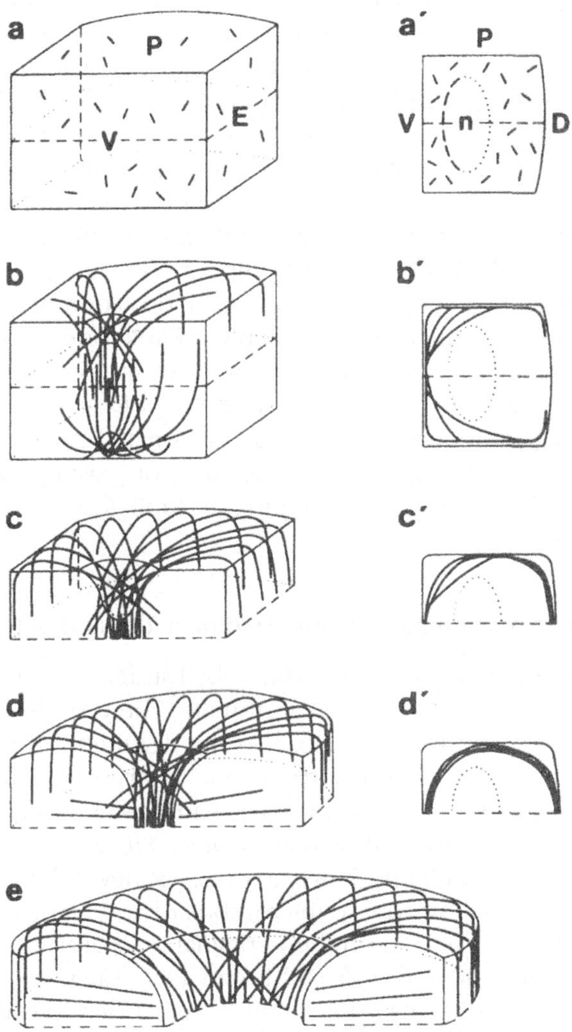

Figure 3.10 Schematic diagrams showing the development of a radial array of microtubules in a guard cell from *Allium* oriented with the ventral wall in face view (a–e) and in corresponding transverse sections (parallel to the end walls) through the central portion of the cell (a'–d'). In c, c'–e only half the array is presented for the sake of simplicity. D dorsal wall. E end wall. P periclinal wall, V ventral wall, n nucleus. (a,a') Early postcytokinetic stage with diffuse cytoplasmic immunofluorescence and a fluorescent signal at the proximal side of the nucleus. (b, b') An incipient array of microtubules (MTs) originating from the mid-region of the central strip of the ventral wall and radiating along the periclinal wall as well as directly through the inner cytoplasm. (c, c') An established radial array of MTs extending distally along the dorsal and end walls; MTs become consolidated into bundles. (d, d') An expanding cell in which the edges between the dorsal and end walls are rounded. The MTs have retracted from the edge where the central strip of the ventral wall joins the periclinal wall, thus creating a space that extends to the median periclinal plane. Periclinally oriented MTs extend from the pore site toward the end walls. (e) An expanded, mature guard cell, flexed into a rounded shape and forming a distinct stomatal pore. From Marc *et al.* (1989a) with permission.

divides twice to give two subsidiary cells and a GMC that subsequently divides symmetrically to form the guard cells. In most respects the ultrastructural changes during development of guard cells of *V. sinensis* parallel those described for *Allium* (see Galatis, 1977; Galatis and Mitrakos, 1979, 1980). However, the divisions are notable for the highly asymmetric and curved PPBs that form prior to division and, in small meristemoids, these are often the only discernible asymmetric feature (Galatis and Mitrakos, 1979). The PPBs still accurately predict the alignment of the developing cell plate, which is also curved. The range of possible locations and orientations of the PPB leading to formation of guard cells is the dominant factor influencing the highly variable orientation of the stomata in the epidermis commonly found in dicot species.

In many species, guard cell growth may still continue after pore formation along with the development of a substomatal cavity. The substomatal cavity may form by natural extension of existing air spaces and/or by some degeneration of developing mesophyll cells underneath a stoma.

3.2.5 Development of graminaceous-type stomata: division of the GMC

The development of graminaceous or dumb-bell stomata is considerably different from that observed in elliptical stomata. The GMC is formed by an asymmetric division positioned close to the distal end of the protodermal cell. In the GMC a broad transverse interphase microtubule band forms (Galatis, 1982; Cleary and Hardham, 1989; Cho and Wick, 1989; Mullinax and Palevitz, 1989) that is not seen at the equivalent stage of development in elliptical stomata (e.g. Mineyuki *et al.*, 1989). The microtubules of the interphase microtubular band gradually appear to shift to a radial array centred on the periclinal walls and subsequently to the longitudinally oriented, but still broad PPB. Palevitz (1991) has argued strongly that the microtubules remain intact during this process and the whole array moves intact to the new location. This is unlike in animal cells where the microtubules usually disassemble and then reassemble in a new location within the cell. The actin cytoskeleton has also been visualized in developing stomatal complexes of Rye (*Secale cereale*) using fluorescent phalloidin (phalloidin is a fungal toxin from *Amanita phalloides* that binds to F-actin) or immunolocalization (Cho and Wick, 1990, 1991). The microfilament arrays are rather different to those associated with symmetric divisions in other tissues (Wick, 1991). In meristematic protodermal cells there is a random cortical meshwork that becomes predominantly transverse in the GMC and thus aligned with the long axis of the cell. The microfilaments are not concentrated with the interphase microtubule band and do not co-align when the PPB forms in the GMC. Initially the PPB is incomplete, with microtubules concentrated on the end walls. As the PPB narrows, microtubules extend to contact the array from the opposite wall and complete formation of the circumferential band. During this time the GMC may be expanding parallel to the long axis of the leaf, i.e. parallel to the forming PPB. The distribution of the microtubule arrays during division of the GMC follows that outlined for the elliptical guard cells. The microfilament distribution is slightly different to other dividing cells, however. Actin staining becomes more diffuse and localized to the cortical region of the cell face adjacent to the spindle poles. There is no evidence for actin associated with the spindle fibres, but both actin and microtubules co-localize to the phragmoplast. Cytochalasin B also disrupts phragmoplast rotation and alignment of the new wall separating the guard cells.

Shortly before division of the GMC, the subsidiary mother cells (SMC) also divide asymmetrically, but in this case, they cut off a lens-shaped cell from the lateral wall adjacent to the GMC. The PPB in the SMC is strongly curved, contacting the wall adjacent to the GMC just above and below the GMC. Striking localization of microfilaments also occurs in the SMC at the pole facing the GMC and extending towards the SMC nucleus. Cytochalasin B inteferes with this nuclear positioning and prevents formation of a curved cell plate in the SMC (Cho and Wick, 1990).

The mechanism involved in the regulation of these events is not understood. However, a protein kinase (known as $p34^{cdc2}$) which is present in all dividing cells so far examined has now been detected in GMC and SMC (Colasanti *et al.*, 1993). They also found the protein to be associated with the PPB of these cells but not the interphase microtubule band. The kinase binds with another protein, cyclin, to form a p34 protein kinase complex which, in turn, regulates a number of critical events in the cell cycle through protein phosphorylation. The protein kinase complex is normally associated with entry of cells into both S phase and M phase of the cell cycle (see Wolniak, 1991). The specific association of $p34^{cdc2}$ with some of the mitotic arrays, but none of the interphase arrays, points to a key role for $p34^{cdc2}$ in the organization of the PPB, although the relevant target substrates in the PPB have not yet been identified.

3.2.6 Development of graminaceous-type stomata: pore formation

The microtubule arrays in the guard cells arise adjacent to the pore site and fan out across the periclinal walls. A similar radial array occurs transiently in the subsidiary cells, suggesting a high degree of co-ordination within the complex (Palevitz and Mullinax, 1989), and subsequently shifts to a transverse orientation. The cortical cytoplasm adjacent to the pore appears to function as a microtubular organizing centre similar to the situation in elliptical stomata (Cleary and Hardham, 1990). A transient swelling of the guard cells then occurs with concomitant elongation. This is followed by a constriction phase which continues until the characteristic dumb-bell-shaped guard cells are formed (Palevitz, 1981a) The guard cell microtubules are normally aligned axially in the central constricted region, but curve in the bulbous ends The subsidiary cell microtubules remain transverse in *Lolium rigidum* (Cleary and Hardham, 1989), although in oat (*Avena sativa*) they shift to an approximate axial orientation (Palevitz and Mullinax, 1989) In Timothy grass (*Phleum pratense*), the individual axial microtubules were up to 4–5 μm long and appeared to nucleate from amorphous or aggregates of membranous material in the cortex (Palevitz, 1981b) Figure 3 11 is a representation of the microtubule patterns which develop in cells and of the changes in cell shape, which occur as the stomatal complex forms in grass species

The cellulose microfibrils follow the microtubule arrangement and are predominantly axially arranged in the central connecting pieces of the guard cells, while the bulbous ends have microfibrils radiating out from the pore (Galatis, 1980) In the Cyperaceae a similar pattern of development occurs to that in the grasses, but the microfibrils are orientated radially along the length of the guard cells (Mishkind *et al* , 1981) The situation is further complicated as the microfibril alignment in *Cyperus esculentum* (yellow nutsedge) appears to be subject to environmental modification greenhouse-grown plants have stomata with axial microfibrils in the central connecting pieces of the guard cells, while plants growing in the field have radially arranged ones (Mishkind *et al* , 1981) This questions the overall significance of the microfibril arrangement in the connecting pieces to the mechanical properties of the guard-cell Stomata with either axial or radial arrangements appear to function normally (Mishkind *et al* , 1981)

Another major difference in the development of grass-type stomata from that of elliptical stomata is that large pores, many microns across, form in the common walls of daughter guard cells of a stoma at each bulbous end (e g Kaufman *et al* , 1970, Srivastava and Singh, 1972, Ziegler *et al* , 1974) (Fig 3 7) Once the dividing wall forming the two guard cells is complete, areas at each end of this wall are reabsorbed to form these pores The pores, first reported by Brown and Johnson (1962) in many grass species, are large enough to allow exchange of organelles between each guard cell pair Thus, the guard cell pair can be regarded as a single binucleate unit and the pressure relations experi-

Figure 3 11 Diagrammatic representation of the major microtubule (MT) patterns observed during development of stomatal complexes in *Lolium* Preprophase band (PPB) in guard mother cell (GMC) precursor cell (1) Interphase MT arrays in GMCs and PPBs before the division of subsidiary mother cells (2–9) Mitotic GMCs and subsidiary cells (SCs) with radial MT arrays (10–15) GCs reinstating cortical MT arrays (16–18) GCs with MTs organized from along the length of the ventral wall (19) Complexes showing elongation of internal periclinal walls with expanding MT wedges (20–24) SCs are

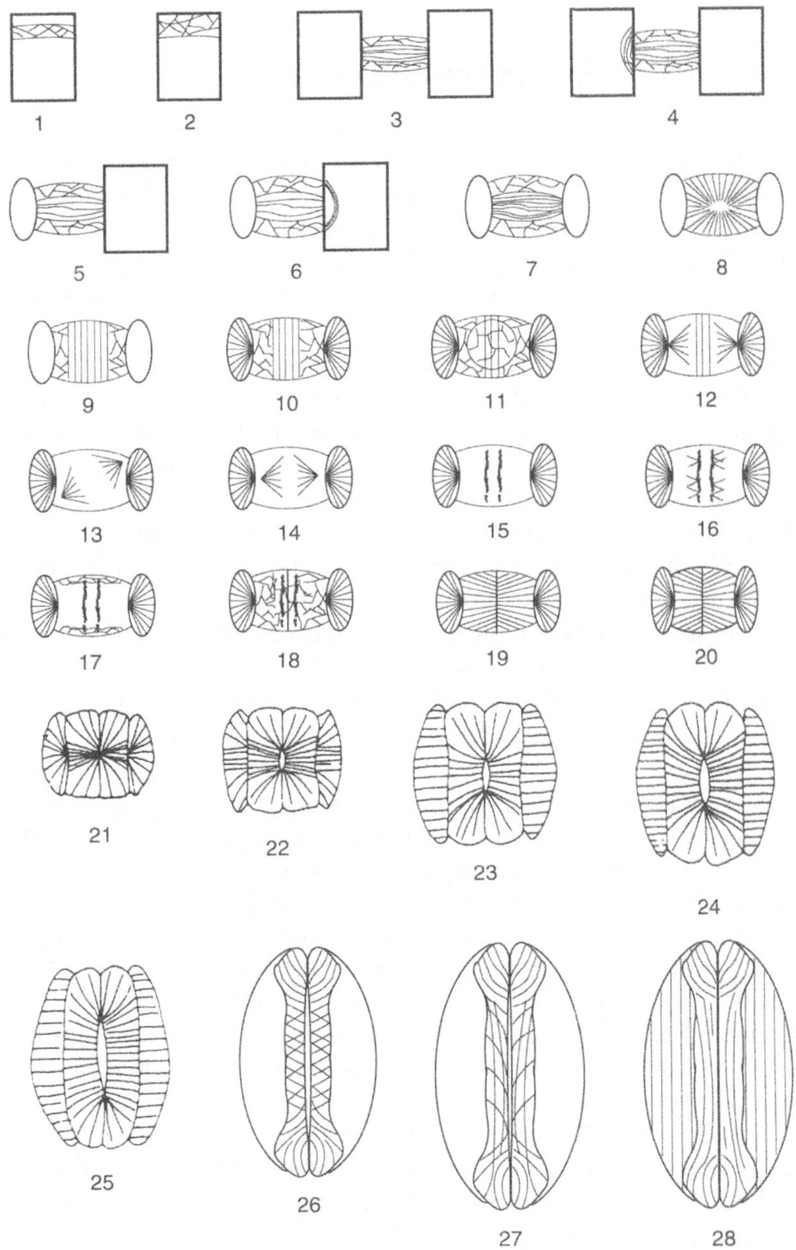

establishing transverse MTs initially at the centre of the internal periclinal walls. Reorganization of MTs on the external periclinal walls of GCs (24). Elongation of GCs and reorientation of MTs from transverse (25) through oblique (26, 27) to axial (28). Redrawn from Cleary and Hardham (1989) with permission.

enced by the contents of one guard cell will also be experienced by the other, unlike in elliptical stomata

3.2.7 Unusual variants in stomatal differentiation

There is considerable variation in the pattern of stomatal differentiation from the two models described above In *Funaria hygrometrica* and just a few other moss genera, each stoma is doughnut shaped consisting of only one guard cell with two nuclei (Schimper, 1848, Sack and Paolillo, 1983a, 1985) (Fig 3 12) The nuclei remain on opposite sides of the small pore, but two large vacuoles span the polar ends of both putative cells (Sack and Paolillo, 1983a) Initially it was considered that a complete dividing wall was formed during division of the GMC and then the ends were reabsorbed to allow cytoplasmic contact between the pair of guard cells as occurs in grass-type stomata However, Sack and Paolillo (1985) discovered that the GMCs of *Funaria* are unusual in that they undergo karyokinesis (nuclear division) but incomplete cytokinesis (cell division) Thus, the developing cell plate never joins the end walls (Fig 3 12) Interestingly, the PPB may also be absent in the GMC They also showed that cuticle formation accompanies development of the pore and may even be involved in separation of the guard cells to form the pore (Sack and Paolillo, 1983b)

Haberlandt (1886) also described a case where binucleate guard cells occurred with incomplete ventral cell walls in *Azolla* (a water fern) which has elliptical stomata Busby and Gunning (1984) made an electron microscopy (EM) study of the development of these stomata and found that the GMCs developed a PPB of microtubules followed by a cell plate which fused with the parental walls at the site of these tubules Subsequently, however, only the central part of the plate was consolidated while the end sections became perforated and may completely disperse, a situation which occurs in the development of grass-type stomata

In *Anemia* the GMC is unusual in that it can form completely within the initial stomatal cell precursor and gives rise to 'floating' guard cells (Galatis *et al* , 1986) The PPB is unconventional, with a ring on the outer periclinal surface and a second internal polar cortical site on the inner periclinal wall adjacent to the developing substomatal air space (Fig 3 13) After division the GMC is encircled by a conical cell plate that fuses with the parent wall at the locations identified by the PPB-like microtubule arrays (Galatis *et al* , 1986)

Of interest concerning the development of stomata is a report by Kubichek (1981) that chlorophyll is synthesized in guard cells of *Tradescantia virginiana* in complete darkness (a red fluorescence in UV light was used as a specific indicator of chlorophyll presence) This observation is unusual in that angiosperms are unable to synthesize

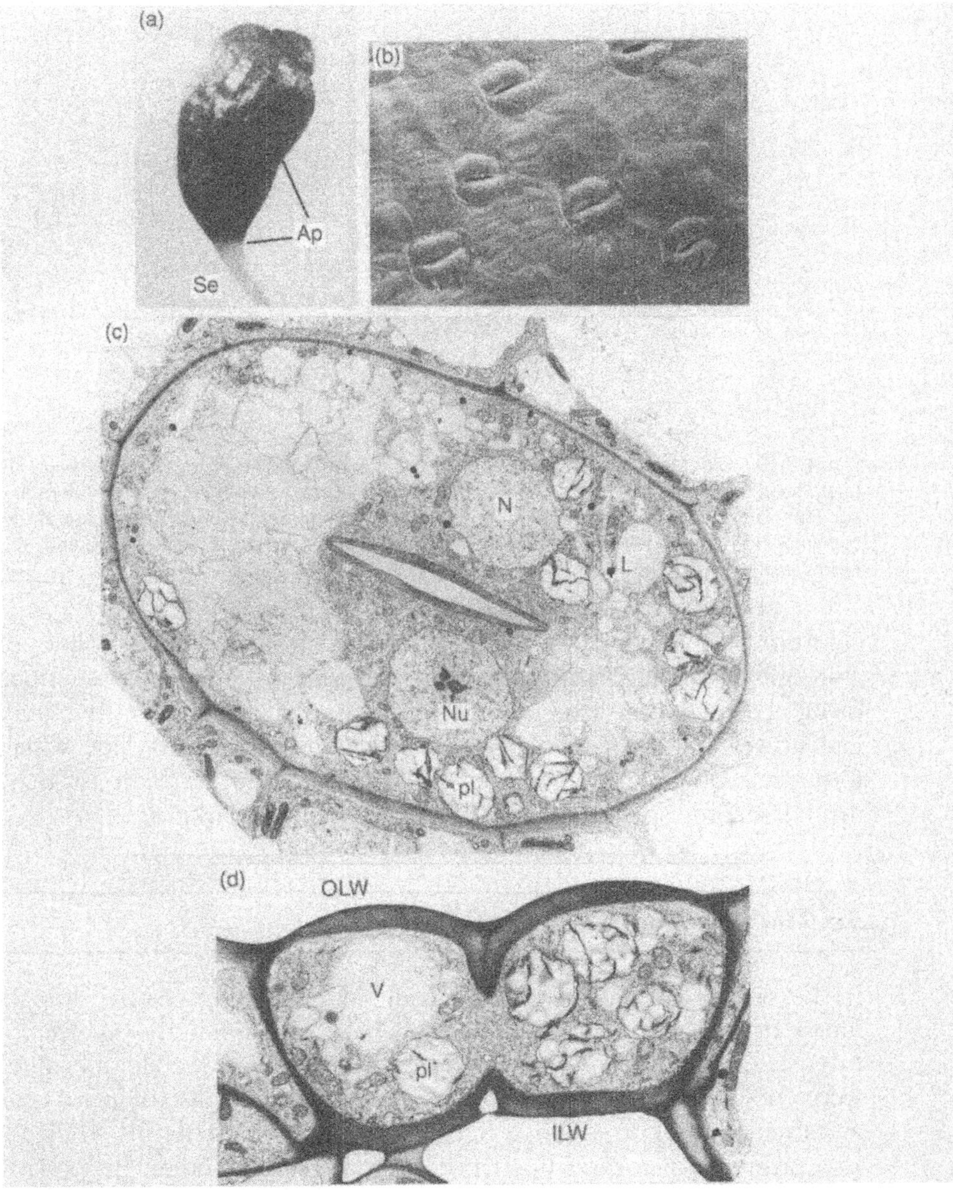

Figure 3.12 Stomata in the capsule of *Funaria hygrometrica*. (a) Side view of a fully expanded capsule. Stomata are distinguishable as white dots on the basal area of the capsule known as the apophysis. Ap, Se, seta. ×10. (b) Scanning electron micrograph of the surface of the capsule showing stomata. ×270. (c) A paradermal section through a mature stoma. N, nucleus; Nu, nucleolus; pl, plastid; L, lipid droplet. ×3000. (d) Cross section through a stoma in a plane just beyond the pore region. Pl, plastid; V, vacuole; OLW, outer lateral wall; ILW, inner lateral wall. ×6600. From Sack and Paolillo (1983) with permission.

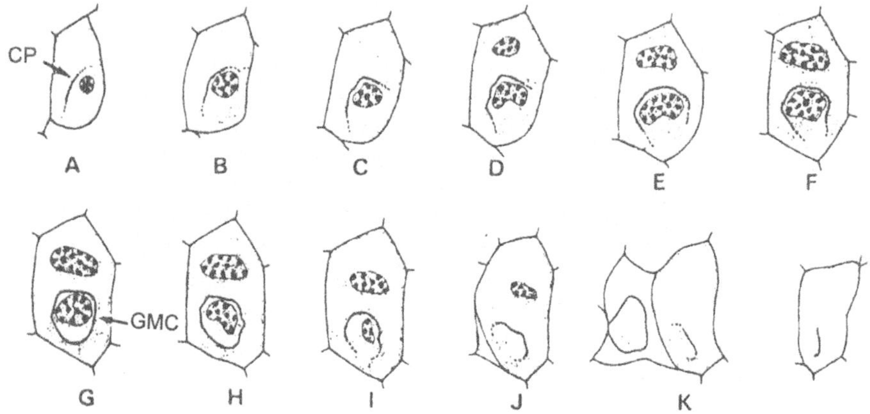

Figure 3.13 Diagrammatic representation of a late cytokinetic stage in the formation of guard mother cells (GMC) in *Anemia mandiocanna* Tracings taken from serial paradermal sections Note the curved growth of the cell plate (CP) in different planes The fusion of the anticlinal cell plate edges has been completed first in a median plane of the cell (G and H) From Galatis *et al*, (1986) with permission

chlorophyll in the dark, the conversion of protochlorophyllide to chlorophyll needing light (some algae and gymnosperms are exceptions being able to synthesize chlorophyll in complete darkness). This finding, however, is difficult to confirm since protochlorophyllide which forms in darkness also fluoresces red in UV light. Furthermore, stomata in at least some species do not develop normally in darkness.

3.3 The ultrastructure of guard cells

In the search for clues to broaden our understanding of how stomata function, fully developed guard cells have also been the subject of extensive electron microscopic studies. Initially, considerable difficulties were encountered with such studies due to the small size of guard cells and their location in, often, a thick, tough layer of epidermis. They are also prone to distortion upon preparation for EM studies, particularly when open, due to their high turgor pressure and the slow penetration of fixatives through the thickened walls. One of the first studies with reasonable preservation was by Brown and Johnson (1962) in a range of grass species. Since then innumerable more studies have been made. Good general reports on guard cell ultrastructure are to be found in Allaway and Milthorpe (1976), Palevitz (1981a) and Sack (1987).

Guard cells are surrounded by a complex and differentially sculpted wall and contain dense cytoplasmic contents with relatively small vacuolar volumes (Fig. 3.14). As well as possessing some fascinating and unusual ultrastructural features guard cells possess all of the usual cell organelles. For comparison, the frequencies of different organelles in guard cells and mesophyll cells of *Allium porrum* (leek) and *V faba* (broad bean) are given in Table 3.2

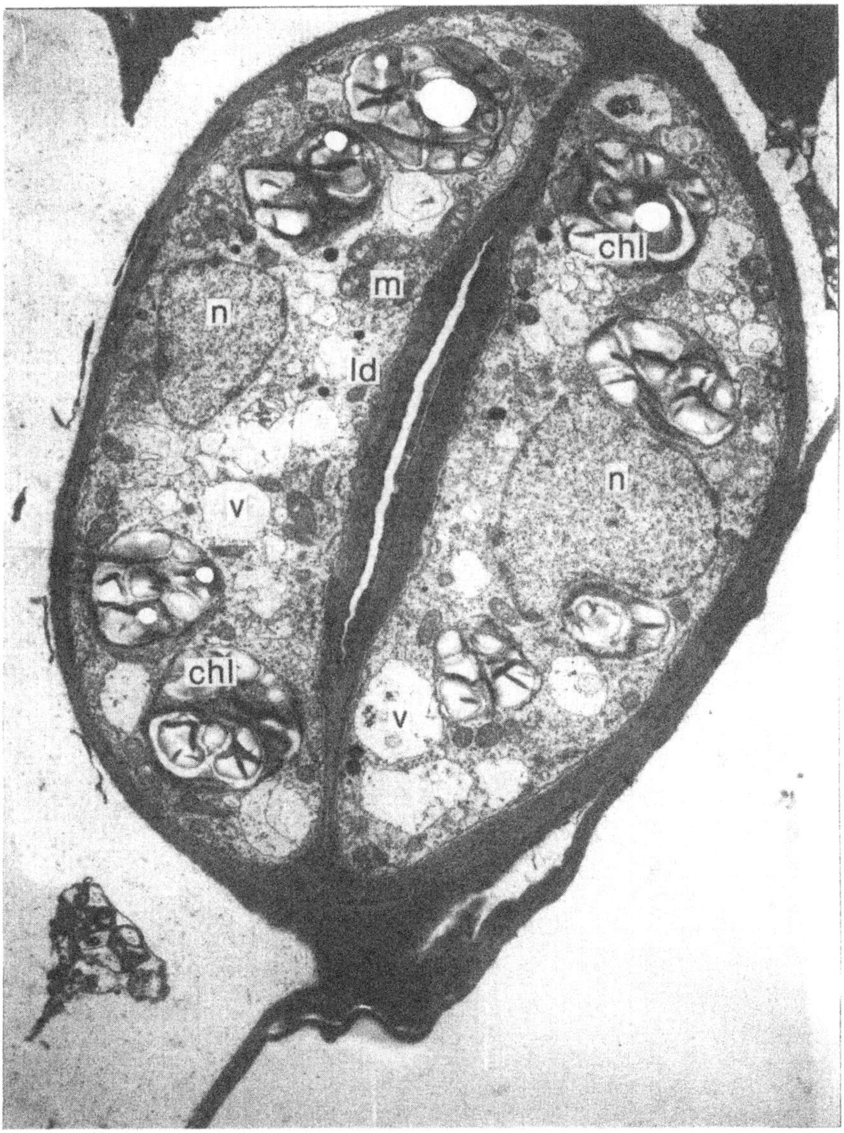

Figure 3.14 Electron micrograph of a stoma of *Phaseolus vulgaris* (chl) chloroplasts containing starch and poor lamellar structure (n) nucleus, (v) vacuole, (m) mitochondria, (ld) lipid droplets From Willmer (1983)

Table 3.2 Numbers of organelles per cell profile in thin section

Cell type	Species	Chloroplasts	Mitochondria	Mitochondria chloroplasts ratio	Microbodies	Spherosomes	No of cells counted
Mesophyll	*Allium porrum*	17 8 (4–29)	16 8 (9–26)	0 94 (0 41–4 3)	4 2 (2–9)	0 6 (0–3)	5
	Vicia faba	15 2 (11–19)	11 0 (7–16)	0 72 (0 37–1 4)	2 0 (1–4)	0	5
Guard cells	*Allium porrum*	5 3 (2–10)	50 8 (36–68)	9 6 (6 0–20)	0 3 (0–2)	12 2 (7–18)	6
	Vicia faba	6 2 (3–11)	22 6 (7–49)	3 6 (1 9–4 5)	0	8 4 (0–31)	5

Values are means with extreme ranges in parentheses obtained by counting all the profiles of organelles in the whole profile of each cell in a thin section (about 600 Å thick) All replicate cells selected at random were from different plants From Allaway and Setterfield (1972)

3.3.1 The guard cell wall

The cell wall of guard cells is specialized and probably uniquely different from walls surrounding other cell types Microscopy with polarized light (Ziegenspeck, 1938, Palevitz, 1981a) and electron microscopy indicate a predominantly radial orientation of the cellulose microfibrils (micellae) in the kidney-shaped guard cells in a manner analogous to the strengthening in a radial tyre (see Chapter 4) In the grasses the micellae radiate out from the pore at the bulbous ends of the guard cells with an axial arrangement predominating in the connecting pieces between the bulbous ends in a similar pattern to that of the microtubules (see Fig 3 11) The differential thickening and orientation of the microfibrils is a major determinant of the direction of cell expansion and the resulting shape changes during stomatal movements (see Chapter 4) The wall of some species is rich in pectins, which impregnate the cellulose microfibrils In some species, such as *C communis*, the junction between the two guard cells stains differentially with ruthenium red and other cationic dyes that bind to pectins The large cation binding capacity of such negatively charged polymers may be important in temporary storage of K^+ during stomatal movements (Chapter 8) The guard cell walls of gymnosperms and some ferns contain lignin, while a β-1,3 glucan (probably callose) has been located next to the plasmalemma in the guard cell walls of species of *Ophioglossum* (a fern) Deposits of silicon (probably existing as silicon dioxide) have also been detected in the middle lamellar region and just below the cuticle of guard cells and subsidiary cells of sugar cane and in the walls of guard cells and subsidiary cells of *Equisetum* and some grass species

Structures, described as osmotically/imbibitionally-induced sacs , located between the walls and the cuticle and positioned on the under surface at each end of the guard cells have been reported (Stevens and Martin, 1977) These were observed in the guard cells of many different species and it was suggested they accumulated K^+ This conclusion was made from transmission and scanning electron microscopy and K^+ histochemical studies However, other investigators have not been able to confirm the above observations

3.3.2 Plastids

For a long time it was considered that all functional stomata possessed chloroplasts However, Nelson and Mayo (1975) reported the absence of chloroplasts, as judged by a lack of chlorophyll autofluorescence, in functional guard cells of certain *Paphiopedilum* (lady slipper orchid) species Later, EM studies by Rutter and Willmer (1979) and D'Amelio and Zeiger (1988) confirmed this unusual feature Additionally, the study of D Amelio and Zeiger (1988) indicated a broad structural diversity in guard cell plastids of the Orchidaceae and of particular interest was an unusual pleomorphism of *Haemaria* plastids Jamieson and Willmer (1984) also found no chloroplasts in functional guard cells of a variegated chimera of *Pelargonium zonale*

The number, size and complexity of guard cell chloroplasts varies greatly depending on the species Two chloroplasts per guard cell are reported in *Anthoceros* while in *Selaginella* three to six chloroplasts per guard cell are found (see Allaway and Milthorpe, 1976) In most species there are usually about 10–15 chloroplasts per guard cell occupying about 8% of the total volume, e g in *Vicia* (Humble and Raschke, 1971), but in ferns chloroplasts are very abundant, with up to 100 reported to pack each guard cell of *Polypodium vulgare* (Stevens and Martin, 1978a) There are also numerous reports which indicate that chloroplast numbers per guard cell increase as levels of ploidy increase (e g Laptev *et al* , 1976 Cardi *et al* , 1993) Thus, Cardi *et al* (1993) found that guard cells of the diploid *Solanum commersonii* Dun contained five to seven chloroplasts while in tetraploid plants guard cells contained 10–13 chloroplasts, an approximate doubling of chloroplast numbers with doubling of ploidy

In most species, guard cell chloroplasts are poorly developed, with little thylakoid structure and scant granal stacking compared with that found in mesophyll cell chloroplasts In some species of *Allium*, chloroplasts are very small and may not be observed with normal light microscopy although fluorescence microscopy and electron microscopy confirm their presence Also, in some *Allium* species the granal stacking is extremely sparse in chloroplasts of the guard cells At the other extreme chloroplasts in guard cells of ferns are large, with considerable thylakoid complexity and an abundance of starch

Starch is lacking in the guard cell chloroplasts of a number of species from various families including the Liliaceae, Iridaceae and Amaryllidaceae In the *Allium* genus starch is lacking in the chloroplasts (Fig 3 15) however although there has been some dispute, recent evidence indicates that small molecular weight fructans (polymerization value of 8–9) occur in the epidermis of *Allium cepa* If fructans occur in the guard cells they could play a similar role to starch, a glucan polymer by supplying carbon skeletons for synthesis of organic anions (see Chapter 9)

Guard cell chloroplasts are not destarched by darkness as are chloroplasts of mesophyll cells, instead, starch tends to accumulate in guard cell chloroplasts in the dark Generally, there is a good negative correlation between guard cell starch content and stomatal aperture, starch levels increasing with decreasing stomatal aperture (but see Chapter 9)

An extensive chloroplast peripheral reticulum (an invagination of the

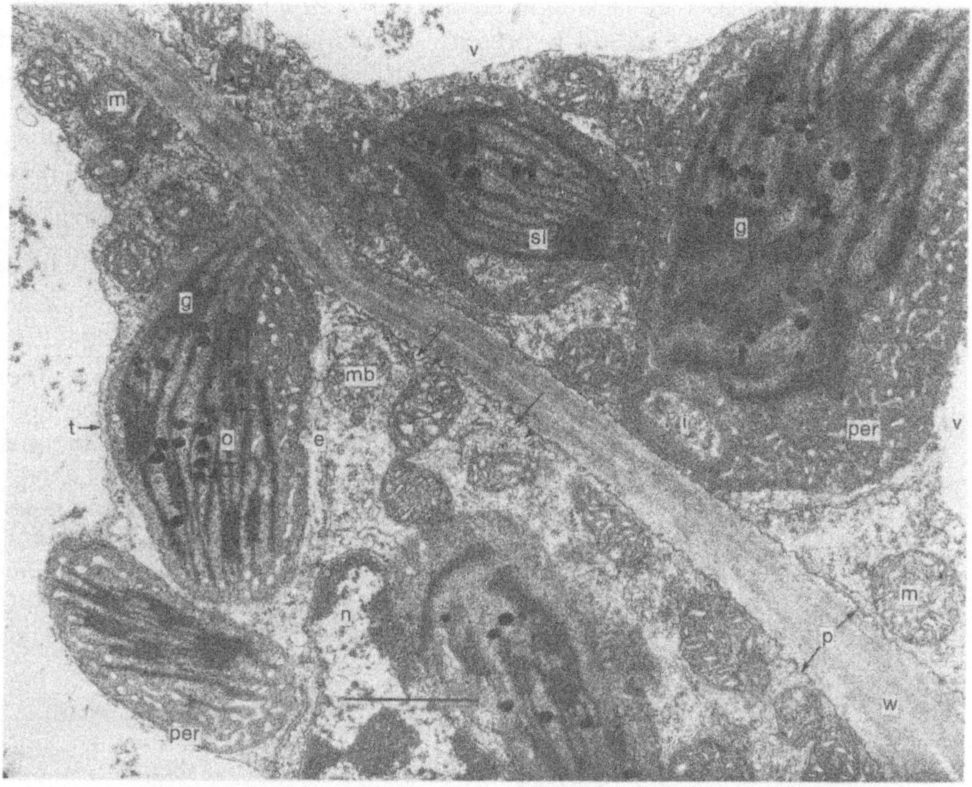

Figure 3.15 Part of a stoma of *Allium porrum* (leek) showing two guard cells separated by their common ventral walls (w). Note the invaginations of the plasmalemma and vesicles apparently fusing with the cell walls (p and arrowed), numerous mitochondria (m), some endoplasmic reticulum (e), a microbody (mb), a nucleus (n), chloroplasts [with grana (g), sparse stromal lamellae (sl), numerous osmiophilic globules (o and arrowed), and a well developed peripheral reticulum (per)], vacuole (v) and tonoplast (t). An invagination (i) of the cytoplasm into a chloroplast is also shown. Bar = 1 μm. From Allaway and Setterfield (1972).

inner chloroplast envelope) has been observed in some guard cells (Fig. 3.15), but this is not a constant feature in all species or within the same species. In C4 plants, where a peripheral reticulum almost always occurs in the mesophyll and bundle sheath chloroplasts, it has been suggested that one purpose of the reticulum is to increase the transfer of metabolites between cytoplasm and chloroplast. However, since the peripheral reticulum is not a constant feature of guard cell chloroplasts, it is not deemed essential for the normal functioning of stomata.

Unusually, microtubule-like structures have also been observed in guard cell chloroplasts (Allaway and Setterfield, 1972) but their functional significance is not understood. Plastoglobuli are also prevalent in guard cell chloroplasts.

Chloroplasts are lacking in guard cells of *Paphiopedilum* species but another type of plastid is present. These plastids are spindle-shaped with fibrous material running down the long axis. They also contain starch and large, numerous plastoglobuli (Fig. 3.16).

Thus, except for a few species, a common feature of guard cell chloroplasts is the abundance of starch, suggesting that the plastids may be acting primarily as amyloplasts (see Chapter 9 for further details on chloroplast function).

3.3.3 Nuclei

The nucleus of a guard cell is similar in size to that found in many other cell types, but it occupies a greater proportion of the cell volume since guard cells are relatively small. For example, in *V faba*, Humble and

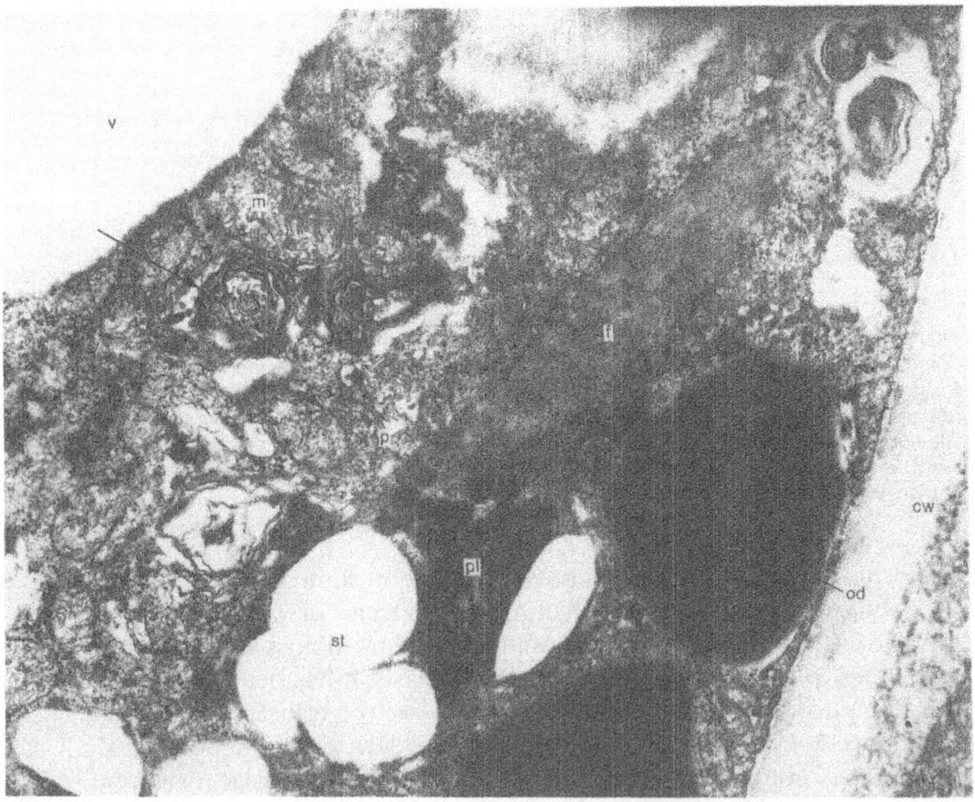

Figure 3.16 Part of a guard cell of *Paphiopedilum insigne* showing one of the curious plastids (p) which contain starch (st), plastoglobuli (pl) and a fibrous material (f) Also shown are the large oil droplets (od), (v) vacuole, (cw) cell wall, (m) mitochondria Arrowed are rarely observed membranous structures consisting of concentric whorls of membranes with a tubular or spherical central inclusion × 75 000 From Rutter and Willmer (1979)

Raschke (1971) give a volume of 0 091 pl, representing 3 5% of the protoplast volume for a closed guard cell while Weyers and Fitzsimons (1982) calculated the nuclear volume of a GCP of *C communis* to be 0 24 pl, equivalent to 6% of the protoplast volume

Nuclei are usually centrally positioned close to the ventral wall in kidney-shaped guard cells The nucleus may change shape during stomatal movements For example, according to Heller and Resch (1967) the nucleus in guard cells of *V faba* is oval when stomata are in the dark and more rounded when stomata are illuminated In *Anemia rotundifolia* the nucleus is rounded when stomata are closed and crenated in open stomata (Guyot and Humbert, 1970) The functional significance of these changes in nuclear form is not known

In dumb-bell-shaped guard cells the nucleus comprises a mass of nuclear material located at each bulbous end connected to each other by a strand of nucleoplasm passing through the central connecting portion of the guard cell, a feature first reported by Flint and Moreland in 1946

3.3.4 Mitochondria

Mitochondria are numerous in guard cells of all species so far examined (see, e g Table 3 2), a characteristic first recorded in grass species by Brown and Johnson (1962) This is a feature common to all cells which have high metabolic activity Their size and shape and appearance of the cristae are not reported to be significantly different from that found in most other cell types However, in *Commelina* mitochondrial profiles in electron micrographs of guard cells are smaller than in neighbouring subsidiary and epidermal cells (Vavasseur, personal communication) The mitochondrial configuration within guard and subsidiary cells of maize has been correlated with stomatal aperture (Lasceve *et al* , 1987) Thus, when stomata were closed and in the dark the mitochondria were partly in the resting (non-respiring) state in the guard cells, while most were in the active state (i e with swollen cristae and a condensed matrix) in the subsidiary cells However, when stomata were opened in the light all the mitochondria were in the resting state in the guard cells and in the active state in subsidiary cells An interesting aspect here is that high metabolic activity in the subsidiary cells may be necessary at all times since there is a shuttle of K and Cl between these cell types during stomatal opening and closing However, subsidiary cells do not contain chloroplasts and therefore all ATP must be derived from oxidative phosphorylation which may reflect the high proportion of active mitochondria in this cell type Conversely, guard cells do contain chloroplasts and in the light it is noticeable that all the mitochondria of these cells are in the resting state indicating that ATP may be derived from photophosphorylation

Miroslavov (1972) also observed differences in the structure of mitochondria when stomata were closed or open, the organelles being swollen with a clear matrix or contracted with a dense matrix, respectively Miroslavov related the differences to whether the organelle accumulated ions or not but surprisingly found the mitochondria swollen when stomata were closed, a situation in which guard cells release ions

3.3.5 Microbodies, spherosomes and lipid droplets

Microbodies are classified as spherical organelles bounded by a single unit membrane They range in diameter from 0 5 to 1 μm They are not usually numerous in the cytoplasm Little is known about the microbodies which have been observed in guard cells of a few species If they prove to be peroxisomes, typical of most other green cells, then some glycollate metabolism may occur in the guard cells since peroxisomes are a central site of glycollate metabolism However, glycollate oxidase as a marker for peroxisomal enzyme activity, is either non-detectable or detectable at only low levels in epidermal tissue and diaminobenzine staining of EM sections of guard cells indicates low catalase (a marker enzyme for peroxisomes) activity (see Chapter 9) Furthermore, the major source of glycollate in green cells is the Calvin cycle which may not be a major pathway in guard cells (see Chapter 9) Thus, it is unlikely that peroxisomes are an important feature of guard cell metabolism or functioning

Spherosomes are about 1 μm in diameter and recent ontogenic studies indicate that they are surrounded by a half-unit membrane rather than a complete unit membrane (Wanner *et al* , 1981) Their function is not clearly understood and may vary depending on their location within a plant In addition to containing lipids, they may contain a variety of enzymes including lipases, phosphatases, endopeptidases, RNase and DNase Spherosomes have also been called oleosomes and lipid bodies although the latter name is misleading because lipid droplets exist in the cytoplasm without being surrounded by a membrane or half-unit membrane

Large numbers of spherosomes have been reported in the guard cells of *Campanula persicifolia*, which contained much phosphatase activity when stomata were closed and little activity when they were open (Sorokin and Sorokin 1968) Fujino (1967) also detected ATPase and/or acid phosphatase activity throughout guard cells of closed stomata, but not open ones Contrary findings were made by Willmer and Mansfield (1970), who detected acid phosphatase and/or ATPase activity distributed throughout guard cells of both open and closed stomata Vacuoles are a site of much acid phosphatase activity in other plant cells and it is possible that the phosphatase activity observed by Fujino (1967) and Willmer and Mansfield (1970) was restricted primarily to the vacuoles of guard cells.

Lipid droplets (apparently without a limiting unit or half-unit membrane) are regularly observed in guard cells and unusually large numbers occur in species of *Paphiopedilum* (Fig. 3.16). The purpose of the oil droplets remains obscure, but one possibility is that they are involved in the synthesis of cuticle and waxes which may be very thick on the upper lateral wall and lips of guard cells.

3.3.6 Vacuoles

Vacuoles perform a diverse range of functions in the plant kingdom, including ion accumulation and turgor generation, as in guard cells; vacuoles are also sites of activity for hydrolytic enzymes similar to that found in the lysosomes of animal cells. There is no evidence for any phagocytic activity in guard cell vacuoles, though they do have a high acid phosphatase activity.

The vacuolar volume of guard cells is small compared with most other cell types and has been observed to change its form and size during stomatal movements. The mature vacuole in *Allium* guard cells arises from aggregation of a reticulate network of tubes and cisternae (Palevitz, 1981a). However, it is unclear whether most guard cells possess a single vacuole or whether it is subdivided. Heller *et al.* (1971) reported that the vacuolar volume increased during stomatal opening, while Guyot and Humbert (1970) concluded that the vacuole of *Anemia rotundifolia* consisted of many small ones when stomata were closed and fewer, larger ones when stomata opened. A three-dimensional analysis involving over 800 serial EM sections of a closed stoma of *C. communis* indicated that the vacuole was a single convoluted organelle and further, that the surface area of the tonoplast was 25% greater than the plasma membrane (Smith *et al.*, quoted in Weyers and Meidner, 1990, p. 189). An alternative approach using confocal laser scanning microscopy (see Box 3.1) has been used to follow changes in the volume and morphology of vacuoles of living guard cells of *C. communis* during stomatal movements (Fig. 3.17A). With such a technique it was estimated that the vacuolar volume of a guard cell increased from about 2.5 pl at zero stomatal apertures to about 6 pl at 16 μm aperture (Fricker and White, 1990). There was rarely more than one vacuole in open guard cells, although small vacuoles occasionally formed on rapid closure induced by ABA. These slowly re-fused in guard cells of a closed stoma and a single vacuole re-formed if the stoma was subsequently opened by fusicoccin (Wood *et al.*, unpublished).

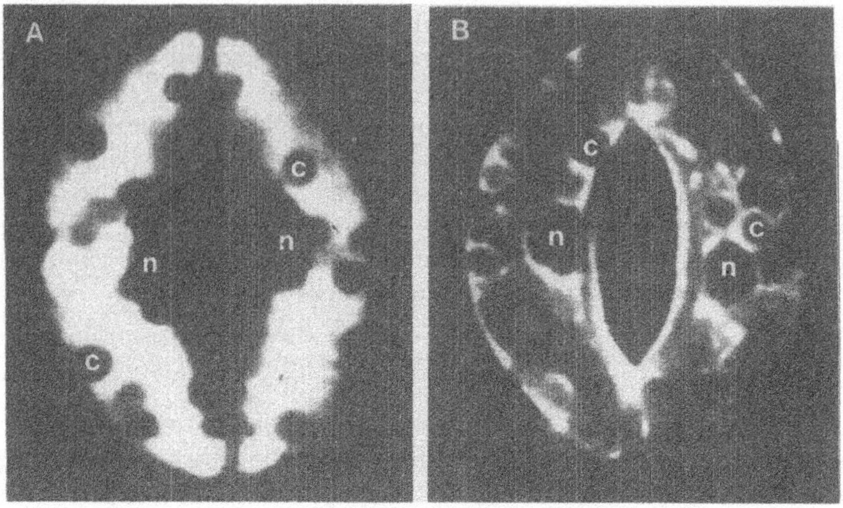

Figure 3.17 Confocal optical sections of a stomatal complex from *Commelina communis* illustrating the clarity that can be achieved using confocal techniques to remove the out-of-focus blur associated with imaging using conventional fluorescence microscopes (A) The cells are stained with the fluorescent dye, acridine orange, which accumulates in the vacuole in response to the prevailing pH gradient The positions of the nucleus (n) and chloroplasts (c) are visible as negatively stained regions (B) The pattern of fluorescence is thought to represent the distribution of ER and mitochondria visualized using the lipophilic fluorescent dye, $DiOC_6$ The ER is concentrated around the nucleus and the chloroplasts The cuticle around the pore lip is also stained

Box 3.1. Confocal scanning optical microscopy (CSOM)

Introduction

Confocal scanning optical microscopy (CSOM) allows non-invasive collection of optical sections through intact biological specimens, virtually free from out-of-focus blur (e.g. Matsumoto, 1993; Mason, 1993). A three-dimensional (3-D) image can be obtained by collecting a set of serial optical sections at progressive focal levels through a specimen. This 3-D image may then be processed digitally for display as a single reconstructed view, a stereo pair or an animation sequence. In addition image analysis techniques can be used to enhance the image contrast or perform specific measurement tasks such as 3-D length, surface area, object counting and volume calculations. A variety of confocal microscope systems are available. Each is a specific compromise between the range of fluorescence excitation wavelengths that are available, the overall sensitivity of the system and the temporal and spatial resolution achievable. These issues are extensively discussed by Pawley (1995). An outline of the basic instrumentation is shown in Fig. 3B.1 and the procedure is given overleaf:

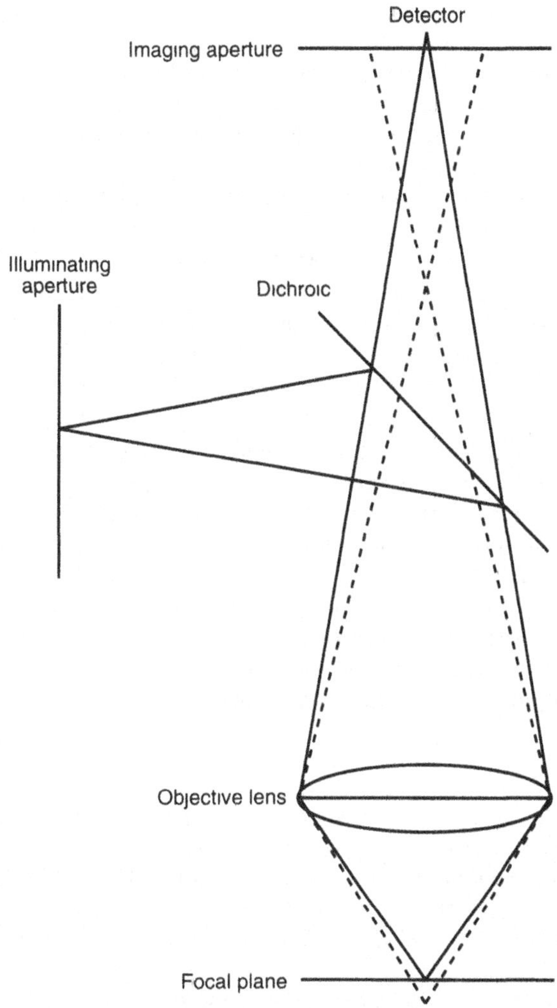

Figure 3B.1 The confocal principle in epi-fluorescence scanning optical microscopy Excitatory laser light from the illuminating aperture passes through an excitation filter (not shown) is reflected by the dichroic mirror and is focused by the microscope objective to a diffraction limited spot at the focal plane within the specimen Fluorescence emissions, excited both within the illuminated in-focus voxel and within the illuminated cones above and below it, are collected by the objective and pass through the dichroic mirror and the emission filter (not shown) However, only those emissions from the in focus voxel (——) are able to pass unimpeded through the imaging aperture to be detected Fluorescence emissions from regions below the focal plane (- - -) and from above it (not shown) have different primary image plane foci and are thus severely attenuated by the imaging aperture, contributing essentially nothing to the final confocal image From Shotton (1989)

Procedure

1 Light from a bright lamp or laser is brought to a focus within the specimen using a conventional microscope objective
2 Light is either reflected or emitted as fluorescence from the specimen from the entire illuminated volume, i e from a small region in the focal plane and from the illuminating cones above and below it
3 A suitable detector such as a photomultiplier, sensitive camera or the human eye, is used to detect the light
4 Light to be detected is passed through an aperture which reduces the contributions from points away from the focal plane
5 To generate a 2-D image at a single focal plane in the specimen, the illumination is scanned in a 2-D pattern Signals from an electronic detector can be digitized and fed into a frame store or computer for subsequent display or analysis Each digitized point in individual sections is termed a voxel
6 To generate a series of 'optical sections' the plane of focus is incremented by a defined amount with an automatic focus control and the scan continued
7 The three-dimensional image can be processed and displayed in a variety of forms ranging from single optical sections in any orientation, reconstructions of the 3-D data from a particular viewpoint, to animated rotating sequences

References

Mason, W T (1993) *Fluorescent and Luminescent Probes for Biological Activity A Practical Guide to Technology for Quantitative Real-Time Analysis*, Academic Press, New York

Matsumoto, B (1993) *Cell biological applications of confocal microscopy (Methods in Cell Biology, Vol 38)*, Academic Press, New York

Pawley, J (1995) *Handbook of Confocal Microscopy*, 2nd edn, Plenum Press, New York

Shotton, D M (1989) Confocal scanning optical microscopy and its applications for biological specimens *J Cell Sci*, **94**, 175–206

3.3.7 Endoplasmic reticulum, dictyosomes and other membranous inclusions

All the structures involved in the synthesis, direction and secretion of wall material and the synthesis of proteins have been observed in guard cells, particularly in the early stages of the development and differentiation of stomata

Rough endoplasmic reticulum (RER) (Fig 3 17B) and polysomes are very frequent during the development of guard cells and are also present in mature guard cells The frequency of these inclusions indicates a high rate of protein synthesis (see Fig 3 3) Smooth ER and dictyosomes (also known as Golgi bodies) are abundant In some cells Golgi bodies appear to be connected and the whole system is known as the Golgi apparatus Dictyosomes are known to be involved in polysaccharide and glycoprotein synthesis and sorting, and therefore increased abundance is to be expected during the massive deposition of wall material characteristic of mature guard cells There is now some evidence that dictyosomes are also involved in lipid synthesis and transport It is interesting to note, therefore, that some guard cells possess an abundance of lipid droplets and all guard cells are surrounded by a well developed cuticle

Plasmalemmasomes (also known as lomasomes or paramural bodies) have been described in a number of plant species and cell types including guard cells of *Verbascum thapsus* L (Miroslavov, 1972), maize (Wille and Lucas, 1984) and *Pelargonium* (Willmer, unpublished, see Fig 3 18) They are possibly concerned in solute transport and are composed of infoldings and inrollings of the plasmalemma into the cytoplasm which may enclose vesicles Their function in guard cells is obscure Such plasma membrane elaborations have been described in *Chara* where they are called charasomes and have been implicated in Cl transport Chloride fluxes across the plasma membrane of guard cells are an important feature of the stomatal mechanism and it is interesting to speculate that plasmalemmasomes in guard cells might serve the same purpose as in *Chara*'

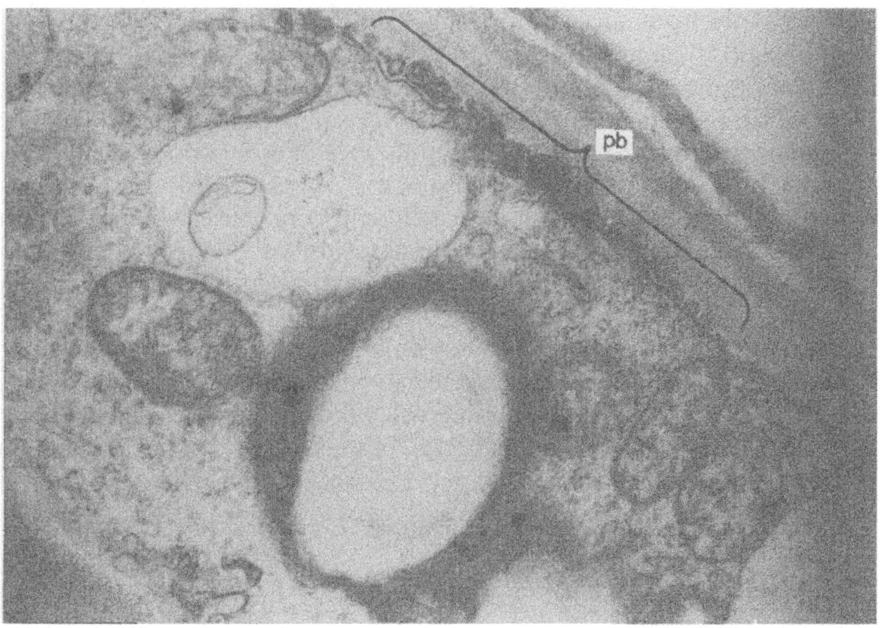

Figure 3 18 Part of a guard cell of *Pelargonium zonale* showing paramural bodies (pb)

Coated vesicles have been described in many plant cell types, including guard cells, where they are reported to be numerous in the cytoplasmic cortex (Palevitz, 1982). The coat of these vesicles consists of three heavy (M_r 180 kDa) and three light (M_r 35–40 kDa) clathrin proteins arranged as a triskelion and then joined together to form hexagons and pentagons. The overall vesicle has a diameter of 50–100 nm and a characteristic basket-like appearance at the EM level. Sanchez (1977) also reported the presence of coated vesicles in guard cells of *Helianthus annuus*, although preservation was not good and identification is difficult. Coated pits in the plasma membrane and coated vesicles were also reported in developing guard cells of the moss, *Funaria hygrometrica* (Sack and Paolillo, 1983a). Interestingly, Doohan and Palevitz (1980) commonly observed coated vesicles in GCPs of onion, but few studies indicate their presence in intact guard cells.

Coated vesicles form from coated pits at the plasma membrane and are associated with endocytosis in animal cells. They are thought to be involved in the recovery of excess plasma membrane and in the recovery of membrane proteins as well as in a general uptake of the extracellular medium (fluid-phase endocytosis). They could be involved in at least two key roles in guard cell functioning, i.e. in the regulation of the surface area of the plasma membrane as the cells swell and contract during stomatal opening and closing, and in the regulation of the density of pumps, channels and receptors in the plasma membrane during stomatal movements. However, attempts to detect fluid-phase endocytosis of fluorescent markers have not been successful (Hillmer *et al.*, 1990; Willmer and Oparka, unpublished). Large plasma membrane enclosed vesicles can be induced to form during rapid osmotic changes of the bathing medium (osmocytosis), which remain intact within the cytoplasm, not fusing with other vesicles (Diekmann *et al.*, 1993). The significance of these observations in the normal functioning of guard cells is unclear.

Other membranous structures including concentric whorls of membrane with tubular or spherical central inclusions (Fig. 3.16) have been occasionally observed in guard cells though, again, their significance is not understood.

The inner surface of the plasma membrane of GCPs and mesophyll cell protoplasts of *V. faba* is also reported to have randomly distributed particles (about 9 nm diameter) and particles distributed in hexagonal arrays (Schnabl *et al.*, 1980). The function(s) of these structures is not clear but it has been suggested that they may be concerned in wall deposition or transport processes.

3.3.8 Plasmodesmata and ectodesmata

Brown and Johnson (1962) were the first to conclude that plasmodesmata did not occur between mature guard cells and subsidiary cells of grass species. This correct conclusion, however, was fortuitous because they did not examine sequential serial sections of individual guard cells Thus, the absence or presence of plasmodesmata connecting mature guard cells with neighbouring cells continued to be a contentious issue for many years Now, the concensus is that complete, functional plasmodesmata do not exist at this location in mature cells However, they are present as unbranched connections, occurring either singly or in small pit fields between guard cells and neighbouring cells as the stomata develop, even up to the stages where the stomatal pore is partly formed Evidence for this view comes from two main sources First, serially sectioned material from a number of different species found that any complete plasmodesmata that occurred between sister guard cells and between guard cells and adjacent epidermal or subsidiary cells were sealed at maturity (truncated)(see Fig 3 19A–E) Plasmodesmata were open on the epidermal or subsidiary cell side but sealed in the middle of the joint cell wall indicating that deposition of wall material from the guard cell side was responsible for the truncation (see Fig 3 19D and E) (Willmer and Sexton, 1979, Wille and Lucas, 1984) Further confirmation of this situation came from studies in which lucifer yellow (a fluorescent vital dye unable to cross membranes) was injected into immature and mature guard cells In the former cell type, dye spread into adjoining cells while in the latter, the dye was retained in the guard cells (Palevitz and Hepler, 1985)

Nevertheless, one should be aware that there are numerous reports at the light microscope level of plasmodesmata existing between mature guard cells and neighbouring cells although, at the electron microscope level, where identification is less equivocal, there are very few reports of their existence in this location and only one study (Pallas and Mollenhauer 1972) allows critical evaluation Pallas and Mollenhauer (1972) reported large numbers (15–20 pores µm) of plasmodesmata in the end walls of guard cells of *V faba* and tobacco, the plasmodesmata were branched and there was no restriction of wall thickening in the pit fields This study remains an isolated case and difficult to reconcile with other EM studies, although perhaps fully mature guard cells were not being observed Observations of complete plasmodesmata in mature guard cells at the light microscope level may be explained if one assumes that the histochemical tests used to detect plasmodesmata give positive reactions to aborted connections or to weakened or structurally different areas of wall material

There are other reasons for expecting plasmodesmata to be absent in mature tissue The electrical coupling, turgor and ionic relations of guard cells indicate they are functionally isolated Distortion, thinning and relative movement of the guard cell wall during stomatal opening might be expected to damage plasmodesmatal connections if they were present Furthermore, it may be argued that guard cells will function best as independent sensors of the environment if they lack cytoplasmic connections to the main plant body

It was believed that fingers of cytoplasm extended from the GCP

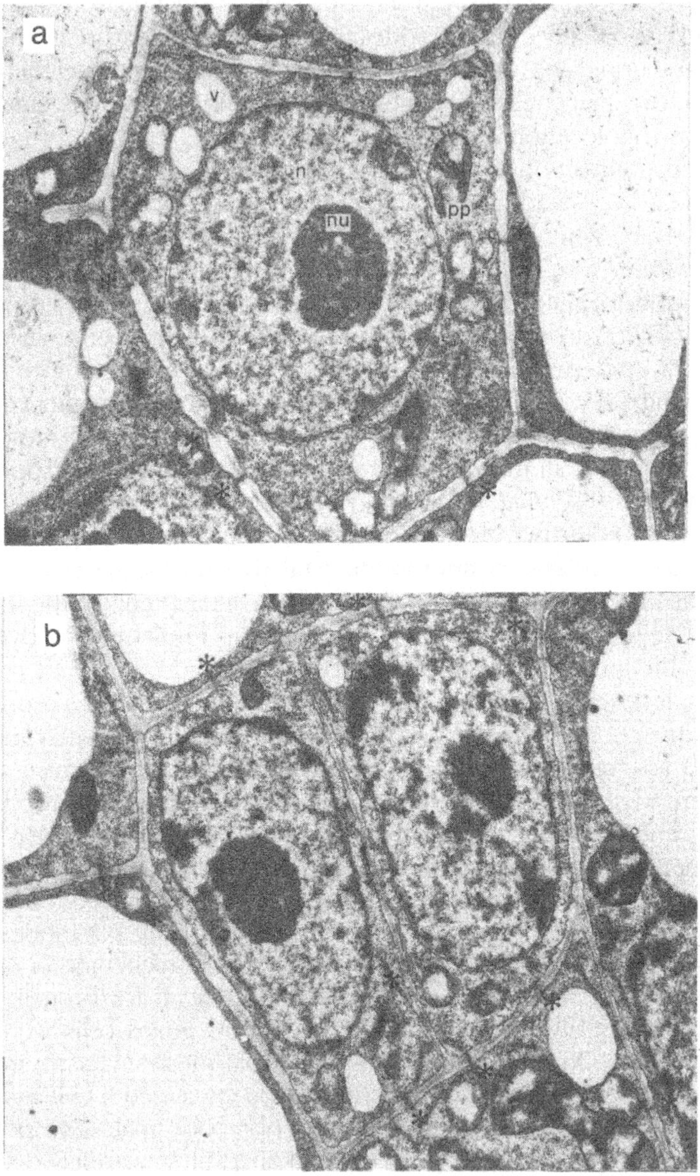

Figure 3 19 Formation of guard cells of *Phaseolus vulgaris* from a guard mother cell (a) Guard mother cell with proplastids (pp) containing starch a few small vacuoles (v) a central nucleus (n) with nucleolus (nu) Plasmodesmata (*) join the guard mother cell with neighbouring epidermal cells which are vacuolated ×15 000 (b) Daughter guard cells connected to each other by plasmodesmata (*) ×15 500 (c) Developing stoma with pore (▶) partly formed Plasmodesmata (*) occur between daughter guard cells in the ventral walls and between guard cells and epidermal cells Rough ER (er) dictyosomes (d) and

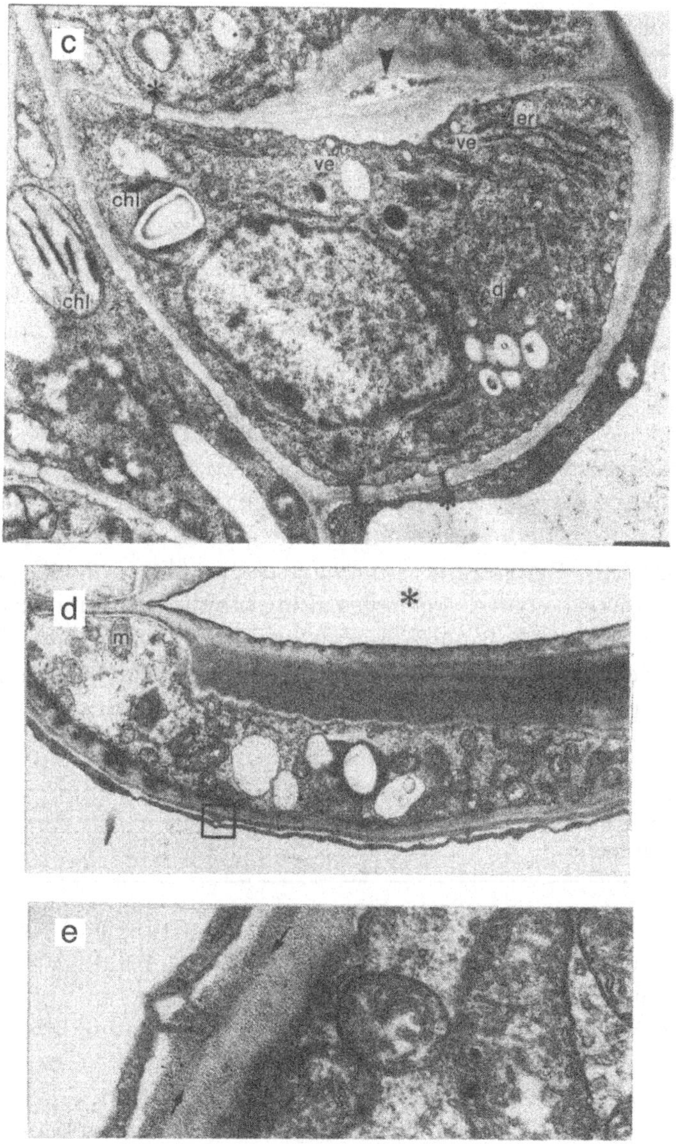

vesicles (ve) probably laden with wall material are present Chloroplasts (chl) with poor lamellar structure but containing starch are also present ×12 500 (d) Part of a fully developed stoma with open pore (*) Mitochondria (m) are abundant The box is enlarged in (e) ×8000 (e) The enlarged area (×50 000) showing incomplete plasmodesmata The arrows indicate the probable line separating the original primary wall from the wall material laid down by the guard cell at a later stage From Willmer and Sexton (1979)

through the outer lateral wall to the cuticular region These structures, known as ectodesmata (Franke 1962), are now considered to be artefacts and may really consist of regions of differently structured or more permeable wall material

3.3.9 The cytoskeleton

The cytoskeleton plays a critical part in the development and differentiation of guard cells and is discussed in detail in Section 3 2 In mature guard cells the cytoskeleton also has a role in, for example, holding the nucleus in a particular position, in cytoplasmic streaming and, possibly, creating a local order or 'structure' to the cytoplasm However, once the cell has acquired its characteristic shape the dynamic activity of the cytoskeleton, particularly of the microtubules, is less evident Cytoplasmic streaming is a feature observed in guard cells, particularly as opening occurs, when stomata are wide open and when they are closing, at least in *C communis*, streaming appears almost absent (Willmer, unpublished) Cytoplasmic streaming is thought to involve the movement of myosin-coated organelles along bundles of unipolar F-actin filaments (F-actin is polymerized actin of a fibrous nature) and therefore both are expected to be present in mature guard cells So far, F-actin has only been reported in developing stomatal complexes (Cho and Wick, 1990, 1991, Mineyuki and Palevitz, 1990)

References

Allaway, WG and Milthorpe, FL (1976) Structure and functioning of stomata, in *Water Deficits and Plant Growth*, Vol IV, Academic Press, New York, pp 57–102

Allaway, W G and Setterfield, G (1972) Ultrastructural observations on guard cells of *Vicia faba* and *Allium porrum* *Can J Bot*, **50**, 1405–1413

Appleby, R F and Davies, WJ (1983) The structure and orientation of guard cells in plants showing stomatal responses to changing vapour pressure difference *Ann Bot*, **52**, 459–468

Baranova, M (1992) Principles of comparative stomatographic studies of flowering plants *Bot Rev*, **58**, 49–99

Birkenhead, K and Willmer, C M (1986) Some biochemical characteristics of guard cell and mesophyll cell protoplasts from *Commelina communis* L *J Exp Bot*, **37**, 119–128

Brown, WV and Johnson, C, Sr (1962) The fine structure of the grass guard cell *Am J Bot*, **49**, 110–115

Bunning, E and Biegert, F (1953) Die Bildung der Spaltoffnungsinitialen bei *Allium cepa* *Z Bot*, **41**, 17–39

Busby, C H and Gunning, B E S (1984) Microtubules and morphogenesis in stomata of the water fern *Azolla* an unusual mode of guard cell and pore development *Protoplasma*, **122**, 108–119

Cardi, T, Innamico, V, D'Ambrosia, F *et al* (1993) *In vitro* regeneration and cytological characterization of shoots from leaf explants of three accessions of *Solanum Commersonii* *Plant Cell, Tissue Organ Culture*, **34**, 107–114

Cho, S-O and Wick, S M (1989) Microtubule orientation during stomatal differentiation in grasses *J Cell Sci*, **92**, 581–594

Cho, S-O and Wick, S M (1990) Distribution and function of actin in the developing stomatal complex of winter rye (*Secale cereale* cv Puma) *Protoplasma*, **157**, 154–164

Cho, S-O and Wick, S M (1991) Actin in the developing stomatal complex of winter rye a comparison of actin antibodies and Rh-phalloidin labeling of control and CB-treated tissues *Cell Motil Cytoskeleton*, **19**, 25–36

Cleary, A L and Hardham, A R (1989) Microtubule orientation during development of stomatal complexes in *Lolium rigidum* *Protoplasma*, **49**, 76–81

Cleary, A L and Hardham, A R (1990) Reinstatement of microtubule arrays from cortical nucleating sites in stomatal complexes of *Lolium rigidum* following depolymerisation of microtubules by oryzalin and high pressure *Plant Cell Physiol*, **31**, 903–915

Colasanti, J, Cho, S-O, Wick S and Sundaresan, V (1993) Localisation of the functional $p34^{cdc}$ homolog of maize in root tip and stomatal complex cells association with predicted division sites *Plant Cell*, **5**, 1101–1111

van Cotthem, W R J (1970) A classification of stomatal types *Bot J Linn Soc*, **63**, 235–46

Cupples, W, Lee, J and Tallman, G (1991) Division of guard cell protoplasts of *Nicotiana glauca* (Graham) in liquid cultures *Plant Cell Environ*, **14**, 691–697

D'Amelio, E D and Zeiger, E (1988) Diversity in guard cell plastids of the Orchidaceae a structural and functional study *Can J Bot*, **66**, 257–271

Dayanandan, P and Kaufman, P B (1973) Stomata in *Equisetum* *Can J Bot*, **51**, 1555–1564

Diekmann, W, Hedrich, R, Raschke, K and Robinson, D G (1993) Osmocytosis and vacuolar fragmentation in guard cell protoplasts their relevance to osmotically-induced volume changes in guard cells *J Exp Bot*, **44**, 1569–1577

Donkin, M E and Martin, E S (1981) Blue light absorption by guard cells of *Commelina communis* and *Allium cepa* *Z Pflanz* , **102**, 345–352

Doohan, M E and Palevitz, B A (1980) Microtubules and coated vesicles in guard-cell protoplasts of *Allium cepa* L *Planta*, **149**, 389–401

Ehleringer, J R (1981) Leaf absorptances of Mohave and Sonoran desert plants *Oecologia*, **49**, 366–370

Esau, K (1977) *Anatomy of Seed Plants*, 2nd edn, Wiley, New York

Fahn, A (1982) *Plant Anatomy*, 3rd edn, Pergamon Press, Oxford

Fitzsimons, PJ and Weyers, J D B (1983) Separation of protoplast types from *Commelina communis* L leaf epidermis *J Exp Bot* , **34**, 55–66

Flint, L H and Moreland C F (1946) A study of stomata in sugar cane *Am J Bot* , **33**, 80–82

Florin, R (1931) Untwersuchungen zur Stammesgeschichte der Coniferales und Cordaitales *K Svensk Vetenskaps Hanl III*, **10**, 1–588

Franke, W (1962) Ectodesmenstudien 1 Mitt Uber pilzformig erscheinende Ektodesmen Kritische Abhandlung uber das Wesen der Ektodesmen *Planta*, **59**, 222–238

Fricker, M D and White, N (1990) Volume measurements of guard cell vacuoles during stomatal movements using confocal microscopy *Trans Roy Microsc Soc* , **1**, 345–348

Fujino, M (1967) Role of adenosine triphosphate and adenosine triphosphatase in stomatal movements *Sci Bull Fac Educ Nagasaki Univ* , **18**, 1–47

Galatis, B (1977) Differentiation of stomatal meristemoids and guard cell mother cells into guard-like cells in *Vigna sinensis* leaves after colchicine treatment *Planta*, **136**, 103–114

Galatis, B (1980) Microtubules and guard-cell morphogenesis in *Zea mays* L *J Cell Sci* , **45**, 211–244

Galatis, B (1982) The organisation of microtubules in guard mother cells of *Zea mays* *Can J Bot* , **60**, 1148–1166

Galatis, B and Mitrakos, K (1979) On the differential divisions and preprophase microtubule bands involved in the development of stomata of *Vigna sinensis* L *J Cell Sci* , **37**, 11–37

Galatis, B and Mitrakos, K (1980) The ultrastructural cytology of the differentiating guard cells of *Vigna sinensis* *Am J Bot* , **67**, 1243–1261

Galatis, B , Apostolakos, P, Katsaros, Chr and Loukari, H (1982) Preprophase microtubule band and local wall thickening in guard cell mother cells of some leguminosae *Ann Bot* , **50**, 779–791

Galatis, B , Apostolakos, P and Katsaros, C (1983) Microtubules and their organizing centres in differentiating guard cells of *Adiantum capillus veneris* *Protoplasma*, **115**, 176–192

Galatis, B , Apostolakos, P and Palafoutas, D (1986) Studies on the formation of floating guard cell mother cells in *Anemia* *J Cell Sci* , **80**, 29–55

Gautier, H , Vavasseur, A , Gans, P and Lasceve, G (1991) Relationship between respiration and photosynthesis in guard cell and mesophyll cell protoplasts of *Commelina communis* L *Plant Physiol* , **95**, 636–641

von Guttenberg, H (1971) Bewegungsgewebe und Perzeptionsorgane, in *Handbuch der Pflanzenanatomie*, Vol V, (eds S Carlquist, W Zimmermann, P Ozenda and H D Wulff), Gebruder Borntraeger, Berlin, pp 203–19

Guyot, M and Humbert, C (1970) Les modifications du vacuome des cellules stomatiques d'*Anemia rotundifolia Schrad C R Acad Sci* , *Ser D*, **270**, 2787–2789

Haberlandt, G (1886) Beitrage zur Anatomie und Physiologie der Laubmoose *Jahrb Wissenschaft Botanik*, **17**, 359–498

Hedrich, R , Raschke, K and Stitt, M (1985) Role for fructose-2,6-bis-phosphatase in regulating carbohydrate metabolism in guard cells *Plant Physiol* , **79**, 977–982

Heller, F O , Kausch, W and Trapp, L (1971) UV-Mikroskopischer Nachweis von Strukturanderungen in Schließzellen von *Vicia faba* *Naturwissenschaften*, **58**, 419

Heller, F O and Resch, A (1967) Funktionell bedingter Strukturwechsel der Zellkerne in den Schlieazellen von *Vicia faba* *Planta*, **75**, 243–252

Herscovich, S , Tallman, G and Zeiger, E (1992) Long-term survival of *Vicia* guard cell protoplasts in cell culture *Plant Sci* , **81**, 237–244

Hillmer, S , Hedrich, R , Robert-Nicoud, M and Robinson, D G (1990) Uptake of Lucifer yellow CH in leaves of *Commelina communis* is mediated by endocytosis *Protoplasma*, **158**, 142–148

Holloway, PJ (1982) Structure and histochemistry of plant cuticular membranes An overview, in *The Plant Cuticle*, (eds D F Cutter, K L Alvin and C E Price), Academic Press, London

Humble, G D and Raschke, K (1971) Stomatal opening quantitatively related to potassium transport Evidence from electron probe analysis *Plant Physiol* , **48**, 447–453

Jamieson, A P and Willmer, C M (1984) Functional stomata in a variegated leaf chimera of *Pelargonium zonale* L without guard cell chloroplasts *J Exp Bot* , **35**, 1053–1059

Jellings, A J and Leech, R M (1982) The importance of quantitative anatomy in the interpretation of whole leaf biochemistry in species of *Triticum*, *Hordeum* and *Avena* *New Phytol* , **92**, 39–48

Jones, H G (1993) *Plants and Microclimate*, 2nd edn, Cambridge University Press, Cambridge

Jones, M G K , Outlaw, WH , Jr and Lowry, O H (1977) Enzyme assays of 10^{-7} to 10^{-14} moles of sucrose in plant tissues *Plant Physiol* , **60**, 379–383

Kaufman, PB, Petering, LB, Yocum, CS and Baic, D (1970) Ultrastructural studies on stomata development in internodes of Avena sativa *Am J Bot*, **57**, 33–49

Kolattukudy, PE (1981) Structure, biosynthesis, and biodegradation of cutin and suberin *Ann Rev Plant Physiol*, **32**, 539–567

Kubichek, SA (1981) Formation of chloroplasts in guard cells of Virginia spiderwort in the dark *Sov Plant Physiol (Fiziol Rast)*, **28**, 556–560

Laptev, YP, Makarov, PP, Glazova, MV *et al* (1976) Stomata and pollen as indicators of the ploidy of plants *Genetika* (translated from Russian), **12**, 47–55

Lasceve, G, Vavasseur, A and Couchat, P (1987) Ultrastructure des mitochondries des stomates de mais, cas d'une ouverture photoactive *Can J Bot*, **65**, 1861–1869

Lin, ZF and Ehleringer, J (1983) Epidermis effects on properties of leaves of four herbaceous species *Physiol Plant*, **59**, 91–94

Marc, J and Palevitz, BA (1990) Regulation of the spatial order of cortical microtubules in developing guard cells of *Allium* *Planta*, **182**, 626–634

Marc, J, Mineyuki, Y and Palevitz, BA (1989a) The generation and consolidation of radial array of microtubules in developing guard cells of *Allium cepa* L *Planta*, **179**, 516–529

Marc, J, Mineyuki, Y and Palevitz, BA (1989b) A planar microtubule-organizing zone in guard cells of *Allium* experimental depolymerization and reassembly of microtubules *Planta*, **179**, 530–540

Martin, ES, Donkin, ME and Stevens, RA (1983) *Stomata*, Edward Arnold, London

Martin, JT and Juniper, BE (1970) *The Cuticle of Plants*, Edward Arnold, London

Mayer, W, Moser, I and Bunning, E (1973) Die Epidermis als Ort der Lichtperzeption fur circadiane Laubblatbewegungen und photoperiodische Induktionen *Z Pflanz*, **70**, 66–73

McDonald, AR, Liu, B, Joshi, HC and Palevitz, BA (1993) γ-tubulin is associated with a cortical-microtubule-organising zone in the developing guard cells of *Allium cepa* L *Planta*, **191**, 357–361

Meidner, H and Mansfield, TA (1968) *Physiology of Stomata*, McGraw-Hill, London

Metcalfe CR and Chalk, I (1950) *Anatomy of the Dicotyledons*, Vol 1, Oxford University Press, London

Mineyuki, Y, Marc, J and Palevitz, BA (1988) Formation of the oblique spindle in dividing guard mother cells of *Allium* *Protoplasma*, **147**, 200–203

Mineyuki, Y, Marc, J and Palevitz, BA (1989) Development of the preprophase band from random cytoplasmic microtubules in guard mother cells of *Allium cepa* L *Planta*, **178**, 291–296

Mineyuki, Y and Palevitz, B A (1990) Relationship between preprophase band organisation, F-actin and the division site in *Allium* Fluorescence and morphometric studies on cytochalasin-treated cells *J Cell Sci*, **97**, 283–295

Miroslavov, E A (1972) Ultrastructural organization of guard cells of open and closing stomata (in Russian) *Dokl Akad Nauk SSR*, **203**, 939–941

Mishkind, M, Palevitz, B A and Raikhel, N (1981) Cell wall architecture normal development and environmental modification of guard cells of the Cyperaceae and related species *Plant Cell Environ*, **4**, 319–328

Mooney, H A, Ehleringer, J and Bjorkman, O (1977) The energy balance of leaves of the evergreen desert shrub *Atriplex hymenelytra* *Oecologia*, **29**, 301–310

Mulroy, J T (1979) Spectral properties of heavily glaucous and non-glaucous leaves of a succulent rosette plant *Oecologia*, **38**, 345–357

Mullinax, J B and Palevitz, B A (1989) Microtubule reorganisation accompanying preprophase band formation in guard mother cells of *Avena sativa* L *Protoplasma*, **149**, 89–94

Nelson, S P and Mayo, J M (1975) The occurrence of functional non-chlorophyllous guard cells in *Paphiopedilum* spp *Can J Bot*, **53**, 1–7

Nobel, PS (1991) *Physicochemical and Environmental Plant Physiology*, Academic Press, San Diego, CA

Nonami, H, Schulze, E-D and Ziegler, H (1990) Mechanisms of stomatal movements in response to air humidity, irradiance and xylem water potential *Planta*, **183**, 57–64

Outlaw, WH Jr Schmuck, L I and Tolbert, N E (1976) Photosynthetic carbon metabolism in the palisade parenchyma and spongy mesophyll of *Vicia faba* L *Plant Physiol*, **58**, 186–188

Outlaw, WH, Jr and Lowry, O H (1977) Organic acid and potassium accumulation in guard cells during stomatal opening *Proc Natl Acad Sci USA*, **74**, 4434–4438

Outlaw, WH, Jr, Mayne, B C Zenger, VE and Manchester, J (1980) Presence of both photosystems in guard cells of *Vicia faba* L implications for environmental signal processing *Plant Physiol*, **67**, 12–16

Outlaw, WH, Jr, Manchester, J and Zenger, VE (1981) The relationship between protein content and dry weight of guard cells and other single cell samples of *Vicia faba* L *Histochem J*, **13**, 329–336

Palevitz, B A (1981a) The structure and development of stomatal cells, in *Stomatal Physiology*, (eds PG Jarvis and TA Mansfield), Cambridge University Press, Cambridge pp 1–24

Palevitz, B A (1981b) Microtubules and possible microtubule nucleation centers in the cortex of stomatal cells as visualised by high voltage electron microscopy *Protoplasma*, **107**, 115–125

Palevitz, B.A. (1982) The stomatal complex as a model of cytoskeletal participation in cell differentiation, in *The Cytoskeleton in Plant Growth and Development*, (ed. C.W. Lloyd), Academic Press, London, pp. 346–376.

Palevitz, B.A. (1986) Division plane determination in guard mother cells of *Allium*: video time-lapse analysis of nuclear movements and phragmoplast rotation in the cortex. *Dev. Biol.*, **117**, 644–654.

Palevitz, B.A. (1991) Potential significance of microtubule rearrangement, translocation and reutilisation in plants, in *The Cytoskeletal Basis of Plant Growth and Form*, (ed. C.W. Lloyd), Academic Press, London, pp. 45–56.

Palevitz, B.A. (1993) Morphological plasticity of the mitotic apparatus in plants and its developmental consequences. *Plant Cell*, **5**, 1001–1009.

Palevitz, B.A. and Hepler, P.K. (1974a) The control of the plane of division during stomatal differentiation in *Allium*. I. Spindle reorientation. *Chromosoma*, **46**, 297–326.

Palevitz, B.A. and Hepler, P.K. (1974b) The control of the plane of division during stomatal differentiation in *Allium*. II. Drug studies. *Chromosoma*, **46**, 327–341.

Palevitz, B.A. and Hepler, P.K. (1976) Cellulose microfibril orientation and cell shaping in developing guard cells of *Allium*: the role of microtubules and ion accumulation. *Planta*, **132**, 71–93.

Palevitz, B.A. and Hepler, P.K. (1985) Changes in dye coupling of stomatal cells of *Allium* and *Commelina* demonstrated by microinjection of Lucifer yellow. *Planta*, **164**, 473–479.

Palevitz, B.A. and Mullinax, J.B. (1989) Developmental changes in the arrangement of cortical microtubules in stomatal cells of oat (*Avena sativa* L.). *Cell Motil. Cytoskeleton*, **13**, 170–180.

Palevitz, B.A., O'Kane, D.J., Kobres, R.E. and Raikhel, N.V. (1981) The vacuole system in stomatal guard cells of *Allium*. Vacuole movements and changes in morphology in differentiating cells as revealed by epifluorescence, video and electron microscopy. *Protoplasma*, **109**, 23–55.

Paliwell, G.S. (1969) Stomatal ontogeny and phylogeny. 1. Monocotyledons. *Acta Botanica Neerlandia*, **18**, 654–668.

Pallas, J.E. and Mollenhauer, H.H. (1972) Physiological implications of *Vicia faba* and *Nicotiana tabacum* guard cell ultrastructure. *Am. J. Bot.*, **59**, 504–514.

Pearson, C.J. and Milthorpe, F.L. (1974) Structure, carbon dioxide fixation and metabolism of stomata. *Aust. J. Plant Physiol.*, **1**, 221–236.

Pickett-Heaps, J.D. and Northcote, D.H. (1966) Cell division in the formation of the stomatal complex of young leaves of wheat. *J. Cell Sci.*, **1**, 121–128.

Raschke, K. (1975) Stomatal action. *Ann. Rev. Plant Physiol.*, **26**, 399–440.

Rasmussen, H. (1981) Terminology and classification of stomata and stomatal development – a critical survey. *Bot. J. Linn. Soc.*, **83**, 199–212.

Rechmann, U., Scheibe, R. and Raschke, K. (1990) Rubisco activity in guard cells compared with the solute requirement for stomatal opening. *Plant Physiol.*, **92**, 246–253.

Robberecht, R. and Caldwell, M.M. (1978) Leaf epidermal transmittance of ultraviolet radiation: its implications for plant sensitivity to ultraviolet-radiation induced injury. *Oecologia*, **32**, 277–287.

Robberecht, R., Caldwell, M.M. and Billings, W.D. (1980) Leaf ultraviolet optical properties along a latitudinal gradient in the arctic-alpine life zone. *Ecology*, **61**, 612–619.

Rutter, J.M. and Willmer, C.M. (1979) A light and electron microscopy study of the epidermis of *Paphiopedilum* spp. with emphasis on stomatal ultrastructure. *Plant Cell Environ.*, **2**, 211–19.

Sack, F.D. (1987) The development and structure of stomata, in *Stomatal Function*, (eds E. Zeiger, G.D. Farquhar and I.R. Cowan), Stanford University Press, Stanford, CA, pp. 59–90.

Sack, F.D. and Paolillo, D.J. (1983a) Protoplasmic changes during stomatal development in *Funaria*. *Can. J. Bot.*, **61**, 2515–2526.

Sack, F.D. and Paolillo, D.J. (1983b) Stomatal pore and cuticle formation in *Funaria*. *Protoplasma*, **116**, 1–13.

Sack, F.D. and Paolillo, D.J. (1985) Incomplete cytokinesis in *Funaria* stomata. *Am. J. Bot.*, **72**, 1325–1333.

Sahgal, P., Martinez, G.V., Roberts, K. and Tallman, G. (1993) Regeneration of plants from cultured guard cell protoplasts of *Nicotiana glauca* (Graham). *Plant Physiol. Suppl.*, **102**, 3i.

Sanchez, S.M. (1977) The fine structure of the guard cells of *Helianthus annuus*. *Am. J. Bot.*, **64**, 814–824.

Sato, N. (1985) Lipid biosynthesis in epidermal, guard and mesophyll protoplasts from leaves of *Vicia faba* L. *Plant Cell Physiol.*, **26**, 805–811.

Schimper, W.P. (1848) Recherches anatomiques et morphologiques sur les mousses, Dissertation, University of Strasbourg.

Schnabl, H., Vienken, J. and Zimmermann, U. (1980) Regular arrays of intramembranous particles in the plasmalemma of guard cell and mesophyll cell protoplasts of *Vicia faba*. *Planta*, **148**, 231–237.

Schwabe, W.W. (1968) Studies on the role of the leaf epidermis in photoperiodic perception in *Kalanchoë blossfeldiana*. *J. Exp. Bot.*, **19**, 108–113.

Schwendener, S. (1881) Über Bau und Mechanik der Spaltöffnungen. *Monatsber. Kgl. Akad. Wiss. Berlin*, **43**, 833–867.

Shimazaki, K.-I. and Zeiger, E. (1985) Cyclic and noncyclic photophosphosphorylation in isolated guard cell chloroplasts from *Vicia faba* L. *Plant Physiol.*, **78**, 211–214.

Shimazaki, K.-I., Igarashi, T. and Kondo, N. (1988) Protection by the epidermis of photosynthesis against UV-C radiation estimated by chlorophyll *a* fluorescence. *Physiol. Plant.*, **74**, 34–38.

Singh, A.P. and Srivastava, L.M. (1973) The fine structure of pea stomata. *Protoplasma*, **76**, 61–82.

Sorokin, H.P. and Sorokin, S. (1968) Fluctuations in the acid phosphatase activity of spherosomes in guard cells of *Campanula persicifolia*. *J. Histochem. Cytochem.*, **16**, 741–802.

Srivastava, L.M. and Singh, A.P. (1972) Stomatal structure in corn leaves. *J. Ultrastruct. Res.*, **39**, 345–363.

Stebbins, G. L. and Khush, G. S. (1961) Variations in the organisation of the stomatal complex in the leaf epidermis of monocotyledons and its bearing on their phylogeny. *Am. J. Bot.*, **48**, 51–59.

Stevens, R.A. and Martin, E.S. (1977) New structure associated with stomatal complex of the fern *Polypodium vulgare*. *Nature*, **265**, 331–334.

Stevens, R. A. and Martin, E. S. (1978a) Structural and functional aspects of stomata. 1. Developmental studies in *Polypodium vulgare*. *Planta*, **142**, 307–316.

Stevens, R.A. and Martin, E.S. (1978b) A new ontogenetic classification of stomatal types. *Bot. J. Linn. Soc.*, **77**, 53–64.

Terryn, N., Arias, M.B., Engler, G. *et al.* (1993) *rha1*, a gene encoding a small GTP binding protein from *Arabidopsis*, is expressed primarily in developing guard cells. *Plant Cell*, **5**, 1761–1769.

Vesque, M.J. (1889) De l'emploi des caractères anatomiques dans la classification des végétaux. *Bull. Soc. Bot. France*, **36**, 41–77.

Wanner, G., Formanek, H. and Theimer, R.R. (1981) The ontogeny of lipid bodies (spherosomes) in plant cells. *Planta*, **151**, 109–123.

Weissenböck, G., Hedrich, R. and Sachs, G. (1986) Secondary phenolic products in isolated guard cell, epidermal cell and mesophyll cell protoplasts from pea (*Pisum sativum* L.) leaves: distribution and determination. *Protoplasma*, **134**, 141–148.

Weyers, J.D.B. and Fitzsimons, P.J. (1982) The non-osmotic volume of *Commelina* guard cells. *Plant Cell Environ.*, **5**, 417–421.

Weyers, J.D.B. and Meidner, H. (1990) *Methods in Stomatal Research*, Longman Scientific and Technical, Harlow, UK.

Wick, S.M. (1991) The preprophase band, in *The Cytoskeletal Basis of Plant Growth and Form*, (ed. C.W. Lloyd), Academic Press, London, pp. 231–244.

Wille, A.C. and Lucas, W.J. (1984) Ultrastructural and histochemical studies on guard cells. *Planta*, **160**, 129–142.

Willmer, C.M. (1983) *Stomata*, Longman, London.

Willmer, C.M. (1993) The evolution, structure and functioning of stomata. *Bot. J. Scot.*, **46**, 433–445.

Willmer, C.M. and Mansfield, T.A. (1970) Further observations of cation-stimulated stomatal opening in isolated epidermis. *New Phytol.*, **69**, 639–645

Willmer, C.M. and Sexton, R. (1979) Stomata and plasmodesmata. *Protoplasma*, **100**, 113–24.

Willmer, C.M., Jamieson, A. and Birkenhead, K. (1987) Leaf epidermal tissue is unsuitable to use for studying biochemical aspects of stomatal functioning. *Plant Sci.*, **52**, 105–110.

Wolniak, S.M. (1991) Patterns of regulation during mitosis, in *The Cytoskeletal Basis of Plant Growth and Form*, (ed. C.W. Lloyd), Academic Press, London, pp. 209–230.

Yang, M. and Sack, F. (1993) An *Arabidopsis* mutant with multiple stomata. *Plant Physiol. Suppl.*, **102**, 122.

Ziegenspeck, H. (1938) Die Micellierung der Turgeszen-mechanismen. Teil 1. Die Spaltöffnungen (mit phylogenetischen Ausblicken). *Bot. Arch.*, **39**, 268–309, 332–372.

Zeiger, E. and Hepler, P.K. (1979) Blue light-induced, intrinsic vacuolar fluorescence in onion guard cells. *J. Cell Sci.*, **37**, 1–10.

Zeiger, E. and Stebbins, G.L. (1972) Developmental genetics in barley: a mutant for stomatal development. *Am. J. Bot.*, **59**, 143–148.

Ziegler, H., Shmueli, E. and Lange, E. (1974) Structure and function of the stomata of *Zea mays*. I. The development. *Cytobiologie*, **9**, 162–168.

4 The mechanics of stomatal movements

4.1 Introduction

Considerable progress has been made in understanding the osmotic changes that generate turgor increases in a guard cell. However, despite the fundamental importance of mechanical interactions between guard cells and neighbouring cells in determining stomatal aperture, much still remains to be elucidated; there is little detailed information on the relevant cell volumes, on wall morphology or on the physico-chemical properties of the wall polymers that translate turgor differences to changes in cell shape during stomatal movements. This chapter deals primarily with the water relations of guard cells and epidermal tissue, and the different modes of deformation of guard cells during stomatal movements. Additionally, the osmotic relations of guard cell protoplasts (GCPs) will be considered, including the dynamics of membrane recycling.

4.2 *Water relations of guard cells*

The water potential of a guard cell (ψ) is essentially equal to the hydrostatic turgor pressure (P) minus the solute pressure (Π_s), and the matric pressure (τ), the latter being commonly included with the solute pressure and called osmotic pressure (Π) (see Nobel, 1991). Thus:

$$\Psi = P - \Pi \quad \text{(Pa)} \tag{4.1}$$

The osmotic pressure (Π) can only be positive, while the turgor pressure (P) is normally positive or zero, although some authorities consider that under special circumstances negative P can occur. (The reader should note that in many texts, the osmotic pressure may be expressed as the osmotic potential [ψ_Π], defined as the negative of the osmotic pressure, and the turgor pressure as the pressure potential, ψ_P.)

Accumulation of solutes in the guard cells (see Chapter 8) increases the osmotic pressure, thus lowering the water potential ψ and causing uptake of water until the turgor pressure increases. The motive force for the opening of stomata is the increased turgor pressure of the guard cells (von Mohl, 1856; Heath, 1938); thus, when guard cells are fully turgid, P should be close to Π. In practice, guard cells have to take up water from the surrounding apoplast. The apoplast has its own water potential, with components due to the matrix pressure of the cell wall polymers, the solute pressure due principally to apoplastic ions and a *negative* hydrostatic pressure arising from surface tension effects at the numerous air–liquid interfaces of the cell wall interstices (see Nobel, 1991).

Meidner and co-workers have measured guard cell turgor pressures directly with pressure probes inserted into the cells (Meidner and Edwards, 1975; Edwards *et al.*, 1976; Edwards and Meidner, 1979). The values they obtained for *Tradescantia virginiana* were 0.1 MPa (0.1 MPa = 1 bar ≈ 1 atmosphere) for guard cells of nearly closed stomata and about 0.8 MPa for guard cells of stomata open to 26 μm (Fig. 4.1). These values are surprisingly low but they may be a feature peculiar to *T. virginiana* since much higher turgor pressures have been obtained for guard cells of *Commelina communis*, increasing from 0.42 MPa at 7 μm to 1.95 MPa at 18 μm stomatal aperture (Meidner, 1982). The measured turgor pressure was significantly lower than expected from the osmotic pressure, however, and was interpreted as evidence for peri-stomatal transpiration (see Section 6.6) against which the GCPs have to compete for water (Edwards and Meidner, 1979).

The increased turgor results from the increased osmotic pressure of the guard cells, although in exceptional cases it has been reported that the osmotic pressure of the guard cells remains fairly constant while the osmotic pressure of epidermal cells increases (Meidner and Mansfield, 1968, p. 21). Early measurements of guard cell osmotic pressure used long incubations in plasmolysing solutions, which are now known to underestimate the real osmotic pressure due to solute leakage from the cells during plasmolysis (Fischer 1973; Willmer and Beattie, 1978; Raschke, 1979). Additional errors may arise from cell volume shrinkage during plasmolysis which is not accounted for, exosmosis phenomena occurring in the unstirred layers of the plasmolyticum surrounding the cells and entry of plasmolytica into the cell. With appropriate controls and short plasmolysis times, osmotic pressures up to about 4 MPa in open stomata have been recorded for *Vicia* (Raschke, 1979). Alternative cryoscopic techniques based on depression of the freezing point as a measure of the solute concentration also give high values of the osmotic pressure. Using such methods, Bearce and Kohl (1970) found much higher values for guard cells than had hitherto been generally obtained,

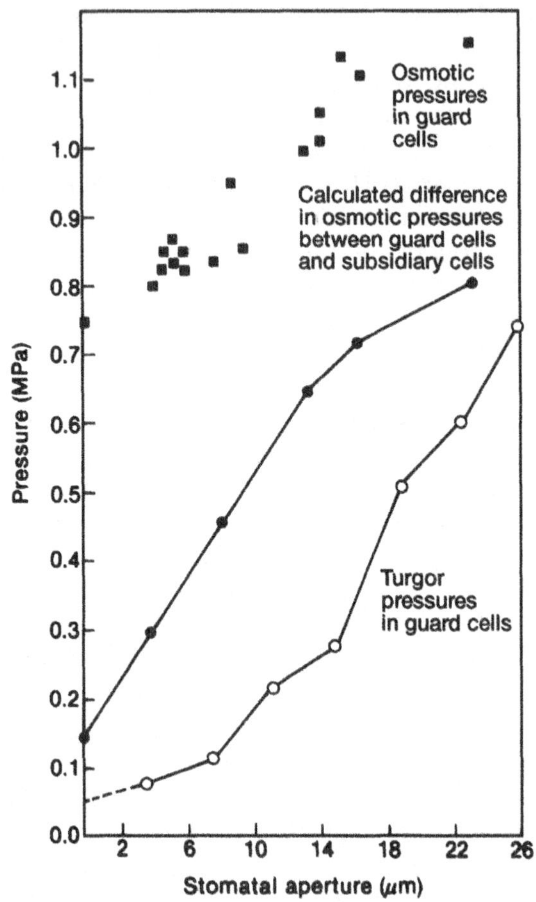

Figure 4.1 The relationship between guard cell osmotic pressure (■), the difference in osmotic pressure between guard cells and subsidiary cells (●), guard cell turgor pressure (○) and stomatal aperture in *Tradescantia virginiana*. The osmotic pressure values were determined at incipient plasmolysis and may therefore be underestimates due to cell shrinkage and ion leakage. Adapted from Edwards and Meidner (1979)

although values varied greatly at each stomatal aperture. Thus, when stomata of *Chrysanthemum* were closed guard cell osmotic pressures ranged from 1.2 to 4.6 MPa, although the relationship between guard cell osmotic pressure and aperture was considered linear. In *Pelargonium* this relationship had a slope of 0.32 MPa μm^{-1}. Slade and Willmer (unpublished) have also used cryoscopic methods to measure osmotic pressures of guard cells of *C. communis* and *T. virginiana* and found much higher osmotic pressures than had previously been obtained by the incipient plasmolysis method. Additionally, for *T. virginiana* and *C. communis* values of osmotic pressure were in the following order: guard cells > inner lateral subsidiary cells > terminal subsidiary cells > outer lateral subsidiary cells (present only in *C. communis*) > epidermal cells. However, further work is needed to fully evaluate the accuracy of the cryoscopic method particularly since there is some evidence that colloids do not behave as dissolved solutes do in terms of depression of freezing points. In addition, cell contents are comprised of a mixture of solutes and therefore freeze or melt over a range of temperature.

MacRobbie (1980) studied the relationship between guard cell osmotic pressure and stomatal aperture using epidermal strips of *C. communis* incubated in plasmolytica for short periods. To 'isolate' guard cells within the epidermal strip the tissue was incubated at pH 3.9 for several hours to kill the epidermal and subsidiary cells. She found that the value of the guard cell osmotic pressure per μm stomatal aperture increased as apertures increased, reaching 0.44 MPa μm^{-1} at apertures above 15 μm (see Table 4.1). With stomatal opening against surrounding cell turgor (subsidiary and epidermal cells remained alive), MacRobbie found that an additional 0.8 MPa μm^{-1} was necessary to achieve the same apertures as those when only guard cells were alive. MacRobbie (1980) concluded that much of the change in osmotic pressure was derived from the accumulation of potassium salts (see Chapter 8), although additional osmotica was required to achieve the measured values of osmotic pressure (there is not complete agreement on this aspect, however, see Fitzsimons and Weyers, 1986). Furthermore, it is considered that the cytoplasm must adjust osmotically to match the osmotic pressure of the vacuole as stomata open and close, since the tonoplast has little mechanical strength. This may involve accumulation or synthesis of considerable amounts of compatible (i.e. non-ionic) organic solutes such as betaines, amino acids or sugars in the cytoplasm. Evidence to date suggests that changing levels of a sucrose pool could be the major cytoplasmic osmoticum.

Table 4.1 Osmotic parameters of *Commelina communis* guard cells at different ranges of stomatal aperture

Guard cell parameter (units)	Stomatal aperture range (μm)		
	5–10	10–15	15–20
Volume at lower and upper end of aperture range (μm^3)	5000–6000	6000–7000	7000–8000
Osmotic pressure requirement for opening (MPa μm^{-1})	0.20	0.31	0.44
Turgor pressure requirement for opening (MPa μm^{-1})	0 18	0.26	0.30
Elastic modulus (MPa)	7	10	11

Data are taken from MacRobbie (1980) and MacRobbie and Lettau (1980a) and refer to guard cells 'isolated' by incubation in MES buffer at pH 3.9 for 2 h. The turgor increase required to open the pore by 1 μm is 1.67 times as great when the pore is 10–15 μm in width as it would be between 5 and 10 μm. (Table adapted from Weyers and Meidner, 1990)

For large stomatal openings and guard cell volumes the guard cell walls will be expected to stiffen and this may explain the increase in osmotic pressure per micron with increasing apertures in *Commelina* as the guard cell approaches its maximum size On the assumption that guard cells behave like ideal elastic bodies (which kidney-shaped guard cells may do over limited aperture changes), the volume change (ΔV) caused by a given change in pressure (ΔP) is related to the flexibility of the cell wall expressed as the elastic modulus (ε), where

$$\varepsilon = \frac{\Delta P}{\Delta V / V} \qquad (4\ 2)$$

In 'isolated' guard cells of *C communis*, ε increased from 7 MPa below 5 µm to 11 MPa above 15 µm (MacRobbie, 1980) (Table 4 1) However, in *Vicia faba*, ε was estimated to be 7 MPa, with little indication of wall stiffening up to apertures of 16 µm (Raschke, 1979) More detailed modelling of cell wall properties has been based on polymer elasticity (Wu and Sharpe, 1979) They found that the guard cell wall initially behaved as an isotropic polymer during expansion, then changed to an anisotropic phase, once a critical volume was exceeded and the micellation pattern restricted radial expansion

Volumes of the GCP and of the cell wall are integral components both in simple determinations of ε or as variables in more complex models (Sharpe and Wu, 1978) Protoplast volumes are also required to convert the number of moles of solutes experimentally measured to concentrations (see Chapter 8) and hence osmotic pressures Estimates of volume are difficult to make, however, and there is considerable variation of values in the literature (e g Fig 4 2) An average value for *V faba* based on microscopic observation indicates that the protoplast volume of a guard cell from a closed stoma is about 2 4 pl and increases linearly by about 0 275 pl µm^{-1} (see also Table 3 1)

4.3 The physico-chemical aspects of stomatal movements

4.3.1 Phases of stomatal opening

Stålfelt (1927) proposed two phases of stomatal opening, i e 'Spannungsphase' (tension phase) and 'Motorische Phase' (motor phase) Another phase may be termed the 'aperture maintenance phase' in which the stomatal aperture reaches a certain level and then remains constant (Rogers *et al* , 1979) 'Spannungsphase' is a period preceding the motor phase in which guard cells ready themselves for stomatal opening; guard cells inflate due to turgor increases but the pore does not open, or opens to only a tiny extent. The motor phase is that phase in which there is an approximately linear relationship between pore opening and guard cell turgor (Fig. 4.2). A mathematical model (Sharpe and Wu, 1978) predicted such a near linear relationship using information from *V. faba* stomata, although an additional pronounced inflection of the line occurred at about 1 μm aperture, possibly representing the transition from Spannungsphase to the motor phase (Fig. 4.2).

4.3.2 The mechanical advantage of neighbouring cells over guard cells

Stomatal movements *in vivo* are determined by the relative turgor pressures within the guard cells acting against the turgor of the neighbouring cells. A common misconception about the pressure relations of

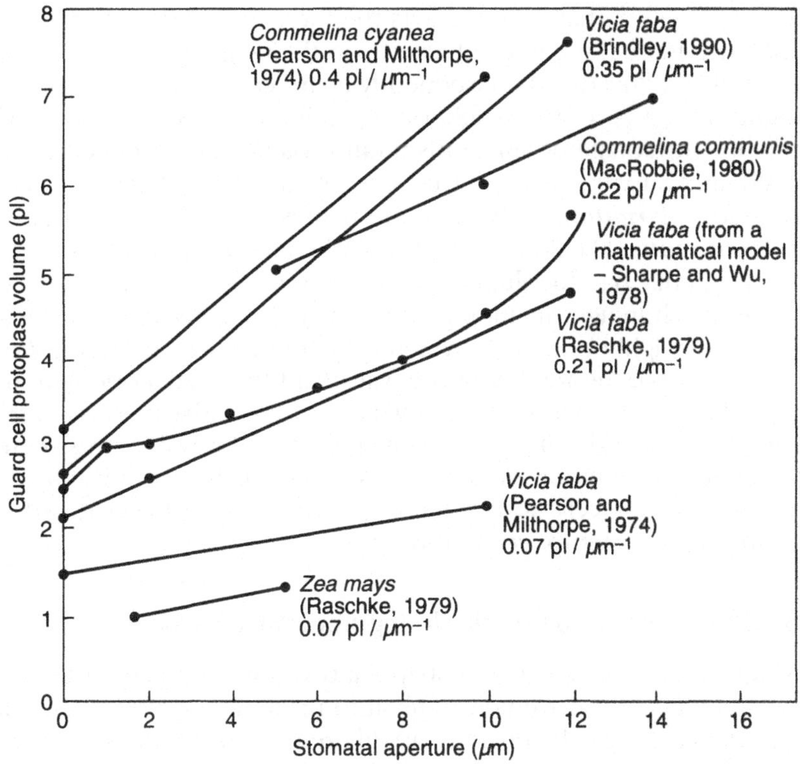

Figure 4.2 The relationship between guard cell protoplast volume with guard cells *in situ* and stomatal aperture. Labels with each line are the names of the species, the source of the information and the increase in volume per μm increase in stomatal aperture. Based on Fig. 4.3; Willmer (1983).

stomatal movements is that the pore aperture is a direct function of the pressure differences between guard cells and neighbouring cells (Stålfelt, 1966). Indeed, von Mohl (1856) and others, during the early days of investigations into the mechanism of stomatal movements, seemed to consider that guard cells have a leverage over neighbouring cells. In fact, the opposite is true; neighbouring epidermal or subsidiary cells have a 'mechanical advantage' over the guard cells (DeMichele and Sharpe, 1973; Edwards *et al.*, 1976; MacRobbie, 1980) and pressures in guard cells must build up to a certain level above pressures in neighbouring cells before the stomatal pore begins to open. The period of build-up of guard cell pressure corresponds to the Spannungsphase of stomatal opening. This mechanical interaction between guard cells and neighbouring cells has also been referred to as the antagonism ratio (Cooke *et al.*, 1976).

The mechanical advantage results from the turgor-generated forces originating in the guard cells and the adjacent cells. The force on the ventral wall of the guard cell (which is facing the pore and has limited retaining forces) partially counteracts the force on the dorsal wall. Thus the effective force the guard cell exerts on the neighbouring cell via the dorsal wall is reduced and consequently guard cell turgor pressure must be higher than that of the neighbouring cell for the pore to open. The mechanical advantage or antagonism ratio has been determined experimentally and varying values have been obtained for different species (e.g. 1.6 for *T. virginiana*, Meidner and Bannister, 1979; greater than 3.1 for *C. communis*, MacRobbie, 1980). Mathematical modelling of guard cell movements has highlighted a number of factors that affect the mechanical advantage, including the contact area between the guard cell and the neighbouring cell (which may not be constant during stomatal movements), the rigidity of the walls, and the spatial orientation of the guard cell and epidermal cells (Wu *et al.*, 1985; Sharpe *et al.*, 1987). A reduction in guard cell size in plants grown under water stress conditions has also been interpreted in terms of the mechanical advantage, as a means for guard cells to maintain open pores with lower guard cell turgor pressures (Spence *et al.*, 1986).

4.3.3 Deformation of guard cells during stomatal movements

Although changes in guard cell shape and volume are central to stomatal movements, it is extremely difficult to make the necessary measurements of guard cell dimensions at different stomatal apertures from fresh or frozen sectioned material. Qualitative descriptions are available for a wide range of potential guard cell movement mechanisms (see Schwendener, 1881; Haberlandt, 1896; Allaway and Milthorpe, 1976; Zeigler, 1987). The four main types are described below, although it may be that a combination of these guard cell deformations may occur.

(1) The simplest type is reported to be in the mosses such as *Mnium cuspidatum* (Haberlandt, 1896) and possibly some ferns (von Guttenberg, 1971) in which the stomatal pore forms as a result of the guard cells becoming narrower and deeper; upon closure the ventral walls become more rounded and the guard cells wider and less deep. Only the ventral walls of the guard cells increase in length as the stomata open and it is this wall that is deformed. Figure 4.3 illustrates these deformations.

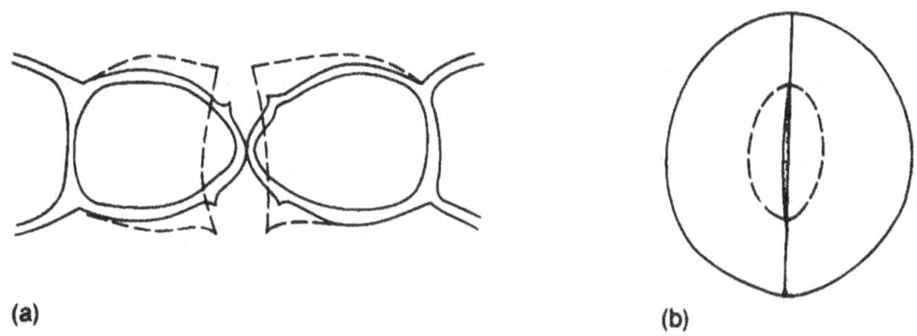

Figure 4.3 Deformation of guard cells of *Mnium cuspidatum*, (a) Side view, (b) plan view (- - -, stoma open, —, stoma closed) From Willmer (1983)

(2) Another type which has been reputed to occur in elliptical stomata is one in which the guard cells tilt outwards above the plane of the epidermis to effect opening. It seems that the outer cuticular ledges of the pair of guard cells separate during this movement resulting in an increase in the eisoidal aperture (the aperture between the lips) rather than a genuine separation of the stomatal throat (Fig. 4.4).

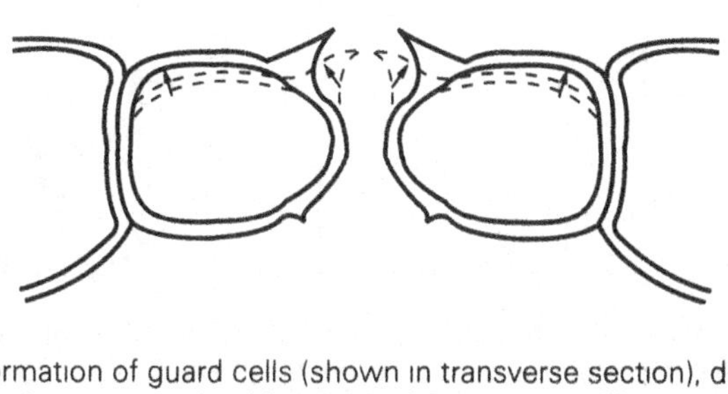

Figure 4.4 Deformation of guard cells (shown in transverse section), during stomatal opening of elliptical stomata with a 'drawbridge'-type section The arrows indicate the direction of movement of the guard cells during opening (- - -, stoma closed, —, stoma open) From Weyers and Meidner (1990)

(3) Guard cell movements in other elliptical-shaped stomata have certain features in common which allow them to be discussed together. The length of a stoma remains virtually constant during opening and closing movements (e.g. Schwendener, 1881; Meidner and Mansfield, 1968), although it has occasionally been questioned. For example, Saxe (1979) found a decrease of 5–10% in pore length (which may be a reflection of stomatal length) during wide opening in *C. communis*. The guard cells, however, increase in length upon stomatal opening: both the ventral and dorsal walls increase by about the same absolute amounts but, since the ventral wall is shorter than the dorsal wall, the former will extend relatively more than the latter (Raschke, 1975; Meidner and Willmer, 1975) (Fig. 4.5). Thus, the thick ventral walls must allow some stretching. The radial orientation of the cellulose microfibrils will allow expansion of the guard cell walls in a lengthwise direction, although it is not known if radial micellation is a common feature of all kidney-shaped guard cells. Such a radial arrangement of the microfibrils will allow the cross-sectional shape to change, but the cross-sectional area of the outer dimensions of the guard cells remains almost constant. Using sausage-shaped rubber balloon models with masking tape wrapped around them to represent the radial micellations, Aylor *et al.* (1973) demonstrated the importance of such a structuring of the guard cell walls. The rubber balloon model, however, has a particular weakness in that the overall length of the model stoma decreases as pressure is increased in the balloons and the pore forms. This is due to the ventral wall being relatively rigid because of the tape but, as indicated above, in reality the ventral wall does stretch in its long axis as a pore opens.

According to Raschke (1975, 1979) opening is effected in the following manner: the ends of the guard cells swell and push each other apart

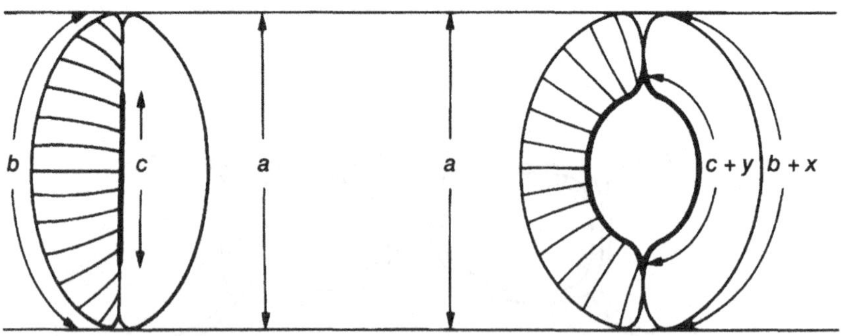

Figure 4.5 Open and closed stomata showing that the overall length, *a*, does not change during stomatal movements but the lengths of the dorsal, *b*, and ventral, *c*, walls of the guard cells increase by distances *x* and *y*, respectively, upon stomatal opening The radial direction of the cellulose microfibrils are also diagrammatically represented in one guard cell of each stoma From Willmer (1983)

and, since the length of the stomatal apparatus remains virtually constant, guard cell expansion occurs outwards into the surrounding epidermal or subsidiary cells (see Fig. 4.6). However, when one guard cell of a pair is punctured the intact guard cell is still able to expand in the usual manner. This tends not to support the view that the guard cells push against each other (Meidner and Willmer, 1975), although it may still be possible that the ends of the guard cells are anchored firmly enough to give sufficient leverage to allow an intact guard cell to swell in the manner described by Raschke.

Although the cross-sectional area of the outer dimensions of the guard cell may not change during stomatal movements in those species having the radial arranged microfibrils, Raschke (1979) considers that the cross-sectional area of the protoplast does alter (Fig. 4.6). According to Raschke, when stomata are closed or nearly closed the cross-sectional shape of the middle portion of the protoplast is triangular in shape but upon stomatal opening becomes more spherical. Changes also occur in the cross-sectional shape of the end regions of the guard cells during stomatal movements: in the bottom of the ventral wall the protoplast exists as a ridge when stomata are closed which is stretched out as turgor increases and stomata open (Fig. 4.6). Such changes in protoplast shape and volume would be associated with considerable decreases in the thickness of the wall of the guard cell.

(4) The wall thickenings of grass-type guard cells are very different from those found in kidney-shaped guard cells. The thin-walled ends of the dumb-bell-shaped guard cells swell as the stoma opens and are able to push the middle portions apart (Fig. 4.7). The middle portions of the guard cells are heavily thickened on the upper and lower surfaces of the lateral walls and these are relatively inflexible. As a result a pore is formed with parallel sides. According to Raschke (1979) the bulbous ends of the guard cells act like bellows pushing the mid-portions apart and, thus, resemble the system that he envisages in elliptical stomata.

Detailed analysis of stomatal mechanics are hampered by the lack of quantitative measurements of guard cell volumes, wall morphology and wall properties. There have been a number of attempts to model guard cell movements and develop a mathematical description of guard cell mechanics based on parameters that are more readily determined (e.g. DeMichele and Sharpe, 1973; Aylor *et al.*, 1973; Cooke *et al.*, 1976). Such models have helped us to obtain a clearer understanding of the mechanics of stomatal movements and highlighted a number of important features summarized by Raschke (1979), including the degree of mechanical interaction between the guard cells and neighbouring cells in determining stomatal apertures. However, there are considerable difficulties in comparing non-living, physical structures, such as steel beams, with the more dynamic biological materials, such as cell walls,

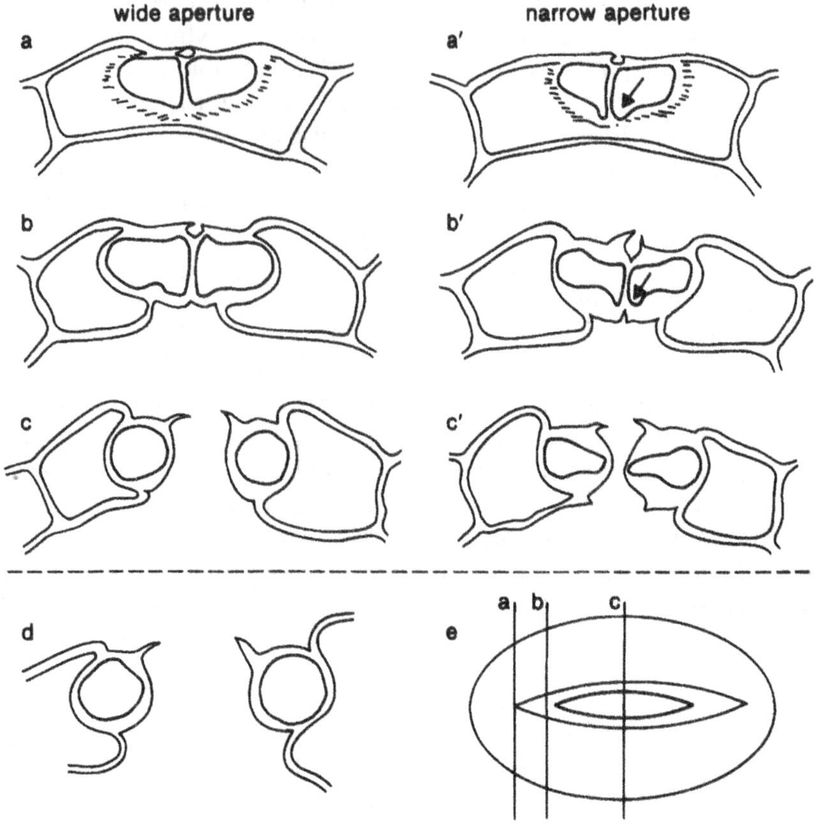

Figure 4.6 Tracings of optical cross-sections through an intact stoma of *Vicia faba* in a wide (10 5 μm) and narrow (1 5 μm) state of opening The stoma was first photographed in the open state and then plasmolysed with a sucrose solution Approximate position of the focal planes is shown in (e) (d) is a median section through another stoma which had opened to an aperture of 21 2 μm Arrows point to the fold that forms in the polar parts of the guard cell when the stoma closes From Raschke (1979)

which surround living cells (reviewed in Sharpe *et al.*, 1987). This is particularly the case for guard cells, where the wall is anisotropic (i.e. the physical properties of the wall, such as stretching and elasticity, are dependent on the direction of measurement). Furthermore, changes in wall thickness occur during stomatal movements and the wall may alter its properties due to binding and release of different ion species as opening and closing occurs. Under such circumstances the cell wall material is unlikely to obey Hooke's law, a fact not allowed for in some of the earlier models. Nevertheless, the most recent models correlate well with experimentally determined data for *V. faba* (see Sharpe *et al.*, 1987), providing a simple geometric interpretation of the mechanical

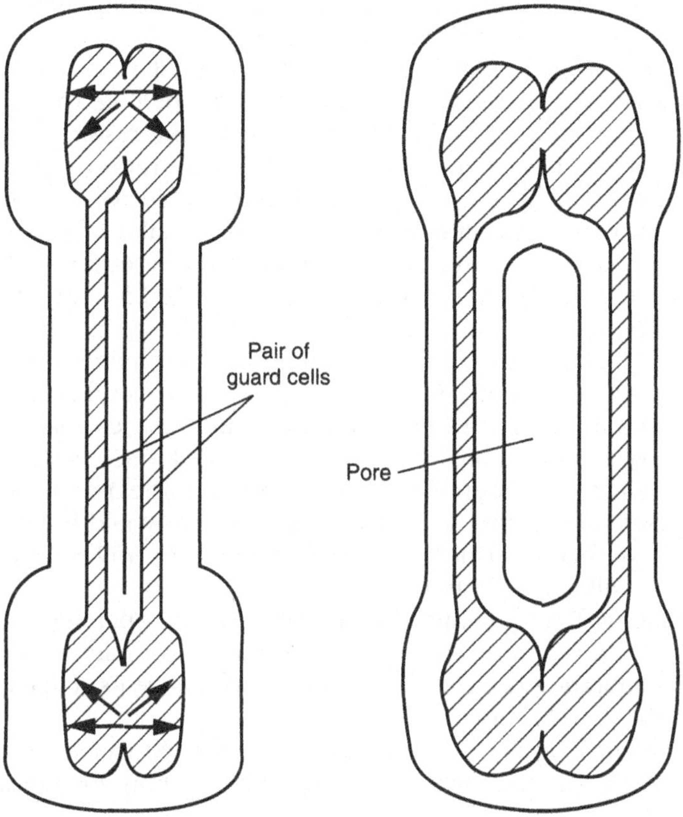

Figure 4.7 Diagrammatic representation of deformation of guard cells in a grass-type stoma during stomatal opening. The cell protoplasts in the bulbous ends of the guard cells, which are connected to each other via pores, swell and push the guard cells apart in the direction indicated by the arrows. After Willmer (1983).

advantage and additional insights into the predicted changes in volume associated with the transition from Spannungsphase to Motorische Phase (Sharpe and Wu, 1978).

4.4 Morphological changes of the plasma membrane and tonoplast of guard cells during stomatal movements

During stomatal opening and closing large changes of the volume and surface area of guard cells occur. For example, it has been estimated that when stomata of *C. communis* open to about 12 μm from the fully closed position the outer dimension (i.e. the perimeter of the cell wall) increases by about 10% (Willmer, unpublished). The plasma membrane surface area may increase by an even greater percentage because, during guard cell swelling, the wall volume remains constant and is stretched in a thinner layer around the protoplast. This extra volume is presumably taken up by additional swelling of the protoplast.

Since plasma membranes only have the physical ability to stretch by 2–3% (Wolfe and Steponkus, 1983), during opening the additional increase in surface area of membrane must be the result of exocytotic incorporation of new membrane material or extension of infoldings of existing membrane. Conversely, upon closure the decrease in protoplast volume must be associated with endocytotic uptake of plasma membrane or the membrane must become more convoluted with infoldings. It is difficult to establish whether the unfolding and folding of corrugations of the plasma membrane take place with the opening and closing of stomata. Electron micrographs of guard cells often depict an undulating topography of the plasma membrane, but such studies are particularly prone to fixation artefacts. Coated pits and vesicles associated with endocytosis are occasionally seen in guard cells (see Section 3.3.7). However, efforts to observe incorporation (in vesicles) of lucifer yellow inside guard cells during stomatal closure, which would indicate endocytosis, have been unsuccessful (Hillmer *et al.*, 1990; Willmer and O'Parka, in Willmer and Pantoja, 1991).

The tonoplast also changes its surface area with opening and closing of stomata, although the situation may be more complex than with the plasma membrane since the vacuole may also change its shape and form considerably (see Section 3.3.6).

It is evident that more investigations are needed to clarify the situation about possible membrane recycling linked to both membranes.

4.5 Physiological and mechanical properties of GCPs

It is appropriate to discuss the physiological and physical properties of GCPs at this point since they can be made to swell and contract osmotically, and in response to physiological stimuli, a property also of intact guard cells. Figure 4.8 shows how the volume, surface area and diameter of guard cells changes as they reach equilibrium in a range of osmotic concentrations of mannitol. GCPs, like all protoplasts, behave almost like perfect osmometers because they have insignificant internal turgor pressure. They, thus obey the Boyle–van't Hoff equation:

$$\Pi(V - NOV) = nRT \tag{4.3}$$

where Π is the solute pressure of the incubation medium (Pa), V is the GCP volume (m^3), NOV is the non-osmotic volume of the protoplasts (m^3), n is the number of solute ions or molecules in the incubation medium (mol), R is the gas constant (8.314 m^3 Pa mol^{-1} K^{-1}) and T is the absolute temperature (293 K = 20°C).

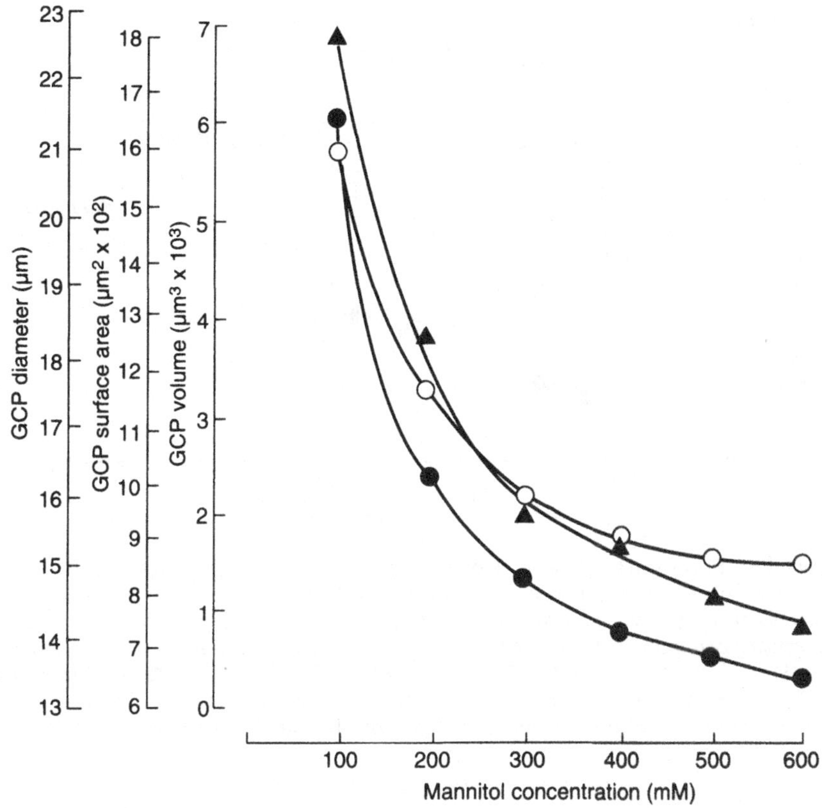

Figure 4.8 Volume (○), surface area (●) and diameter (▲) changes of guard cell protoplasts of *Commelina communis* in response to varying concentrations of osmoticum (mannitol).

If the equation is rearranged so that V is expressed in terms of Π then, for a perfect osmotic system, a plot of V against the reciprocal of Π is linear with a slope of nRT which cuts the y-axis at a value equal to the non-osmotic volume. Figure 4.9 (from Weyers and Meidner, 1990, p. 169) shows some Boyle–van't Hoff plots for GCPs from a number of species. Exactly what the non-osmotic volume consists of is not clear, but Fitzsimons and Weyers speculate that it is mainly chloroplast starch. If this is true, then the non-osmotic volume will change with stomatal aperture and possibly as GCPs change their volumes. Weyers and Fitzsimons (1982, 1985) found that the *NOV* of GCPs varied according to species and that the volume was relatively large being 10–40% of the total protoplast volume in *C. communis*. Hence studies of the water relations of guard cells may have significant error if this factor is not taken into consideration.

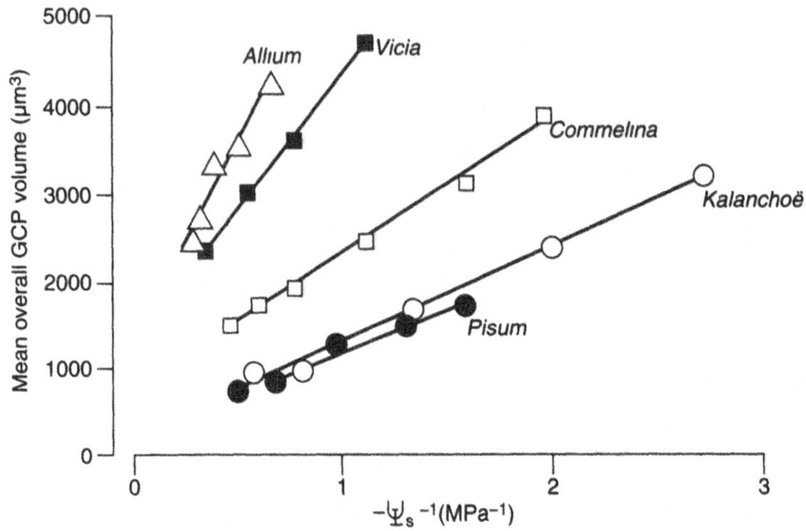

Figure 4.9 Boyle–van't Hoff analysis of guard cell protoplasts (GCPs) from five species The mean solute contents (perfect solute basis) and *NOV*s estimated from the least squares linear regressions are as follows *Allium porrum* (△), 1780 fmol and 1231 μm^3, *Commelina communis* (□), 590 fmol and 732 μm^3, *Kalanchoe daigremontiana* (○), 430 fmol and 185 μm^3, *Pisum sativum* (●), 360 fmol and 234 μm^3, and *Vicia faba* (■), 1250 fmol and 1052 μm^3 The high degree of linearity for each data set shows that the GCPs were acting as osmometers within the concentration ranges used Data of Weyers and Fitzsimons (1985)

Figure 4.8 shows that the surface area of GCPs increases by about 40% as they increase in diameter from 14.3 to 21.2 μm. As indicated above, membranes can only stretch by about 2–3% and GCPs would burst unless new membrane material was inserted into the existing membrane. Conversely upon shrinkage membrane material must be reabsorbed to maintain the spherical shape of the protoplasts. Diekmann *et al.* (1993) showed that when GCPs from pea shrank rapidly under hypertonic conditions, large numbers of endocytotic vesicles, termed osmocytotic vacuoles by the investigators, were formed by invagination of the plasma membrane. However, these vacuoles did not fuse with existing vacuoles nor with the plasma membrane upon subsequent reswelling of the GCP under hypotonic conditions.

References

Allaway, W.G. and Milthorpe, F.L. (1976) Structure and functioning of stomata, in *Water Deficits and Plant Growth*, Vol. IV, Academic Press. New York.

Aylor, D.E., Parlange, J.Y. and Krikorian, A.D. (1973) Stomatal mechanics. *Am. J. Bot.*, **60**, 163–171.

Bearce, B.C. and Kohl, H.C., Jr (1970) Measuring osmotic pressure of sap within live cells by means of a visual melting point apparatus. *Plant Physiol.*, **46**, 515–519.

Brindley, H.M. (1990) Fluxes of $^{86}Rb^+$ in 'isolated' guard cells of *Vicia faba* L. *Planta*, **181**, 432–439.

Cooke, J.R., DeBaerdemaeker, J.G., Rand, R.H. and Mang, H.A. (1976) A finite element shell analysis of guard cell deformation. *Trans. Am. Soc. Agric. Eng.*, **19**, 1107–1121.

DeMichele, D.W. and Sharpe, P.I.H. (1973) An analysis of the mechanics of guard cell motion. *J. Theor. Biol.*, **41**, 77–96.

Diekmann, W., Hedrich, R., Raschke, K. and Robinson, D.G. (1993) Osmocytosis and vacuolar fragmentation in guard cell protoplasts: their relevance to osmotically-induced volume changes in guard cells. *J. Exp. Bot.*, **44**, 1569–1577.

Edwards, M. and Meidner, H. (1979) Direct measurements of turgor pressure potentials. *J. Exp. Bot.*, **30**, 829–837.

Edwards, M., Meidner, H. and Sheriff, D.W. (1976) Direct measurements of turgor pressure potentials of guard cells. II. The mechanical advantage of subsidiary cells, the Spannungsphase and the optimum leaf water deficit. *J. Exp. Bot.*, **27**, 163–171.

Fischer, R.A. (1973) The relationship of stomatal aperture and guard cell turgor pressure in *Vicia faba*. *J. Exp. Bot.*, **24**, 387–399.

Fitzsimons, P.J. and Weyers, D.J.B. (1986) Volume changes of *Commelina communis* L. guard cell protoplasts in response to K^+, light and CO_2. *Physiol. Plant.*, **60**, 463–468.

von Guttenberg, H. (1971) Bewegungsgewebe und Perzeptionsorgane, in *Handbuch der Pflanzenanatomie*, Vol. V, (eds S. Carlquist, W. Zimmermann, P. Ozenda and H.D. Wulff), Gebrüder Borntraeger, Berlin, pp. 203–219.

Haberlandt, G. (1896) *Physiologische Pfanzenanatomie*, 2nd edn, Englemann, Leipzig.

Heath, O.V.S. (1938) An experimental investigation of the mechanism of stomatal movement with some preliminary observations upon the response of the guard cells to 'shock'. *New Phytol.*, **37**, 385–395.

Hillmer, S., Hedrich, R., Robert-Nicoud, M. and Robinson, D.G. (1990) Uptake of Lucifer yellow CH in leaves of *Commelina communis* is mediated by endocytosis. *Protoplasma*, **158**, 142–148.

MacRobbie, E.A.C. (1980) Osmotic measurements on stomatal cells of *Commelina communis* L. *J. Membr. Biol.*, **53**, 189–198.

MacRobbie, E.A.C. and Lettau, J. (1980) Ion content and aperture in isolated guard cells of *Commelina communis* L. *J. Membr. Biol.*, **53**, 199–205.

Meidner, H. (1982) Guard cell pressures and wall properties during stomatal opening. *J. Exp. Bot.*, **33**, 355–359.

Meidner, H. and Edwards, M. (1975) Direct measurements of turgor pressure potentials of guard cells. I. *J. Exp. Bot.*, **26**, 319–330.

Meidner, H. and Mansfield, T.A. (1968) *Physiology of Stomata*, McGraw-Hill, New York.

Meidner, H. and Bannister, P. (1979) Pressure and solute potentials in stomatal cells of *Tradescantia virginiana*. *J. Exp. Bot.*, **30**, 255–265.

Meidner, H. and Willmer, C.M. (1975) Mechanics and metabolism of guard cells. *Curr. Adv. Plant Sci.*, **17**, 1–15.

von Mohl, H. (1856) Welche Ursachen bewirken Erweiterung und Verengung der Spaltoffnungen? *Bot. Z.*, **14**, 697–704, 713–721.

Nobel, P.S. (1991) *Physicochemical and Environmental Plant Physiology*, Academic Press, San Diego, CA.

Raschke, K. (1975) Stomatal action. *Ann. Rev. Plant Physiol.*, **26**, 309–340.

Raschke, K. (1979) Movements of stomata, in *Encyclopedia of Plant Physiology. Vol. 7, Physiology of Movements*, (eds W. Hampt and M.E. Feinleib), Springer, Berlin.

Rogers, C.A., Powell, R.D. and Sharpe, P.J.M. (1979) The relationship of temperature on stomatal aperture and potassium accumulation in guard cells. *Plant Physiol.*, **63**, 388–391.

Saxe, H. (1979) A structural and functional study of the coordinated reactions of individual *Commelina communis* L. stomata (Commelinaceae). *Am. J. Bot.*, **66**, 1044–1052.

Schwendener, S. (1881) Über Bau und Mechanik der Spaltöffnungen. *Monatsber. Kgl. Akad. Wiss. Berlin.*, **43**, 833–867.

Sharpe, P.J.H. and Wu, H.-I. (1978) Stomatal mechanics: volume changes during opening. *Plant Cell Environ.*, **1**, 259–268.

Sharpe, P.J.H., Wu, H.-I. and Spence, R.D. (1987) Stomatal mechanics, in *Stomatal Function*, (eds E. Zeiger, G.D. Farquhar and I.R. Cowan), Stanford University Press, Stanford, CA, pp. 91–114.

Spence, R.D., Wu, H., Sharpe, P.J.H. and Clark, K.G. (1986) Water stress effects on guard cell anatomy and the mechanical advantage of the epidermal cells. *Plant Cell Environ.*, **9**, 197–202.

Stålfelt, M.G. (1927) Die photischen Reaktionen im Spaltöffnungsmechanismus. *Flora*, **121**, 236–272.

Stålfelt, M.G. (1966) The role of the epidermal cells in the stomatal movements. *Physiol. Plant.*, **49**, 241–256.

Weyers, D.J.B. and Fitzsimons, P.J. (1982) The non-osmotic volume of *Commelina* guard cells. *Plant Cell Environ.*, **5**, 417–421.

Weyers, D.J.B. and Fitzsimons, P.J. (1985) Properties of some enzymes used for protoplast isolation, in *The Physiological Properties of Plant Protoplasts*, (ed. P.-E. Pilet), Springer, Berlin, pp. 152–161.

Weyers, J.D.B. and Meidner, H. (1990) *Methods in Stomatal Research*, Longman Scientific and Technical, Harlow, UK.

Willmer, C.M. (1983) *Stomata*, Longman, London.

Willmer, C.M. and Beattie, L.N. (1978) Cellular osmotic phenomena during stomatal movements of *Commelina communis*. 1. Limitations of the incipient plasmolysis technique for determining osmotic pressures. *Protoplasma*, **95**, 321–332.

Willmer, C.M. and Pantoja, O. (1991) The plasma membrane and tonoplast of guard cells, in *The Plasma Membrane; A Biophysical Approach*, (ed. Y.Y. Leshem), Kluwer, Dordrecht.

Wolfe, J. and Steponkus, P.L. (1983) Mechanical properties of the plasma membrane of isolated plant protoplasts. *Plant Physiol.*, **71**, 276–285.

Wu, H.-I. and Sharpe, P.J.H. (1979) Stomatal mechanics II: material properties of guard cell walls. *Plant Cell Environ.*, **2**, 235–244.

Wu, H.-I, Sharpe, P.J.H. and Spence, R.D. (1985) Stomatal mechanics. III. Geometric interpretation of the mechanical advantage. *Plant Cell Environ.*, **8**, 269–274.

Ziegler, H. (1987) The evolution of stomata, in *Stomatal Function*, (eds E. Zeiger, G.D. Farquhar and I.R. Cowan), Stanford University Press, Stanford, CA, pp. 29–58.

5 The theory of gas diffusion through stomata

5.1 Introduction

Over the last century our understanding of gas exchange between a leaf and its environment has increased greatly from the use of simple concepts of resistance to diffusion (e.g. Brown and Escombe, 1900) to the development of complex models and theory (see, e.g. Penman and Schofield, 1951; Monteith, 1983). Although the general concepts are now well established (see, e.g. Weyers and Meidner, 1990; Nobel, 1991; Jones, 1992) some controversial aspects, such as the diffusion of gases inside leaves (see, e.g. Parkhurst, 1994), remain. In this section, emphasis will be on the theory of diffusion of gases through stomata. The various techniques for measuring stomatal resistance, or its reciprocal, stomatal conductance, have been extensively covered elsewhere (see, e.g. Sestak *et al.*, 1971; Weyers and Meidner, 1990).

Leaves are the major organs of gas exchange in plants (with the exception of many submerged aquatic ones) and their surfaces are cuticularized to varying degrees. The cuticle is relatively impermeable to gases including water vapour and CO_2. Thus gas exchange is mainly governed by stomatal aperture and the distribution of stomata in the surface of the leaf. Normally when a leaf is photosynthesizing, the stomata are open and there will be a net movement of CO_2 into and water vapour and oxygen out of a leaf. Partial closure of stomata will have a larger relative effect on water vapour loss than CO_2 uptake, mainly because the diffusion pathway for CO_2 is longer than that for water vapour. Stomatal restriction is therefore a smaller proportion of the total impedence for carbon dioxide fluxes. However, at very small apertures a relatively high efflux of water vapour may retard CO_2 influx.

5.2 Fick's law and the driving forces for diffusion

The flux of CO_2 into or water vapour out of a leaf depends largely on two quantities, i.e. the driving forces for movement and the resistances encountered *en route* The driving forces for diffusion through the stomata are determined by the difference in free energy of the molecules in the atmosphere and the substomatal cavity For gases at a constant temperature, the free energy is related primarily to the number of molecules present per unit volume (molar concentration) or the mass of gas per unit volume (mass concentration) Thus Fick's law states that the rate of mass transfer (*J*) is directly related to the concentration gradient ($\partial c/\partial x$) by a constant termed the diffusion coefficient (*D*)

$$J = -D\partial c / \partial x \qquad (5\ 1)$$

Where J is the rate of mass transfer (g m^{-2} s^{-1}), D is the diffusion coefficient (m^2 s^{-1}) and $\partial c/\partial x$ is the concentration (or density) gradient of the gas (g m^{-4}) The minus sign associated with D in equation (5 1) indicates diffusion occurs towards the region of lower concentration The diffusion coefficient, D, is not strictly constant, but is affected by gas concentration, temperature, pressure and the molecular weight of the gas (from Graham's law which states that the relative rates of diffusion of two gases are inversely proportional to the square root of their densities or molecular weights) The diffusion coefficient is effectively constant at a fixed temperature and pressure if the changes of gas concentration across the tube (pore) are negligible, which is the case for CO_2 Thus at 20°C and atmospheric pressure (0 1013 MPa), D_{CO_2} in air is 15 1 mm^2 s^{-1} and D_{wv} for saturated water vapour is 24 2 mm^2 s^{-1} Cowan and Milthorpe (1968) suggest that the diffusion coefficient for water vapour in free air should be reduced when considering flux through a stomatal pore to take into account molecular interactions of water with the pore walls, thus the value of D_{wv} should be reduced as the pores become smaller

In the case of water vapour, the various measurements of concentration have particular terms Absolute humidity corresponds to the mass or number of molecules per unit volume (i e equivalent to c in equation 5 1) The maximum value of the absolute humidity for water vapour in equilibrium with pure water is termed the saturation concentration of water vapour in air, which increases markedly with temperature (Fig 5 1) As the air is often not saturated with water vapour, relative humidity expresses the absolute humidity as a percentage of this maximum saturation value for a given temperature However, as the leaf temperature is often higher than the air temperature due to absorption of radiant energy, the same concentration of water vapour would correspond to a different relative humidity at each location Thus it is usual to express concentrations in terms of partial pressures or mole fractions which explicitly include a temperature term The absolute humidity is related to the partial pressure of water vapour (e) as follows

$$c = eM / RT \qquad (5\ 2)$$

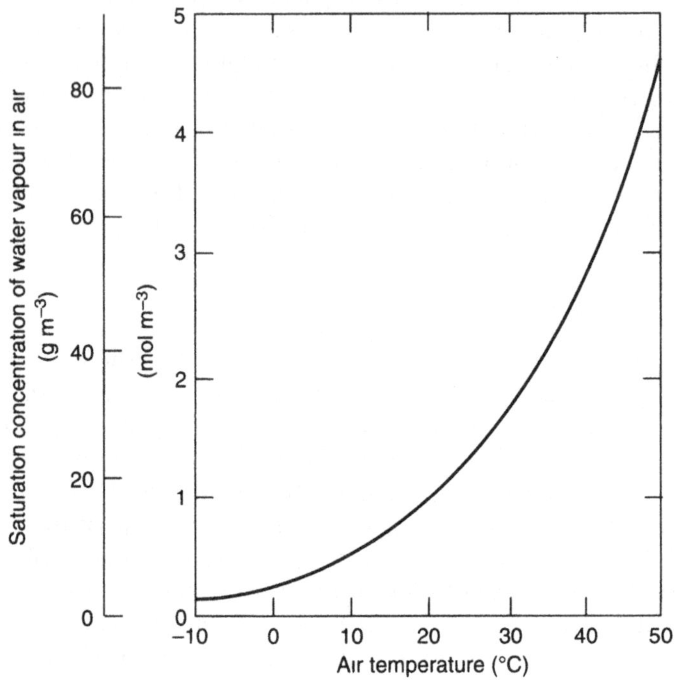

Figure 5 1 Variation in saturation values of water vapour concentration with temperature From Nobel (1991)

Where c is the concentration (g m^{-3}), e is the partial pressure of water vapour (Pa), M is the molecular weight (g mol^{-1}), R is the universal gas constant (m^3 Pa mol^{-1} K^{-1}) and T is the absolute temperature (K)

The relative humidity (RH) can also be expressed as

$$RH = 100e / e_s \tag{5 3}$$

where e_s is the saturated vapour pressure at that temperature

Another commonly used expression is the vapour pressure deficit (VPD), where

$$\mathrm{VPD} = e_s - e \tag{5 4}$$

When considering transpiration, it is useful to relate humidity measurements in the intercellular air spaces or the atmosphere to the water potential of the tissues If the air is saturated, the water potentials in the liquid phase and the gaseous phase are equal As the relative humidity drops, however, the water potential of the air (ψ_{wv}) also becomes more negative according to the equation

$$\Psi_{wv} = \frac{RT}{V_w} \ln \frac{RH}{100} \tag{5 5}$$

Where ψ_{wv} is the water potential of water vapour in air (Pa), R is the universal gas constant (8 314 m^3 Pa mol^{1} K^{1}), T is the absolute temperature (293 K = 20°C) and V_w is the partial molar volume of water (1 8 × 10^{5} m^3 mol^{1} at 20°C)

This highlights the strong driving force for evaporation at anything less than 100% relative humidity, as the term RT/V_w has a value of 135 MPa at 20°C Thus even at 80% relative humidity, the water potential of the air is still –300 atmospheres!

In characterizing diffusion of gases, it is usually more convenient to replace the gradient at a single point ($\partial c/\partial x$) in equation (5 1) with the difference in concentration between two locations separated by a finite distance, termed the path length for diffusion (l) Thus for example, the flux of CO_2 from the air to the leaf using equation (5 1) becomes

$$J = -D_{CO_2}(c_{air} - c_{leaf}) / l \tag{5 6}$$

where J is the mass transfer (g m s ^{1}), D_{CO} is the diffusion coefficient for CO_2 (m^2 s^{1}), c_{air} is the concentration of CO_2 in air (g m^{3}), c_{leaf} is the concentration of CO_2 in the intercellular air spaces (g m^{3}) and l is the path length for diffusion (m) The nominal constant D_{CO}/l is termed the conductance and given the symbol g The reciprocal of the conductance is also commonly used and is given the symbol r, for resistance

5.3 The analogy between electrical circuits and leaf resistances

During the exchange of gases between a leaf and its environment the gas molecules meet a number of resistances besides the stomatal resistance (see Fig 5 2) Such resistances can be treated in a comparable manner to resistances in electrical circuits and equation (5 6) has the same form as Ohm's law which states that the current is equal to the potential divided by the resistance Resistances in series are additive, whilst those in parallel can be combined according to the general equation

$$\frac{1}{r_{tot}} = \frac{1}{r_1} + \frac{1}{r_2} + \frac{1}{r_3} \tag{5 7}$$

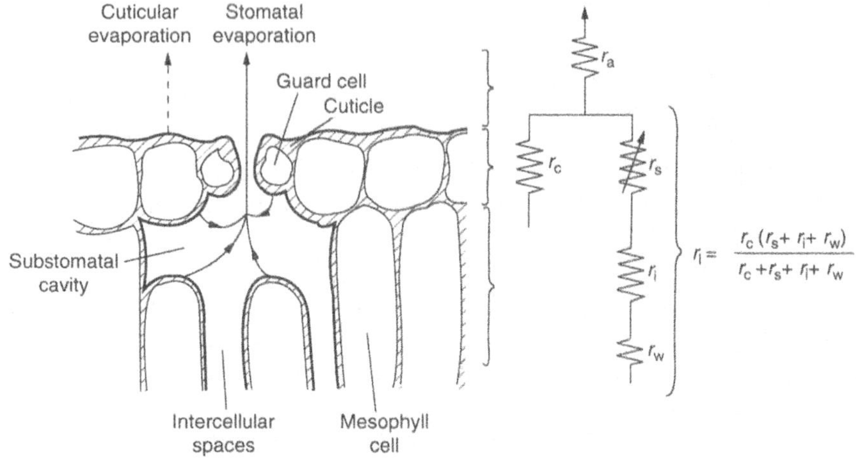

Figure 5.2 Pathways for water loss from one surface of a leaf showing the boundary layer (r_a) cuticular (r_c) variable stomatal (r_s) intercellular space (r_i) wall (r_w) and leaf (r_l) resistances. The total leaf resistance is the parallel sum of r_l for upper and lower surfaces. From Jones (1992)

Figure 5.2 shows the resistances encountered as water vapour is transpired from one surface of a leaf. Evaporation of water from cell walls is considered by some to be the location of the first resistance in the pathway of water loss. In series with this resistance (r_w) is the transfer resistance within the intercellular spaces (r_i) followed by the stomatal resistance (r_s) and the boundary layer resistance (r_a). The cuticular transfer pathway which has very high resistances (r_c) is in parallel with the stomatal resistance (r_s). The stomatal resistance is variable according to the extent of stomatal opening. Figure 5.2 indicates that r_l is the leaf resistance from one leaf surface only and the total leaf resistance is the parallel sum of r_l for the upper and lower surfaces using equation (5.7). Typical values for the resistances (and conductances) for each part of the diffusion pathway are given in Table 5.1.

The pathway for CO_2 diffusion from the surrounding air into the leaf shares some components with the pathway for water vapour moving in the opposite direction (Fig. 5.3). The CO_2 concentration in the atmosphere is currently 0.036%, but may be locally different due to the effects of respiration increasing CO_2 levels in canopies during the night and previous photosynthetic activity decreasing CO_2 concentrations during the day. The inward diffusion of CO_2 from outside the leaf to the carboxylation site within a chloroplast encounters an additional resistance, i.e. the mesophyll resistance, r_m, which is a complex comprising the resistance of CO_2 movement across the plasma membrane, the cytosol,

Table 5.1 Summary of representative values of conductances and resistances for water vapour diffusing out of leaves From Nobel (1991)

Component condition	Conductance		Resistance	
	mm s^{1}	mmol m^{2} s^{1}	s m^{1}	m^{2} s mol^{1}
Boundary layer				
thin	80	3200	13	0 3
thick	8	320	130	3
Stomata				
large area – open	19	760	53	1 3
small area – open	1 7	70	600	14
closed	0	0	∞	∞
mesophytes – open	4–20	160–800	50–250	1 3–6
xerophytes and trees – open	1 4	40–160	250–1000	6–25
Cuticle				
crops	0 1–0 4	4–16	2500–10 000	60–250
many trees	0 05–0 2	2–8	5000–20 000	125–500
many xerophytes	0 01–0 1	0 4–4	10 000–100 000	250–2500
Intercellular air spaces				
calculation	24–240	1000–10 000	4 2 42	0 1–1
waxy layer				
typical	50–200	2000–8000	5–20	0 1–0 5
certain xerophytes	10	400	100	2 5
typical	40–100	1600–4000	10–25	0 2–0 6
Leaf (lower surface)				
crops – open stomata	2 10	80–400	100–500	2 5 13
trees – open stomata	0 5–3	20–120	300–2000	8–50

and the chloroplast envelope plus a biochemical resistance of CO_2 fixation (Fig 5 3) CO_2 fluxes are therefore not good indicators of stomatal behaviour because the fluxes are affected by many factors in addition to passage through stomatal pores

Figure 5 4 presents some typical values of leaf conductances (g_l = $1/r_l$) found in plants which grow in different environments Generally succulents which normally grow in arid environments have the lowest conductances, while plants from wet habitats have the highest conductances This is to be expected if a primary aim of stomata is to optimize carbon fixation under prevailing conditions of water availability

5.4 Diffusion through a single pore and multipores

Within a leaf the pore area of 'open' stomata is normally between 0 5 and 5 0% of the total leaf area and yet leaves often evaporate 50% as much water as would evaporate from an equivalent leaf area of wet blot-

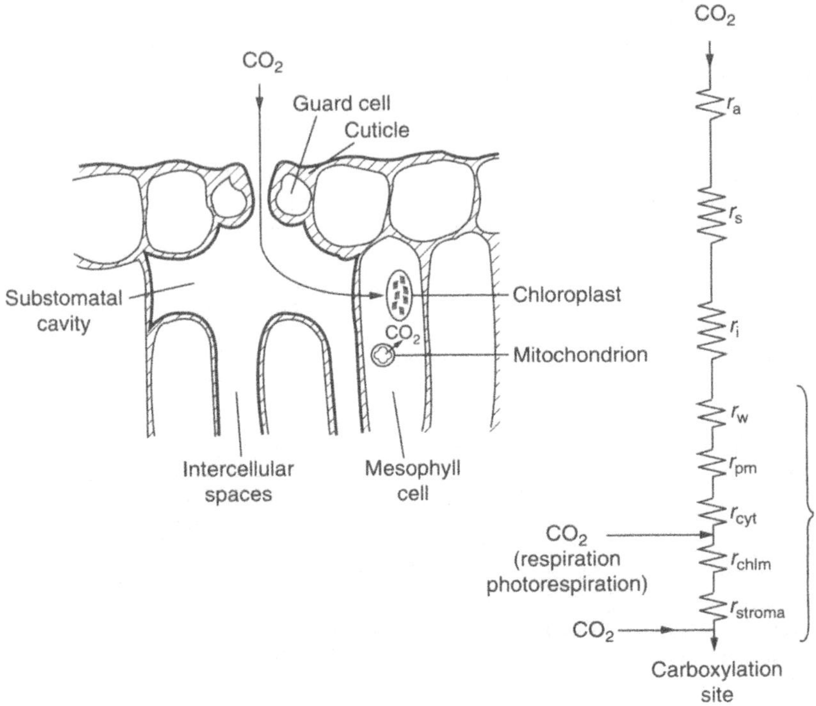

Figure 5.3 Pathway for CO_2 entry into a leaf showing the various resistances encountered by CO_2 en route to a chloroplast in a mesophyll cell Key r_a = boundary layer resistance, r_s = stomatal resistance, r_i = intercellular space resistance, r_w = wall resistance, r_{pm} = plasma membrane resistance of mesophyll cell, r_{cyt} = cytoplasmic resistance, r_{chlm} = chloroplast envelope resistance, r_{stroma} = stromal resistance The sum of r_w, r_{pm}, r_{chlm}, r_{stroma} and the enzyme carboxylation resistance is known as r_m, the mesophyll resistance

ting paper. The main reason for this is that water molecules from a free water surface are part of a relatively dense column of molecules extending above the water surface, while water molecules diffusing from a pore can go in any direction within an imaginary hemisphere centred above the pore. Thus the water vapour gradient will be much steeper in the latter situation than the former, resulting in a higher water loss. Essentially water molecules are diffusing more rapidly from the edges than the centre of the pores and hence the phenomenon is also known as the 'edge-effect'. The spacing of stomata may also assist in the efficient loss of water for reasons indicated below. In the same manner, stomata are efficient at allowing entry of CO_2 from the atmosphere into the leaf for photosynthesis; however, it is important to appreciate that plants are generally attempting to conserve water.

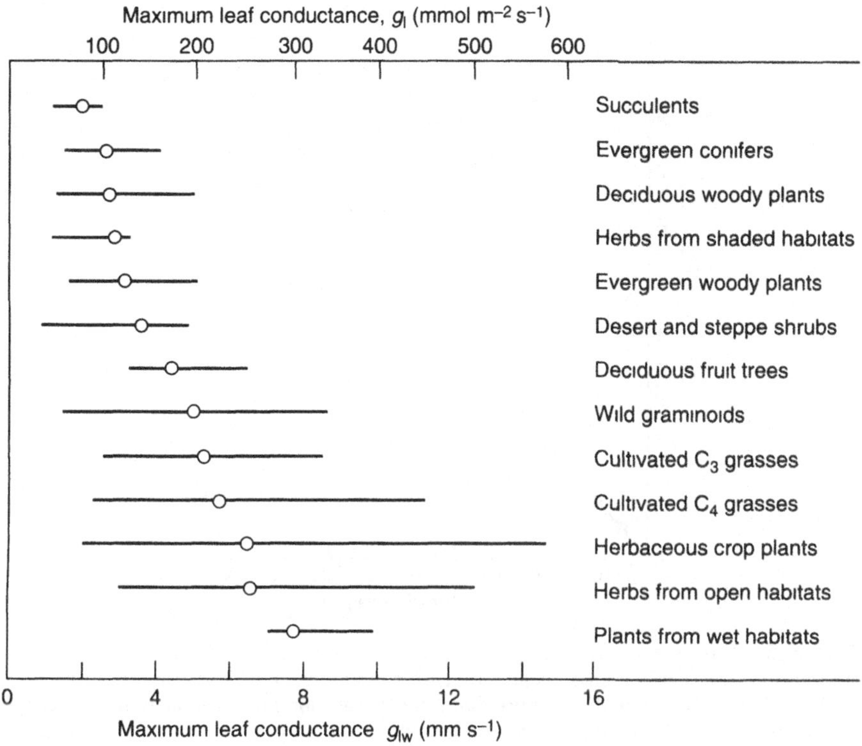

Figure 5.4 Maximum leaf conductance (g_{lw}) in different groups of plants The lines cover about 90% of individual values reported The open circles represent group average conductances From Jones 1993 adapted from Korner *et al* (1979)

It was Brown and Escombe (1900) who first reported that diffusion through a single pore of stomatal dimensions was more nearly proportional to the diameter than the cross-sectional area of the pore (Fig 5 5) This is in accordance with Stefan's law which states that diffusion through small apertures is proportional to the diameter (*d*) of the aperture and the difference in density $\Delta\rho$ measured between the underside of the pore and a point some distance away from it Thus

$$J\alpha 2D\Delta\rho d \qquad (5\,8)$$

The diameter relation holds provided that the diffusion path length to cross-sectional area of the tube is small However, it has been suggested that as stomata close this ratio increases so that transpiration is more proportional to the pore area than pore diameter (e g Lee, 1967) Nevertheless, the efficiency of small pores is partly due to the 'edge-effect'

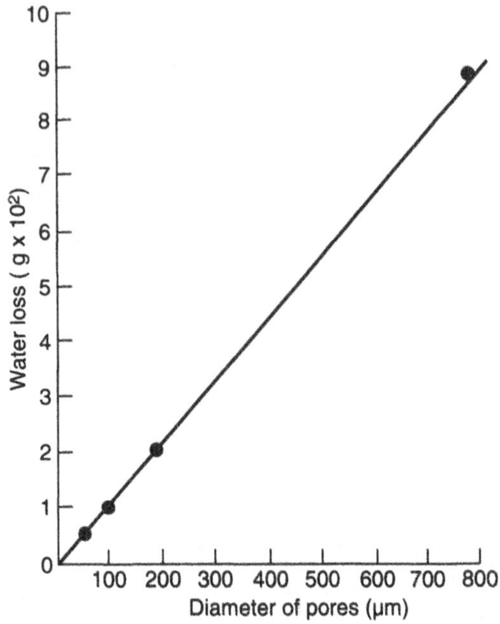

Figure 5.5 The relationship between diffusion through a small pore and the diameter of the pore. Water loss increases linearly with an increase in pore diameter. This relationship of proportionality to pore diameter rather than proportionality to pore area is a consequence of the large source region above the pore. From Ting and Loomis (1963).

Additionally two 'end correction' terms (equivalent to $2\pi r/4$) are generally added to take into account that the pore area is much less than the leaf area and the lines of flux of a gas converge on entering and leaving the stomatal pore making diffusion less effective. In other words there is an increase in the effective length of the pore. However, there are differences of opinion about the formulation of the 'end correction'. Some authorities consider only one 'end correction' is needed since diffusion will be slowed as influx lines converge and increase as efflux line separate. Also, for convenience the pores are considered to be circular, but in fact the pores are better described as ellipsoidal. Thus, Parlange and Waggoner (1970) make a double end correction for elliptical pores:

$$\textbf{end correction} = b.\ln(4a/b) \tag{5.9}$$

where a is the long axis of the pore (m) and b is the short axis of the pore (m). Values for circular and elliptical geometries are most markedly different when narrow stomatal apertures occur, but the overall contribution of the end correction is greatest at wide stomatal apertures.

Stomatal distribution is also an important factor in determining diffusion rates (see Ting and Loomis, 1963; Ting and Loomis, 1965). If the pores are far enough apart the water vapour shells do not overlap and the pores act independently. However, if the pores are close together the shells coalesce to form a single vapour shell over a surface, representing the boundary layer (Fig. 5.6). Thus, as pore diameters increase so, eventually, water vapour shells may overlap. Figure 5.7 illustrates diffusion through a multipore surface analogous to leaf epidermis; at small pore apertures diffusion rates increase linearly (the pores are acting independently) while at the larger pore apertures the diffusion rate is nearly constant (the shells overlap acting as one unit).

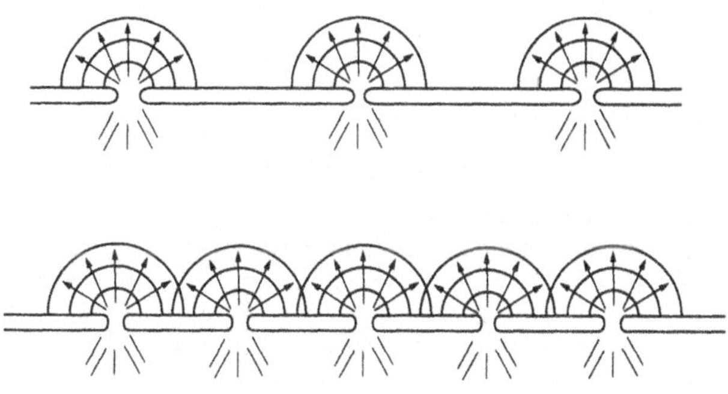

Figure 5.6 Drawing to illustrate diffusion stream-lines and vapour shells that build up over pores. The top diagram depicts three pores that are far enough apart to act independently (there is no overlap of diffusion shells). The lower diagram showing five pores illustrates what happens when the pores are so close that the diffusion shells overlap. Coalescence forms a single vapour shell over the entire surface that represents the boundary layer. Thus a multipored surface such as this will act similarly to a free surface of equal area. From Ting and Loomis (1965).

Thus, on a leaf area basis,

$$\underset{(\mathrm{m\ s^{-1}})}{\text{stomatal conductance}} = \frac{\underset{\text{coefficient }(\mathrm{m^2\ s^{-1}})}{\text{effective diffusion}} \times \underset{(\mathrm{m^{-2}})}{\text{stomatal frequency}} \times \underset{(\mathrm{m^2})}{\text{pore area}}}{\text{pore depth (m)} + 2 \text{ end corrections (m)}} \tag{5.10}$$

Stomatal diffusive resistance (r_s) is the reciprocal of stomatal conductivity (g_s) and therefore has units of s m^{-1}.

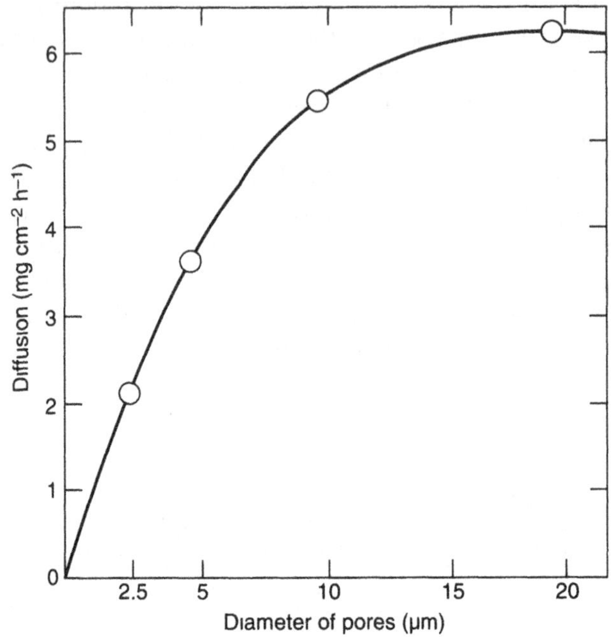

Figure 5.7 Diffusion of water vapour through a model multipore surface analogous to leaf epidermis The diameter of the pores ranges from 2 5 to 20 μm, which is over the range of stomatal dimensions There are 2500 pores cm^{-2} The graph shows the expected linear increase in diffusion at small apertures and the nearly constant diffusion rate at wide pore apertures The shape of the curve is predictable from the principles of small-pore diffusion outlined in the text From Ting and Loomis (1965)

5.5 Cuticular conductance

The cuticle allows a limited amount of evaporation in parallel to water vapour fluxes through the stomatal pores. Measurements of cuticular conductance are typically made either on hypostomatous leaves by sealing the epidermis containing stomata to prevent stomatal transpiration, or under conditions that should close all the stomata. Typical values are given in Table 5.1. The cuticular conductance also varies greatly between species and within species depending on the environmental conditions during leaf development. In addition, cuticular conductance may change over much shorter time periods in response to changes in variables such as the degree of hydration of the leaf, temperature and light intensity. Thus, Meidner (1986) found that cuticular conductance in *Prunus* and other species was high in the early morning when the leaf and cuticle water content was high, but then declined during the day as the leaf water content dropped. In addition, van Gardingen and Grace (1992) found that the cuticular conductance of *Fagus sylvatica* decreased as the water vapour pressure deficit was increased in steps, but no response was observed when vapour pressure deficit was decreased from an initially high value. They suggested that the cuticular conductance reflected the degree of hydration of the cuticle in response to increasing VPD. However, the time-constant for the process of cuticle rehydration was too long in intact leaves to see responses to decreasing VPD.

The level of cuticular conductance is also affected by factors that damage the cuticle integrity, such as wind and particle abrasion (Pitcairn *et al.*, 1986). Such localized regions of cuticle damage are also likely to be the major penetration site for pollutants, such as acid mists (e.g. Hoad *et al.*, 1994).

5.6 Boundary layer conductance

The boundary layer refers to the air layer adjacent to the leaf surface where the surface friction reduces the wind speed in comparison to the bulk air flow. The air layer immediately adjacent to the leaf surface effectively remains stationary and heat and mass transfer of molecules, such as water vapour, occurs by diffusion alone. The diffusion function described in Section 5.4 will apply in still air but, as Bange (1953) demonstrated, in moving air, the vapour shells are removed or diminished and vapour loss is more linearly related to pore diameter (Fig. 5.8). Hence, in still air the low boundary layer conductance will have a significant effect on the flux of water vapour out of a leaf unless the stomata are closed or virtually so, while in moving air, efflux is regulated more by the stomata than the boundary layer.

In reality, however, still air is virtually never experienced under field conditions and even very slight breezes will reduce the boundary layer so that vapour shells associated with each pore do not overlap (e.g. Lee, 1967). The air flow moving away from the leaf surface will be laminar until a particular velocity threshold is exceeded and turbulence sets in. The thickness of the laminar boundary layer (δ) is related to the kinematic viscosity of air (v) and the wind speed (u) at a distance, x, downwind from the leading edge of a leaf by the semi-empirical expression:

$$\delta \approx \sqrt{(vx / u)} \quad (5.11)$$

Where δ is the thickness of the boundary layer (m), v is the kinematic viscosity of air (1.51×10^{-5} m^2 s^{-1} for dry air at 20°C), x is the distance downwind from the leading edge of the leaf (m) and u is the wind speed (m s^{-1}).

Thus the thickness of the boundary layer varies across the surface of a leaf and may be further modified by the leaf shape and surface topography (see review by Schuepp, 1993). Local leaf temperature is also influenced by the thickness of the boundary layer controlling heat dissipation

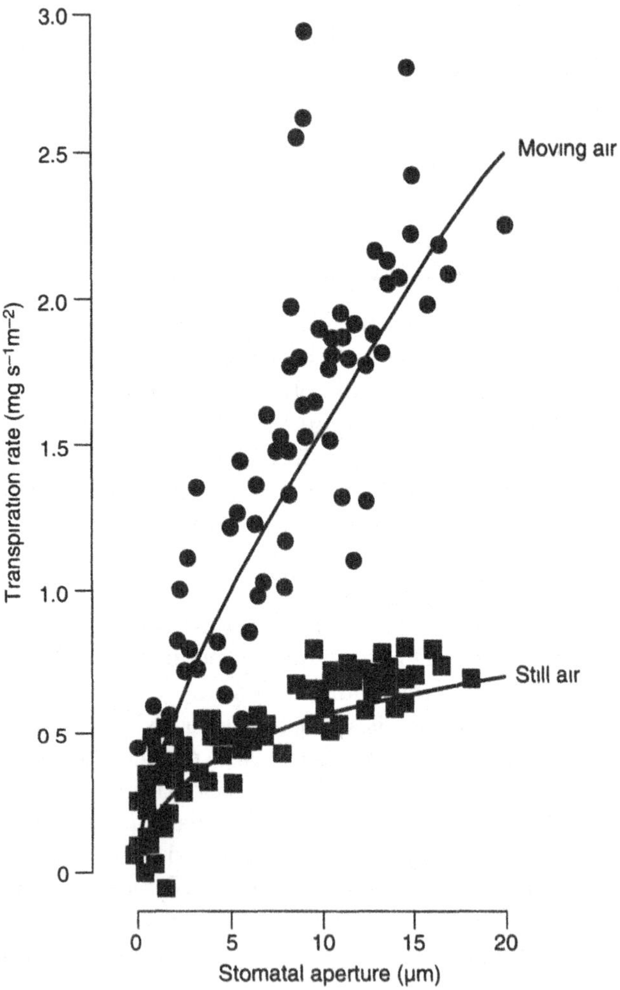

Figure 5.8 Transpiration rate of *Tradescantia zebrina* leaves at different stomatal apertures in still and windy conditions Adapted from Bange (1953) The data are based on gravimetric determinations of water loss over 1–5 min from leaf discs placed in a special holder to eliminate water loss from the edges and minimize changes in disc temperature The VPD was 1% and the temperature 23 ± 2°C (i e RH about 90%) Stomatal apertures were estimated from 25 light microscope measurements and transpiration rates were corrected for stomatal frequency and cuticular water loss Lines are predictions based on a theoretical analysis From Weyer and Meidner (1990)

by conduction. As air is a relatively poor conductor of heat the surface temperature of leaves with a large boundary layer may be significantly higher than the surrounding air by as much as 15°C, although for most smaller leaves values of 1–5°C are more common (Nobel, 1991). Leaf temperature also varies across the surface of the leaf as the boundary layer increases in thickness (Fig. 5.9). This variation in leaf temperature will also affect the local value of the saturated water vapour pressure (see Section 5.2).

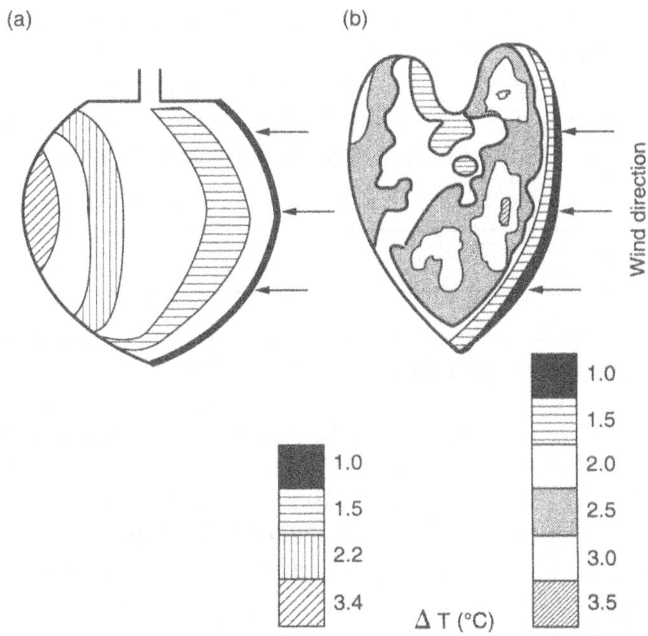

Figure 5.9 Temperature differences observed between surface of artificial (a) and natural (b) leaf, and ambient air. From Wigley and Clark (1974)

5.7 *Water use efficiency*

The water use efficiency (WUE) of a plant is an indicator of how effective the plant is at fixing CO_2 relative to the amount of water lost. The reciprocal of water use efficiency, the transpiration ratio (TR) is also commonly used to describe this feature. Thus

$$\text{TR} = \text{moles of water transpired} / \text{moles of } CO_2 \text{ fixed} \quad (5.12)$$

Typically, C3 plants have transpiration ratios of about 500, while C4 plants transpire less water per mole of CO_2 fixed, having transpiration ratios of around 250 For CAM plants exhibiting night opening of stomata, values may be nearer 50

The reasons why there is much more water lost than CO_2 fixed are firstly that the driving forces for CO_2 and water vapour fluxes are greatly different, being 50 times larger for the latter gas This is because the gradient of CO_2 from inside to outside the leaf is relatively low (internal CO_2 concentration is about 0 01– 0 02% while atmospheric CO_2 concentration is about 0 036%) while it is high for water vapour (leaf air spaces are saturated while the atmosphere ranges from typically 50–90% saturated Secondly, CO_2 diffuses more slowly through air than water vapour as it is a larger molecule with a lower diffusion coefficient (see Section 5 2) Thirdly, CO_2 has a longer diffusion pathway than water vapour adding to the resistance encountered in this pathway (see Section 5 3)

References

Bange, G G (1953) On the quantitative explanation of stomatal transpiration *Acta Botanica Neerlandia*, **2**, 255–296

Brown, H and Escombe, F (1900) Static diffusion of gases and liquids in relation to the assimilation of carbon and translocation in plants *Phil Trans Roy Soc Lond , Ser B*, **193**, 233–291

Cowan, I R and Milthorpe, F L (1968) Plant factors influencing the water status of plant tissues, in *Water Deficits and Plant Structure*, Vol 1, (ed T T Kozlowski), Academic Press, New York, pp 137–193

van Gardingen, PR and Grace, J (1992) Vapour pressure deficit response of cuticular conductance in intact leaves of *Fagus sylvatica* L *J Exp Bot* , **43**, 1293–1299

Hoad, S P, Jeffree, C E and Grace, J (1994) Effects of wind and simulated acid mist on leaf cuticles, in *Air Pollution and the Leaf Cuticle*, (eds K E Percy, N Cape and R Jagels), Springer, Heidelberg, pp 225–238

Jones, H G (1993) *Plants and Microclimate*, 2nd edn, Cambridge University Press, Cambridge

Lee, R (1967) The hydrologic importance of transpiration control by stomata *Water Res* , **3**, 737–752

Meidner, H (1986) Cuticular conductance and the humidity response of stomata *J Exp Bot* , **37**, 517–525

Monteith, J L (1983) *Principles of Environmental Physics*, Arnold, London

Nobel, P (1991) *Physicochemical and Environmental Plant Physiology*, Academic Press, San Diego, CA

Parkhurst, D.F. (1994) Diffusion of CO_2 and other gases inside leaves. *New Phytol.*, **126**, 449–479.

Parlange, J.Y. and Waggoner, P.E. (1970) Stomatal dimensions and resistance to diffusion. *Plant Physiol.*, **46**, 337–342.

Penman, H.L. and Schofield, R.K. (1951) Some physical aspects of assimilation and transpiration, in *Fixation of Carbon Dioxide*, (eds J.F. Danielli and R. Brown). *Symp. Soc. Exp. Biol.*, **5**, 115–129.

Pitcairn, C.E.R., Jeffree, C.E. and Grace, J. (1986) Influence of polishing and abrasion on the diffusive conductance of leaf surface of *Festuca arundinacea* Schreb. *Plant Cell Environ.*, **9**, 191–196.

Schuepp, P.H. (1993) Leaf boundary layers. *New Phytol.*, **125**, 477–507.

Sestak, Z., Catsky, J. and Jarvis, P.G. (1971) *Plant Photosynthetic Production: Manual of Methods*, Junk, The Hague.

Ting, I.P. and Loomis, W.E. (1963) Diffusion through stomates. *Am. J. Bot.*, **50**, 866–872.

Ting, I.P. and Loomis, W.E. (1965) Further studies concerning stomatal diffusion. *Plant Physiol.*, **40**, 220–228.

Weyers, J. and Meidner, H. (1990) *Methods in Stomatal Research*, Longman Scientific and Technical. Harlow, UK.

Wigley, G. and Clark, J.A. (1974) Heat transport coefficients for constant energy flux models of broad leaves. *Boundary Layer Meteorology*, **7**, 139–150.

6 Stomatal responses to environmental factors

6.1 Introduction

Stomata are situated in the leaf surface where they are best positioned to control the influx and efflux of gases between the interior of a leaf and its environment Furthermore, guard cells are usually only connected to neighbouring cells via their dorsal walls, and, at maturity, these walls do not possess functional plasmodesmata (see Chapter 3) Thus, because of their relative isolation from the rest of the plant body, stomata are ideally suited for sensing and responding to environmental factors

Over the years there have been disagreements about the effects of environmental factors on stomatal behaviour For the first half of this century, for example, the consequences of changing CO_2 concentrations on stomata were not fully appreciated and the interpretation of many early experiments must remain dubious because of this It was mainly the work of Heath and his colleagues which established the importance of controlling CO_2 concentrations surrounding leaves when stomatal behaviour was being monitored Even in recent times the view that light and CO_2 have direct effects on stomatal movements has been occasionally questioned

Stomata are influenced indirectly or directly by a wide range of environmental variables Moreover, guard cells themselves directly perceive, transduce and respond to many environmental stimuli including light (quality and quantity), CO_2 concentration, temperature, probably humidity and or vapour pressure deficit gradients and pollutants Additionally, availability of mineral nutrients can influence guard cell responses and the biological clock regulating stomatal circadian rhythms probably resides in the guard cells All of these incoming signals are integrated to produce a net stomatal response

6.2 Interaction of environmental factors which influence stomatal behaviour

Stomatal behaviour can be influenced by environmental factors in a direct or indirect manner Factors which impinge upon the guard cells to affect their water relations or metabolism thereby influencing stomatal aperture are direct and can be unequivocally demonstrated in guard cell protoplasts (GCPs, which contract and swell in an analogous manner to closing and opening of stomata) or in epidermal strips (though the effects could still be indirect if aperture changes are a result of responses in the epidermal and or subsidiary cells) Indirect effects are those which influence photosynthesis of the mesophyll or water relations of tissues other than the guard cells and which subsequently influence stomatal behaviour Originally Stålfelt introduced the terms 'hydropassive and hydroactive to describe the effects of tissue and cell water relations on stomatal movements Hydropassive effects are those that do not involve metabolic activities of the guard cells, but result from the availability of tissue water to change the turgor relations of the stomatal complexes Hydroactive effects are those that influence stomatal movements via metabolic processes of the guard cells For example, a fall in leaf turgor may activate abscisic acid (ABA) synthesis in mesophyll cells from where it is transported to the guard cells to set in motion metabolic events which ultimately bring about hydroactive closure (see Chapter 8)

The environmental stimuli will interact so that stomatal aperture is a resultant of all these factors Superimposed on the interactions of these environmental factors are modifications of movements brought about by circadian rhythms and possibly a variety of other, shorter-term oscillations

Studies of the effects of one factor on stomatal movements in isolation is difficult because of this interaction between environmental stimuli For example, increasing leaf irradiance will tend to increase the leaf temperature which, in turn, will increase evaporation and lower the leaf water potential, the increased temperature may also change the intercellular CO_2 concentrations (C_i) by changing rates of photosynthesis, respiration and photorespiration Thus, all factors may be in play at one time so that stomatal aperture is a resultant of all these factors (e g Fig 6 1, which shows the interaction of just two factors, i e CO_2 concentration and light quantity, on stomatal conductance in wheat leaves)

The major consequences of stomatal movements are to change the rate of diffusion of CO_2 into a leaf for photosynthesis and loss of water vapour which evaporates from the cell walls in contact with air spaces within the leaf Some of the water lost serves for evaporative cooling of the leaf when exposed to high temperatures, but much is unavoidably lost when the stomata are open It has been estimated by Raschke (1979) that several hundred molecules of water are lost from a plant for each molecule of CO_2 taken up Usually this means water must be conserved by a plant but there is considerable debate about the precise role

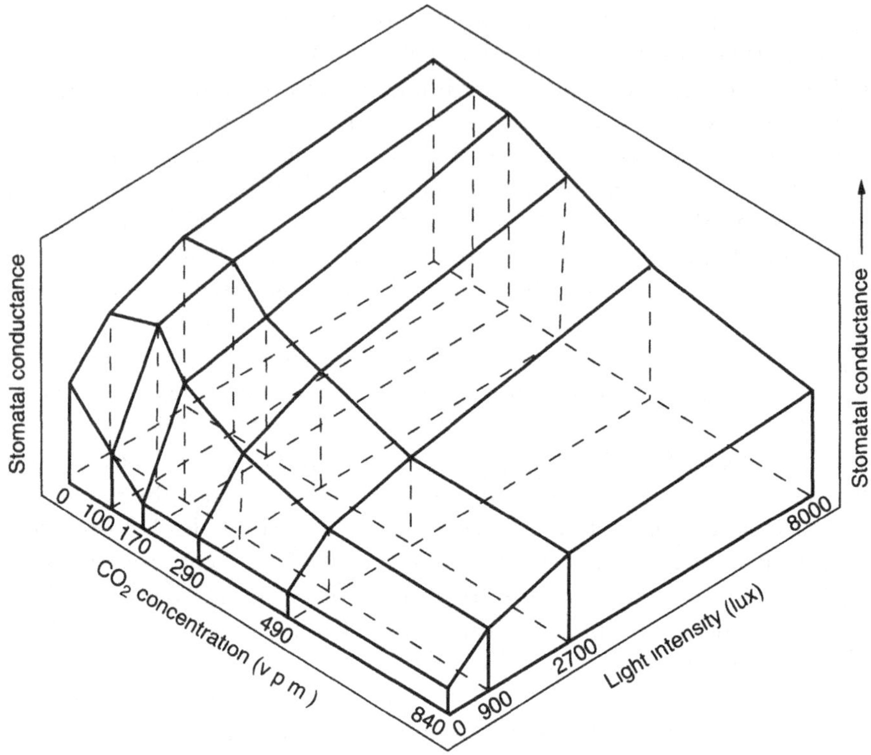

Figure 6 1 The effect of photon flux density and CO_2 concentration on stomatal opening in wheat leaves From Heath and Russell (1954)

of stomata in regulating the compromise between water loss and CO_2 assimilation The transpiration stream which is largely controlled by stomata, will also aid in the transport of ions from the roots, although whether this is an essential role is debatable (see Chapter 1) The uptake and transport of ions in the transpiration stream may however, be less critical to the survival of the plant and need less fine control by stomata since active ion uptake and relocation of ions within a plant will occur, to some extent, independently of transpiration It seems, therefore, that it would be of advantage to a plant if stomata could respond to environmental stimuli by maximizing photosynthesis while minimizing excessive water loss The relative constancy of C_i, the internal leaf CO_2 concentration, is considered to be a consequence of a control system linking photosynthesis and transpiration to stomatal movements which ensures that the maximum amount of CO_2 is fixed per unit of water transpired, i e stomata optimize gas exchange by maintaining a constant ratio of the 'cost of water' against the 'benefit of carbon' (Cowan and Farquhar, 1977) Thus, stomatal aperture is a compromise between the need to conserve water and to maintain assimilation at a rate dependent on the capacity of the mesophyll to fix CO_2 In efforts to understand these opposing priorities Cowan (1977) and Raschke (1975, 1979), using a control system analogous to electrical control systems, modelled the interaction between factors affecting gas exchange between a leaf and its environment. They envisage 'sensors' of environmental factors coupled with positive and negative feedback (closed loops) and feedforward (open loops) which establish a relationship between CO_2 uptake and water loss. Figure 6.2 summarizes the ways in which the different environmental factors interact to produce a net effect on stomatal aperture using such control systems.

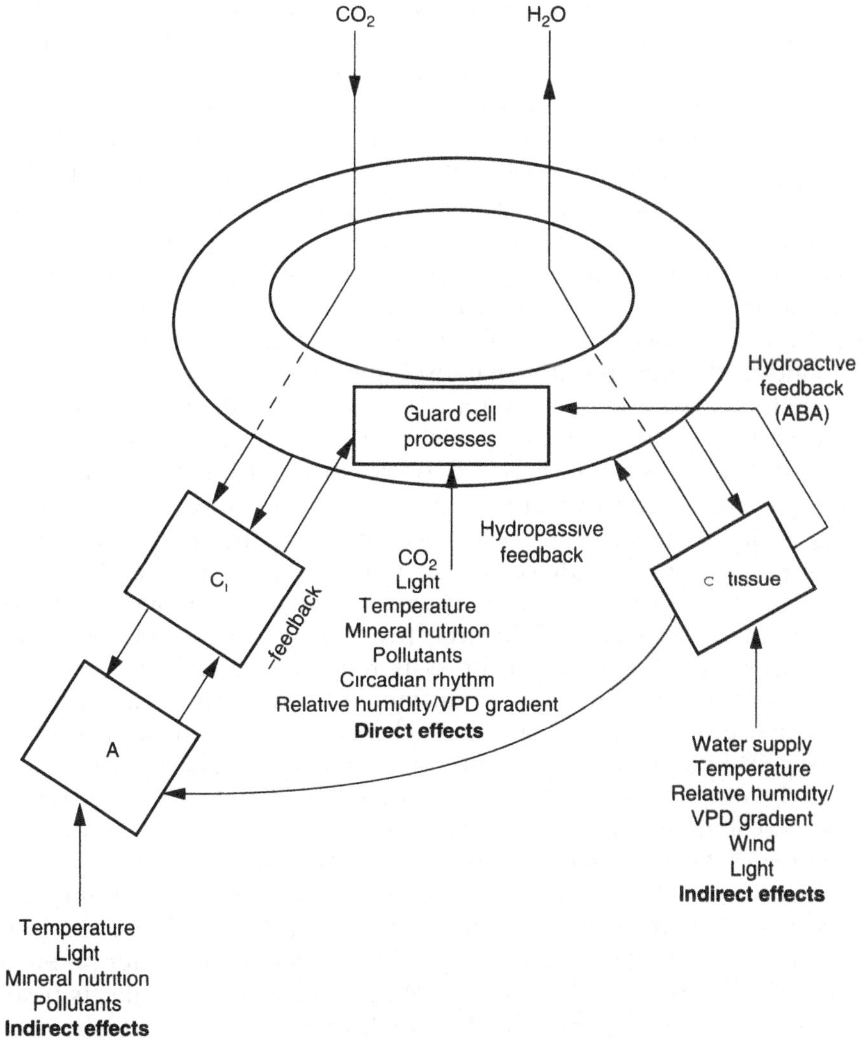

Figure 6.2 Direct and indirect effects of environmental factors, and their interactions, on stomata A, Photosynthetic assimilation of CO_2 by mesophyll cells, C_i, intercellular space CO_2 concentration From Willmer (1988)

Two major feedback loops are envisaged, one regulating CO_2 concentrations in the leaf and the other regulating leaf water potentials or leaf turgor In the 'CO_2 loop', CO_2 enters the leaf via stomata where it is assimilated primarily in the mesophyll (A) The internal leaf CO_2 concentration is a resultant of photosynthetic activity of the mesophyll cells decreasing C_i, and photorespiration and respiration of cells plus diffusion of CO_2 into the leaf increasing C_i Guard cell CO_2 metabolism could also affect CO_2 concentrations at the surface or inside the guard cells since they have a high respiration rate relative to mesophyll cells and also the ability to fix considerable amounts of CO_2 via phosphoenolpyruvate carboxylase activity This is an important consideration because it is probably the CO_2 concentration at the surface of the guard cell plasma membrane or inside the guard cell which is ultimately 'sensed' and sets in motion opening or closing movements influencing C_i values Light indirectly affects stomata via mesophyll photosynthesis if photon flux density increases, C_i will decrease due to stimulation of photosynthesis and the guard cells will open further allowing entry of more CO_2 into the leaf, conversely, if photon flux density decreases, C_i will tend to increase due to a drop in photosynthesis and the stomata will close to some extent (negative feedback) Temperature, mineral nutrition, and pollutants can all affect photosynthetic activity and therefore indirectly affect the stomatal response via C_i

The loop involving tissue water potential is a resultant of the amount of water lost and the availability of water from other tissues and the soil If the leaf water potential falls due to excessive transpiration caused, for example, by an increased atmospheric vapour pressure deficit gradient (it is assumed the internal leaf airspaces including the substomatal cavity are saturated), this condition will be transmitted to the guard cells via physical (hydropassive) and/or chemical signals triggering metabolic processes in the guard cell (hydroactive) and stomata will close (negative feedback) As the water stress is relieved, tissue water potential increases and this condition is transmitted to the guard cells, again via hydroactive and/or hydropassive mechanisms and stomata open With hydroactive processes ABA can be considered as a 'signal' which is transported to the guard cells from the mesophyll as tissue water potential falls and initiates processes which bring about stomatal closure (see Chapter 7) As the water stress is relieved ABA levels in or at the surface of guard cells will fall and stomata begin to open again Vapour pressure deficit, temperature, wind and light (which tends to increase leaf temperature) can all affect transpiration and therefore tissue water potential thus affecting this feedback loop Soil water potential also affects tissue water potential via its effect on the supply of water to the leaf The occasional observation of overshoot' when stomata open and oscillations of aperture which occur before equilibrium is reached are typical responses of such opposing loops

In addition there may be complex interactions between the water and CO_2 loops as water stress is known to inhibit photochemical and enzymatic processes of photosynthesis (e.g. Keck and Boyer, 1974; Boyer, 1976).

There are also direct effects of environmental factors on guard cell processes which influence stomatal movements. Light, temperature, availability of mineral ions, pollutants, circadian rhythms and possibly atmospheric humidity and or water vapour pressure gradients between the guard cell and the atmosphere may also directly affect stomata and these are discussed in the appropriate sections below.

It must be appreciated that the sensors and signals are largely unknown. Wong *et al.* (1979) believe that some product or products of mesophyll photosynthesis can control stomatal movements via positive feedback; it is envisaged that such metabolites, which have some controlling influence on the rate of photosynthetic processes, are transported to the guard cells. Interestingly, Lee and Bowling (1992) concluded that a product or products of photosynthesis in the mesophyll of *Commelina communis* diffused to the guard cells and stimulated stomatal opening. More results, however, are needed to clearly establish the significance of their findings. Also, as indicated above, ABA may be considered a 'signal' and cell turgor or water potential a 'sensor' in the negative feedback loop, since ABA accumulates in leaves under water stress and induces stomata to close. Such an analysis as that described above tends to over-simplify the complex biological system that exists. For example, the simple feedback loop involving ABA in reality describes a whole series of biological events. If our present understanding of the situation regarding the control of stomata by ABA is correct, then a lowered water potential (or possibly turgor potential) of mesophyll cells triggers ABA synthesis in the cell cytoplasm. Then, in a very directional and efficient diffusional or transport system, the ABA must find its way to the epidermal layer where it eventually reaches the guard cells (see Chapter 7). Up to the point of reaching the cell neighbouring the guard cells, ABA may travel either symplastically or apoplastically. For the ABA to reach the plasma membrane of the guard cell it must diffuse through the common cell wall since plasmodesmata are absent between guard cells and neighbouring cells.

6.3 Stomatal responses to light quality and quantity

Plants respond to a wide range of incident radiation including the visible spectrum (about 400–700 nm), UV and IR radiation. Moreover, some of the wavebands of radiation are major environmental signals to which stomata respond directly.

Stomata of most plants open in response to light and close in response to darkness (e.g. Fig. 6.3). Stomata in CAM plants are an exception to this rule, opening in the dark to varying extents according to the degree of CAM being exhibited (see Section 6.4.1). Stomata may also close during late evening hours in daylight and open during darkness of early morning because of an endogenous rhythm (see Section 6.8), and some species under certain conditions exhibit midday stomatal closure (see Section 6.6.1). Generally, closing is faster than opening in response to changing light levels (e.g. Fig. 6.3).

The effects of light on mesophyll photosynthesis and stomata can be separated by observing stomatal responses in epidermal strips. Such studies indicate that very low photon flux densities are required to saturate stomatal opening relative to values needed to saturate photosynthesis. For example, in epidermal strips of *C. communis* less than 100 $\mu mol\ m^{-2}\ s^{-1}$ (about 0.05% of full sunlight) in CO_2-free air or normal air saturates stomatal opening (see Fig. 6.4). These values may vary to some extent, however, according to species, pretreatment of plants and environmental conditions such as the temperature and water potential of the incubating medium.

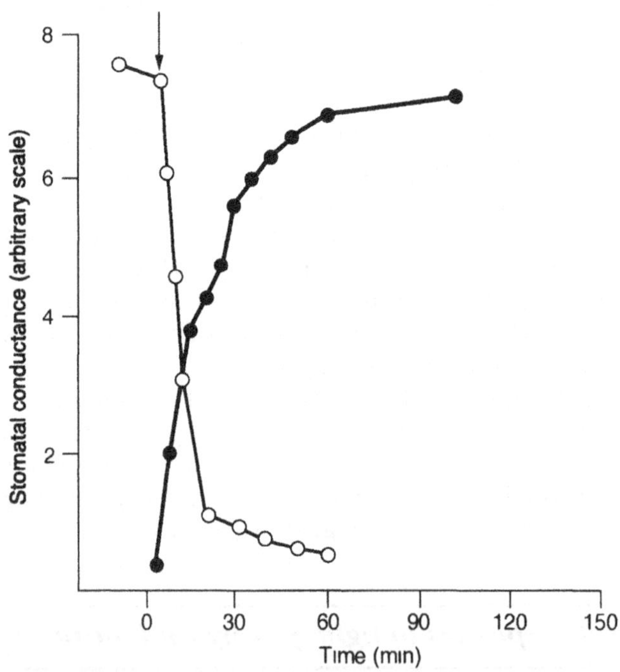

Figure 6.3 Stomatal opening (●) and closing (○) in *Xanthium pennsylvanicum* leaves The arrow indicates where the plants were exposed to a high light level (PFD) preceding an 8 h dark period, or to darkness preceding exposure to high light level After Mansfield and Heath (1963) and Mansfield and Meidner (1966)

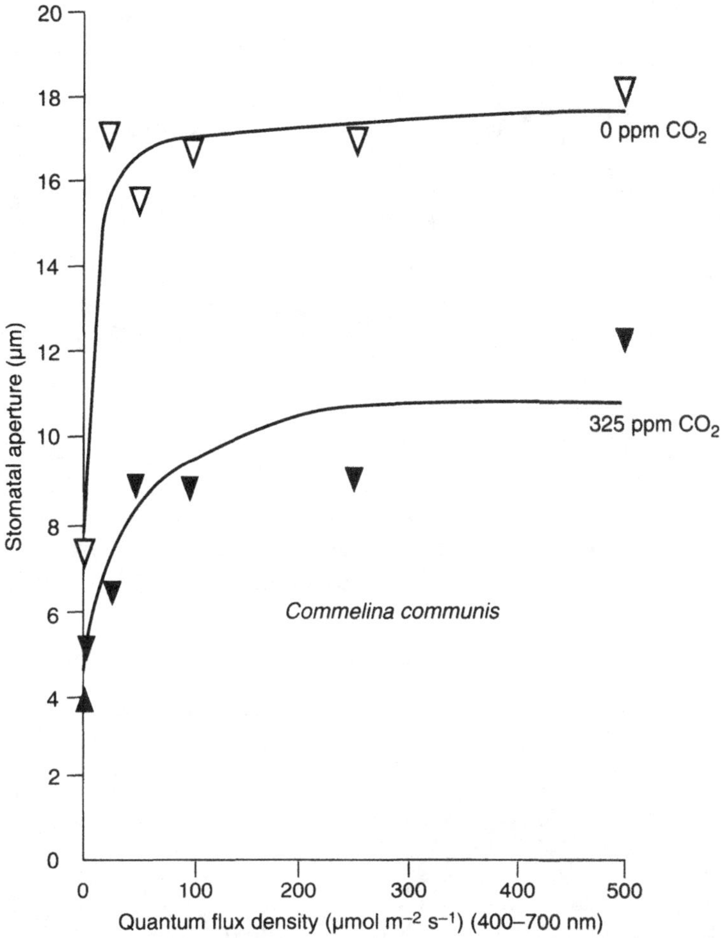

Figure 6.4 The effect of quantum flux density on stomatal opening in epidermal strips of *Commelina communis*; ▼ and ▽ represent opening in air (325 ppm CO_2) and CO_2-free air, respectively, with epidermal strips incubated in 10 mM MES buffer, pH 6.15, and 50 mM KCl at 20°C. From Travis and Mansfield (1981).

Light quality also directly affects guard cells. At equal energy levels, light of different wavebands promotes stomatal opening in epidermal strips in the following order of effectiveness: blue > red > green (e.g. Figs 6.5a and b, and Fig. 6.6). Most investigators have found that blue light is two to 20 times more effective at causing stomatal opening than red light, while other wavelengths have little or no effect. However, at equal photon flux densities white light (and blue plus red light) is more effective at stimulating stomatal opening than blue light alone, indicating there is an interaction between the different wavebands of light (e.g. Fig. 6.6). An action spectrum for maintenance of stomatal opening in

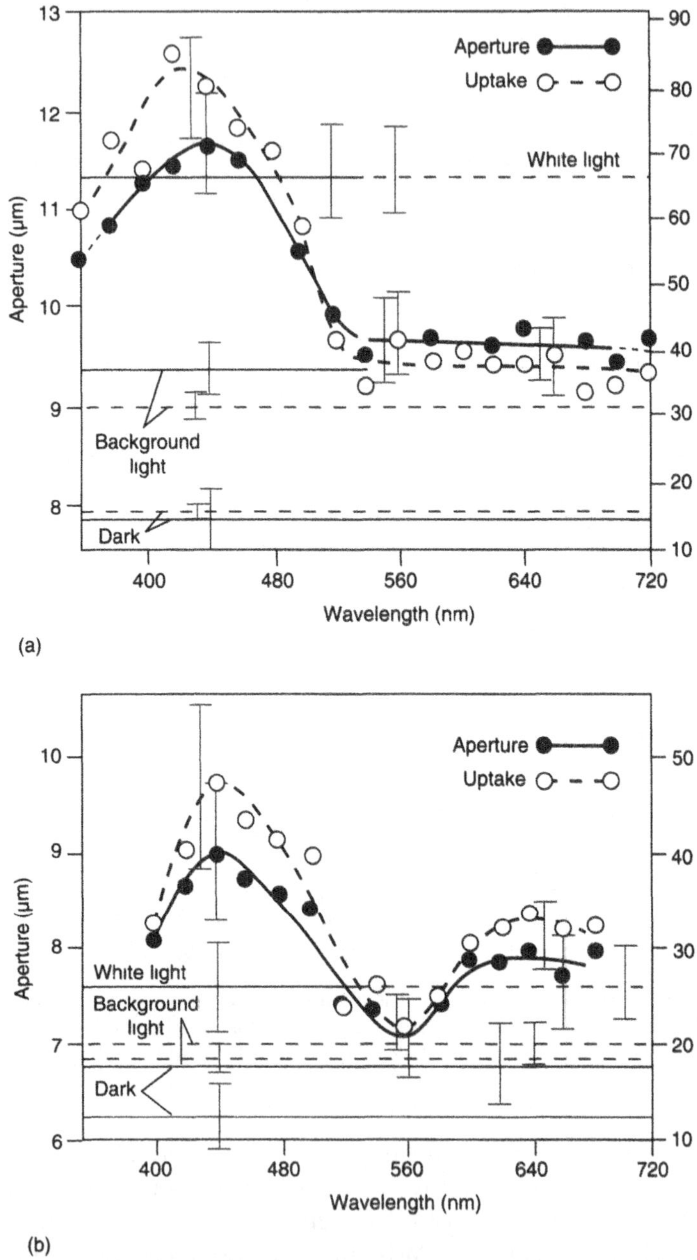

Figure 6.5 Effect of light at different wavelengths and equal quantum flux density on Rb^+ uptake and stomatal opening in epidermal strips of *Vicia faba*. Exposure was for 3 h in CO_2-free air. Rb^+ uptake is based on a unit area basis (mm^2) of epidermal tissue. Vertical bars represent twice the standard error of the mean for uptake and aperture. (a) Low quantum flux density (7.8×10^{14} quanta cm^{-2} s^{-1}) except for 360 and 720 nm which received 5.9×10^{14} and 6.6×10^{14} quanta cm^{-2} s^{-1}, respectively. (b) High quantum flux density (38×10^{14} quanta cm^{-2} s^{-1}). From Hsaio *et al.* (1973).

epidermal strips of *Senecio odoris* has also been obtained and blue light was more than twice as effective as red light (Kuiper, 1964). Such responses have been observed both in intact leaves and in epidermal strips (see below). GCPs also respond to light signals. Thus, light causes GCPs of all species so far tested to swell, with blue light being particularly effective, and to contract in response to darkness, high CO_2 or ABA in an analogous manner to stomatal opening and closing, respectively.

There is considerable evidence that two guard cell photoreceptors are involved in stomatal opening, i.e. chlorophyll and a flavin or carotenoid. Although chlorophyll is considered to be the pigment mediating red light-stimulated stomatal opening, it absorbs both blue and red light. Therefore chlorophyll may also be involved in at least part of the blue light-stimulated opening. In the case of chlorophyll, light is acting primarily as an energy source for cellular metabolic processes (rather than as a specific signal to initiate some cellular process). Most species possess chlorophyll in their guard cells although the chloroplasts are usually poorly developed, but some electron transport and

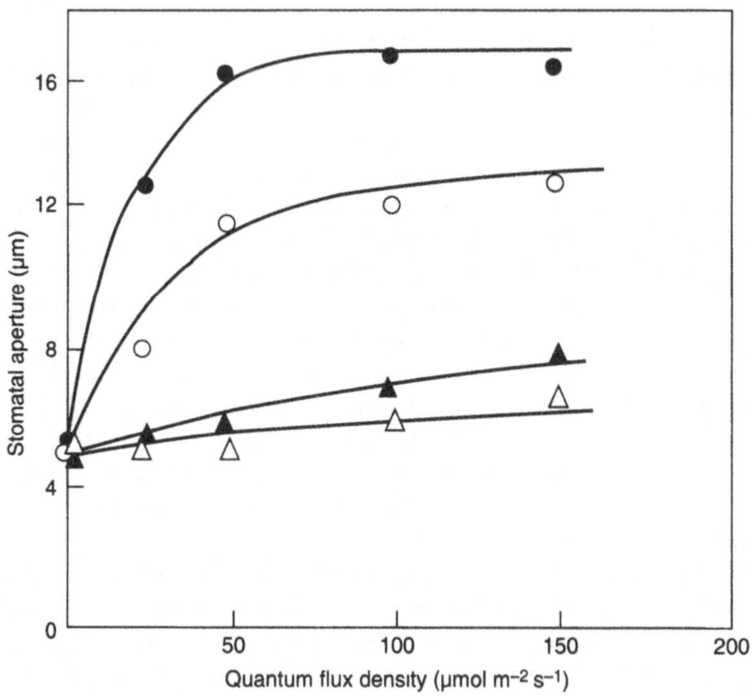

Figure 6.6. Effect of red (▲), blue (○), green (△) and white (●) light at different energy levels, and darkness on stomatal opening in epidermal strips of *Commelina communis*. The epidermis was incubated in 10 mol m^{-3} MES buffer, pH 6.15, containing 50 mol m^{-3} KCl at 25°C with CO_2-free air bubbling through the medium. Apertures were measured after 3 h incubation. Data adapted from Pemadasa (1982).

photophosphorylation activity will occur (see Chapter 9) The ATP and reducing power generated from such activity could be used for bringing about stomatal movements

Phytochrome is also a red/far-red absorbing pigment and has been implicated in the regulation of stomatal movements in some species Although several investigators conclude that it stimulates wider stomatal openings or increases rates of opening, the magnitude of the responses were relatively small and the significance of the results remains inconclusive Thus, Habermann (1977) concluded that stomatal movements of *Helianthus annuus* were phytochrome-mediated, but changes in stomatal aperture were only of the order of 1–2 μm Likewise, Roth-Bejerano and co-workers (Roth-Bejerano and Itai, 1981, Roth-Bejerano *et al* , 1985, Roth-Bejerano and Nejidat, 1987) concluded that there was phytochrome control of stomatal opening in epidermal strips of *C communis*, but aperture changes in response to red and far-red light were also very small According to Evans and Allaway (1972) phytochrome is not involved in stomatal closure of *Vicia faba*

Phytochrome may be involved in the development of stomata (see Chapter 3) and guard cells probably have the capacity to synthesize phytochrome, like most if not all chloroplast containing cells Thus phytochrome has been detected in guard cells in the epicotyls of dark grown pea (*Pisum sativum* L) seedlings using immunofluorescence techniques (Saunders *et al* , 1983) but this does not necessarily mean that the photoreceptor is involved in stomatal control Nevertheless, there is some evidence that phytochrome may be involved with the entrainment of circadian rhythms of stomatal movements (see, e g Holmes and Klein, 1985, Karlsson, 1988, Deitzer and Frosch, 1990)

There is much evidence that another, blue light absorbing photoreceptor is involved in stomatal opening Although the action spectra of stomatal opening and photosynthesis are similar, the greater quantum efficiency of blue light at stimulating stomatal opening, particularly at low photon flux densities (PFD), suggests there is a specific blue light receptor Furthermore, blue light stimulates stomatal opening in *Paphiopedilum harrisseanum*, an orchid without guard cell chloroplasts (Nelson and Mayo, 1975)

There is still debate about whether the blue light photoreceptor is a flavin or a carotenoid and about its location in the guard cell Karlsson and colleagues (Karlsson, 1986, Karlsson *et al* , 1992) observed that some guard cell carotenoids had very similar absorption spectra to the action spectrum for blue light-stimulated stomatal opening, although the same group (Karlsson *et al* , 1983) found that plants treated with an inhibitor of carotenoid synthesis still responded to blue light Recent work has suggested that a blue light absorbing carotenoid pigment, zeaxanthin, within the guard cell chloroplast may be involved in the blue light responses (e g Quinones *et al* , 1993) (see Chapter 9) It has previously been considered, however, that the blue light photoreceptor resided in the plasma membrane or possibly the tonoplast of guard cells

Other than the obvious effect of light on electron transport in chloroplasts, the transduction of red and blue light signals are uncertain but are considered in more detail in Chapter 8 Some basic information will, however, be presented here Most electrophysiological studies of guard cells indicate that light hyperpolarizes the guard cell plasma membrane, a phenomenon associated with proton extrusion and potassium uptake by the guard cell Moreover, Hsiao *et al* (1973) showed that potassium uptake and concomitant stomatal opening in 'rolled' epidermal strips (epidermis is rolled with a glass rod which preferentially bursts the epidermal cells, leaving intact guard cells) of *V faba* occur in blue light at low irradiance, while, at higher irradiance, there is also some stimulation of these processes in red light (Fig 6 5a and b) The maximum efficiency for K^+ uptake and stomatal opening is usually between 420 and 460 nm (e g Fig 6 5a and b) The action spectrum for malate formation in sonicated epidermal strips (this disrupts the epidermal cells and is assumed to empty these cells of their contents, while leaving the guard cells intact) also indicates that blue light (peak at 433 nm) is about eight times more effective than red light (670–680 nm) (Ogawa *et al* , 1978) In the same study it was also shown that at low irradiance, blue light was much more effective with background red light than either red or blue light alone It was postulated that two pigment systems were involved, one mediated by chlorophyll accounting for the red light and part of the blue light response and another pigment accounting for the remaining blue light response

Blue and red light effects are also observed with GCPs Thus, in high energy red light GCP alkalinize the medium in which they are suspended, probably as a result of CO_2 fixation (Calvin cycle activity may be very low but PEP carboxylase activity is very high in guard cells, see Chapter 9) If a pulse of low energy blue light is superimposed on the red light which saturates photosynthesis, the suspending medium acidifies (Fig 6 7) This suggests that blue light promotes H^+ efflux and the process is independent of guard cell photosynthesis Potassium uptake is also associated with the blue light-induced swelling of GCP and concomitant H^+ efflux (e g Shimazaki *et al* , 1986) (see Fig 6 5a and b)

Blue light also causes a decrease in the starch levels within guard cells, although the starch decrease may be an effect of stomatal opening brought about by blue light rather than a direct effect of the light It is interesting to speculate that at least in part, blue light may be affecting stomatal movements by altering the CO_2 levels within guard cells since it is known that blue light enhances respiration and PEP carboxylase activity

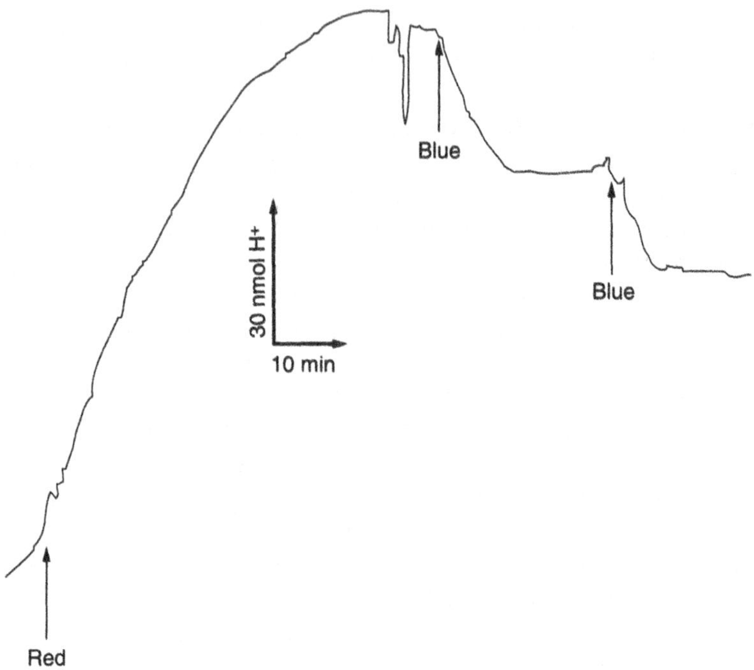

Figure 6.7 Changes of pH of a medium containing guard cell protoplasts of *Commelina communis*. In red light (800 μmol m^{-2} s^{-1}) the medium alkalinizes before the pH stabilizes; when a pulse (1 min) of blue light (50 μmol m^{-2} s^{-1}) is superimposed on the red light the medium acidifies and then stabilizes. A second pulse of blue light triggers off another period of acidification of the medium. From Willmer and Pantoja (1992).

How light signals are perceived and transduced by guard cells to bring about opening or closing stomatal responses is discussed further in Chapter 8.

Light has direct and indirect effects on stomata as indicated in Section 6.2. Preliminary evidence of the direct effect came from Heath and Russell (1954), who demonstrated a positive correlation in leaves of wheat between stomatal aperture and light quantity when intercellular CO_2 concentration at a particular level between zero and 840 ppm was kept constant using a viscous flow porometer (see Fig. 6.1).

An indirect effect of light can occur in the following manner. As the light quantity changes, so the photosynthetic rate of a leaf may change which, in turn, will change the intercellular CO_2 levels. Stomata will, therefore, respond to these changing internal levels of CO_2 within the leaf. This type of response may be particularly important in C4 plants where the CO_2 gradient across stomata may be large, but it will be less so in C3 plants where the CO_2 concentration drop across the stomata maybe 30 ppm or less.

Thus, in whole leaves stomatal responses to light vary according to species, whether C4 or C3, the age of the leaf or plant, the pretreatment of the leaf or plant and other accompanying environmental factors. In general, stomatal conductance saturates at photon flux densities above 400 μmol m^{-2} s^{-1} and less than 1000 μmol m^{-2} s^{-1} in normal air (Fig. 6.8). Also stomata in shade-adapted leaves open at lower light levels than stomata in sun-adapted leaves.

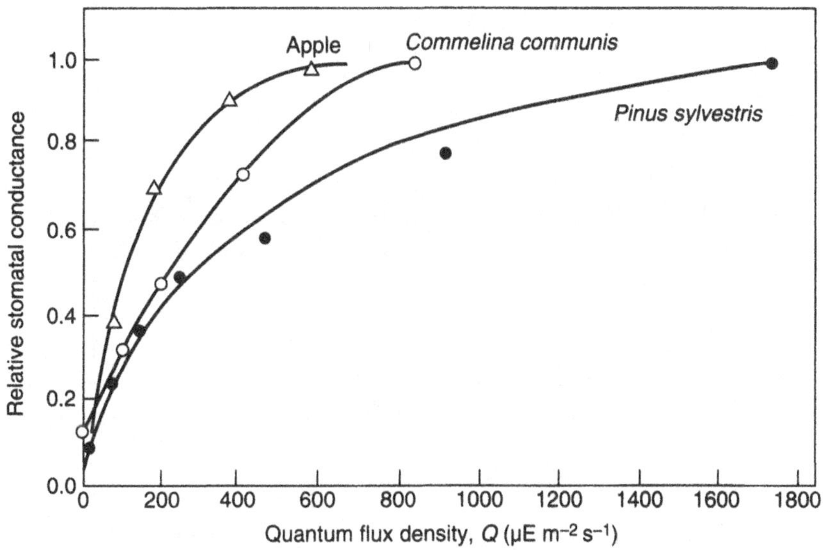

Figure 6.8 The effect of quantum flux density on stomatal opening in leaves of various species in air containing about 340 ppm CO_2. Data for *Commelina communis* and *Pinus sylvestris* adapted from Morrison and Jarvis (1987) and for apple from Warrit *et al.* (1980).

Stomatal responses to light quality are also observed in leaves (e.g. Fig. 6.9a and b). However, the effects of periods or pulses of blue or red light on stomatal conductances have been of particular interest. A 20 min period of blue light caused a rapid increase in transpiration rate before falling (overshoot) in species of Graminae and Cyperaceae, while in other species only a slow increase in transpiration rate was observed (Johnsson *et al.*, 1976). A 20 min period of red light, however, caused a slow, prolonged increase in transpiration in all plants they tested. Johnsson *et al.* (1976) concluded that the different stomatal responses of the two groups of plants to red and blue light may be due, in part, to differences in stomatal anatomy, i.e. the grass-type stomata behaving differently to the elliptical stomata.

A different effect is observed if a pulse (of the order of 30 s) of low PFD blue light is applied to a background of continuous, high PFD red

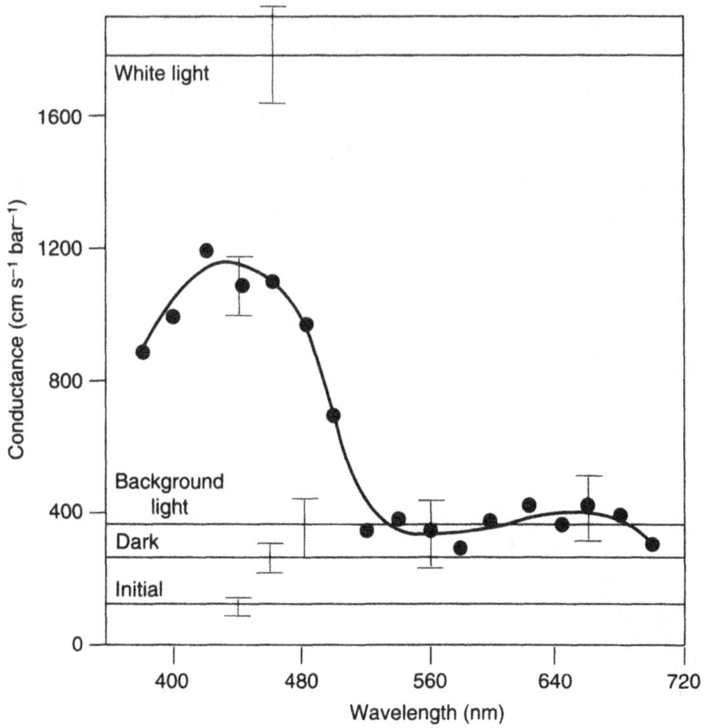
White light
1600
1200
800
400
0
Conductance (cm s⁻¹ bar⁻¹)
Background
light
Dark
Initial
400
480
560
640
720
Wavelength (nm)

(a)

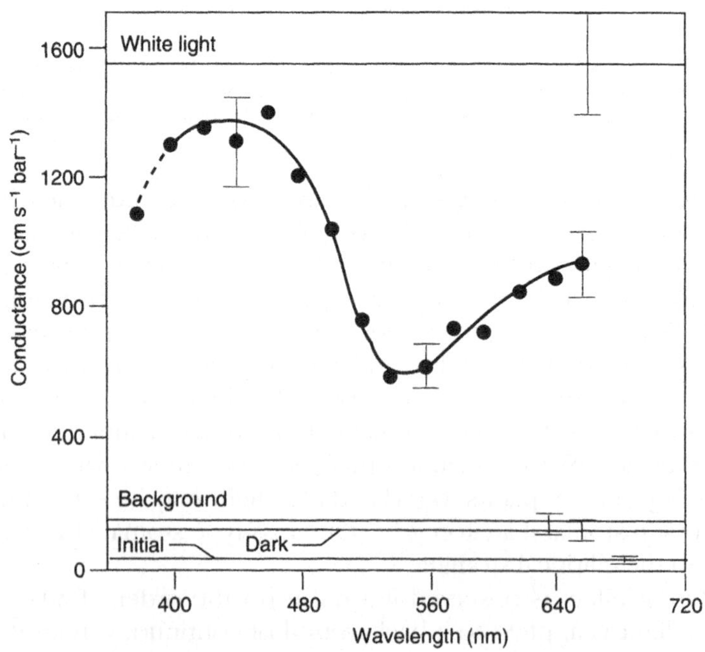
White light
1600
1200
800
400
0
Conductance (cm s⁻¹ bar⁻¹)
Background
Initial
Dark
400
480
560
640
720
Wavelength (nm)

(b)

light Thus, in a number of species including *C communis*, *V faba*, *Phaseolus vulgaris*, *Paphiopedilum harrissianum* (see Zeiger *et al* , 1987) and *Hedera helix* (Karlsson and Assmann, 1990) there is a sharp increase in stomatal conductance almost immediately following the blue light pulse In *C communis* stomatal conductance reaches a maximum about 15 min after the pulse and then conductances slowly return to original values 50 or 60 min later, a pulse of red light on the high PFD background red light had no effect on the stomata (see Zeiger *et al* , 1987) However, a blue light pulse (in the order of seconds to 1 min) of low PFD in darkness produces only very slight increases in stomatal conductance (Lasceve *et al* , 1993) Low or CO_2-free atmospheres, background red light (Assmann, 1988, Lasceve *et al* , 1993) or low vapour pressure differences (Assmann, 1988) enhance the blue light stimulated conductances Thus, it appears that directly or indirectly leaf water status and carbon metabolism of leaves influence the blue light stomatal response Lasceve *et al* (1993) attempted to relate the blue light effects to redox reactions occurring at the guard cell plasma membrane (see Chapter 8) In contrast to this pattern, in *Zea mays* leaves, either a pulse of red light or a pulse of blue light on a high PFD (1000 μmol $m^{-2} s^{-1}$) red light gave rapid increases of conductances (Assmann *et al* , 1992) Assuming photosynthesis of the maize leaf was saturated by 1000 μmol $m^{-2} s^{-1}$ red light, the significance of this finding is uncertain

The blue light response of stomata is thought to be involved in the rapid opening of stomata to sun-flecks and dawn-opening of stomata Leaves growing in deep shade of dense canopies or the under canopy may only be exposed to high PFD for brief periods as sun-flecks Short pulses of red light fail to elicit opening while short pulses of blue light result in a rapid and substantial opening (e g Iino *et al* , 1985) sufficient to support significant photosynthesis Moreover, the time course of stomatal opening to sun-flecks (Kirschbaum *et al* , 1988) is similar to the blue light response rather than the red-light response of stomata With dawn opening of stomata blue light will be very depleted, but because the blue light response is saturated at such low levels it could still be responsible for the light on signal (see, e g Zeiger *et al* , 1981)

6.3.1 Abaxial and adaxial stomatal responses to light

Considerable differences have been found between stomatal responses on the abaxial (lower) and adaxial (upper) leaf surfaces, particularly in their responses to light This is partly due to differences in microclimate

Figure 6 9 Spectral dependence at equal quantum flux density of stomatal opening in *Vicia faba* leaf discs Exposure was for 3 h under normal air and opening in the abaxial epidermis was measured with porometers Vertical bars represent twice the standard error of the mean (a) Low quantum flux density (7×10^{14} quanta $cm^{-2} s^{-1}$) (b) High quantum flux density (38×10^{14} quanta $cm^{-2} s^{-1}$ except for 380 nm which received 32.5×10^{14} quanta $cm^{-2} s^{-1}$) From Hsiao *et al* (1973)

within the leaf (between the upper and lower epidermal surfaces) and partly an inherent difference between the two epidermes Even with patchy stomatal opening in leaves the stomata in the upper surface may have considerably different apertures to those opposite in the lower surface

In most species stomata on the abaxial leaf surface are more sensitive to light than the adaxial stomata they open at lower light levels and have wider open stomata at all light levels than adaxial stomata There are exceptions to this For example, in peanut (Pallas, 1980) and cotton (Lu *et al* , 1993a) the stomatal resistance of adaxial stomata is less than that of the abaxial ones at the same light level Experiments with isolated epidermis of *C communis*, a situation in which the effects of mesophyll tissue are avoided, indicate that stomata on the abaxial leaf surface are much more sensitive to light than those on the adaxial surface (Travis and Mansfield, 1981) This finding supports the view that differences in photosensitivity between stomata on the two leaf surfaces are at least partly inherent

In some species (e g maize, cotton, sunflower), when light is incident on the adaxial leaf surface, the abaxial stomata open faster than the adaxial ones However, if a leaf of sunflower is inverted so that light is incident on the abaxial surface then adaxial stomata open faster than the abaxial ones

6.3.2 Effects of UV radiation on stomata

The effects of UV radiation on stomatal behaviour are becoming more relevant as ozone depletion in the stratosphere is increasing, because ozone absorbs shorter, damaging wavelength ultraviolet radiation (UV-C, < 280 nm, and portions of UV-B, 280–315 nm) Most of UV-B radiation appears to cause stress and damage in plants and, since UV radiation is absorbed by vital cell components such as proteins and nucleic acids, plants have evolved mechanisms to counter such problems Whether plants can adapt rapidly enough, however, to counter the increasing UV loading as the ozone layer is diminished, remains uncertain

The epidermal layer of leaves is a major site of UV absorption (see Section 2 1) due both to the presence of phenolic compounds in cells of the epidermis and to the cuticle comprised of a complex layer of cutin and waxes An absorption spectrum (with maxima at 265 and 330 nm, and a minimum at 280 nm) of guard cells from *V faba* suggested that the UV absorbing pigments were phenolic compounds, possibly flavonoids (Heller *et al* , 1971) Zeiger and Hepler (1979) considered that a vacuolar autofluorescence in guard cells of onion was caused by a flavin, while Vierstra *et al* (1982) concluded that guard cells of *V faba* contained an abundance of the flavonol, kaempferol 3-*O*-galactoside 7-*O*-rhamnoside. Contour-Ansel and Louguet (1985) reported high levels of tannins and simple phenolic compounds (particularly gallic acid, protocatechuic acid, gentisic acid and vanillic acid) in the epidermis of *Pelargonium zonale* and the amounts varied according to whether the epidermis was illuminated or darkened before analysis. The simple phenolics may, however, be hydrolysis products of the more complex flavonoids since such compounds have been detected in guard and epidermal cells of a variety of plant species including *V. faba* (Weissenbock *et al.*, 1984), *Allium cepa* (Weissenbock *et al.*, 1987), *P. sativum* (Weissenbock *et al.*, 1987), *Tradescantia virginiana* (Takahama, 1988) and *Z. mays* (Schnabl *et al.*, 1989). In a detailed study Weissenbock *et al.* (1984) estimated the total flavonol content of guard cells of *V. faba* to be 85 fmol per protoplast (about the same as in epidermal cells but six times greater than in mesophyll cells). The flavonols were kaempferol glycosides and at least 12 different ones were detected residing in the vacuole. Different flavonol glycosides dominated in guard cells and epidermal cells of *P. sativum* and their relative contents were also different (Weissenbock *et al.*, 1986). Thus, quercitin 3-triglucoside and its *p*-coumaric ester were much more abundant than the kaempferol glycosides, and 80, 300 and 25 fmol total flavonoid was estimated to be in guard, epidermal and mesophyll cells, respectively. Weissenbock *et al.* (1987) also detected high amounts of flavonol glycosides, particularly kaempferol and quercitin aglycones substituted with sulphate and glucuronate, in guard cells of *A. cepa*. The role of the phenolics in guard cells remains uncertain though it may be that they are part of the protective layer of the epidermis against UV damage to photosynthetic activity of the mesophyll (e.g. Shimazaki *et al.*, 1988) or they may have an antioxidative function (e.g. Takahama, 1988).

There are a number of reports that stomatal behaviour is influenced by UV radiation. Generally, such radiation is found to close stomata (e.g. Murali and Saxe, 1984) but Teramura *et al.* (1983) found that UV-B radiation (280–315 nm) decreased stomatal resistances, especially under water stress, and that cucumber was particularly sensitive to UV-B. In the presence of white light (PFD = 50 μmol m^{-2} s^{-1}) Negash and Bjorn (1986) obtained a UV (PFD = 2 μmol m^{-2} s^{-1}) action spectrum of stomatal closure in leaves of *Eragrostis tef* in which 285 nm or shorter wavelengths were very effective at causing closure, but the effectiveness thereafter decreased sharply until at wavelengths of 313 nm there was no significant effect. They found that increased irradiance with white light overcame the effects of UV though, subsequently, there was a tendency for stomata to gradually close. Generally, it appears that UV radiation causes leakiness of ions from tissues presumably due to membrane or cell damage, but effects on stomata appear to be more difficult to define. For example, UV-C irradiation of guard cells of *V. faba* resulted in irreversible cell damage as gauged by membrane leakiness, while such observations were not observed with UV-B or UV-A (315–400 nm) irradiation (Negash *et al.*, 1987). Curiously, Ogawa *et al.* (1978) found that UV-A radiation of guard cells of *V. faba* stimulated malate formation.

6.4 Carbon dioxide

It is probable that guard cells can perceive and respond to the CO_2 concentration at the external surface of the plasma membrane or within guard cells, since GCPs and stomata in epidermal strips are influenced by changing CO_2 concentrations. Thus, GCPs expand in the presence of zero or low CO_2 concentration and contract in high CO_2 concentrations (e.g. Gotow *et al.*, 1982, Fitzsimons and Weyers, 1986) (Fig. 6.10). In epidermal strips stomata open wider in CO_2-free air than ambient air and in epidermal strips of *C. communis* there is an almost linear stomatal response to CO_2 concentration in the range 0–360 ppm (Fig. 6.11). Little is known about the transduction mechanism and this aspect is discussed further in Chapter 8.

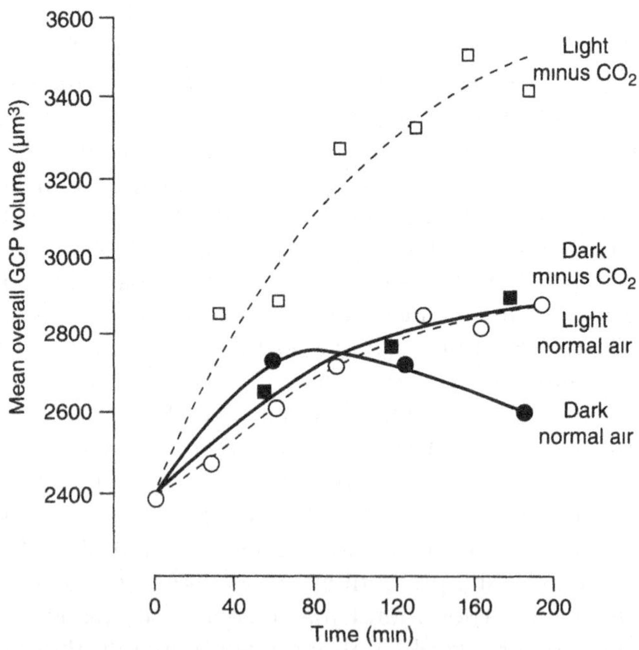

Figure 6.10 Effects of CO_2 and light on volume changes of *Commelina communis*. Guard cell protoplasts were suspended in 400 mM mannitol buffered to pH 6.2 with MES–Tris buffer containing 1 mM KCl. The medium was aerated with normal air or CO_2-free air and the protoplasts illuminated with 'white' light (10 W m^{-2} PAR from white fluorescent tubes) or kept in darkness. From Fitzsimons and Weyers (1986)

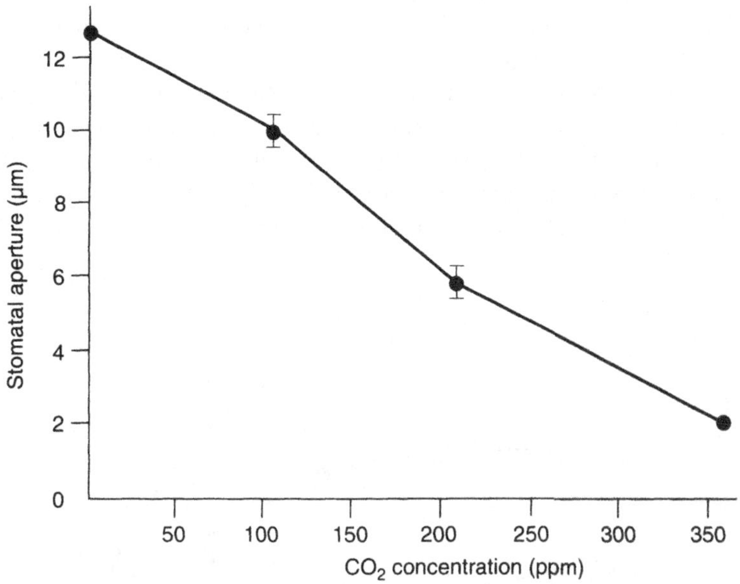

Figure 6.11 Effect of CO_2 concentration on stomatal opening in epidermal strips of *Commelina communis*. Epidermis was incubated in 10 mol m^{-3} MES buffer, pH 6.5, containing 20 mol m^{-3} KCl at 26°C and a quantum flux density of 170 μmol m^{-2} s^{-1}. Air containing the different CO_2 concentrations was bubbled through the medium at a rate of 100 cm^3 min^{-1} and apertures were measured after 2 h incubation. From Willmer (1988).

In intact leaves stomata normally close as ambient CO_2 concentrations increase over the physiological range, while stomata open as CO_2 concentrations decrease over the same range. Thus, high CO_2 concentrations can close stomata, even in the light, while CO_2-free air can open stomata, even in the dark. However, atmospheric CO_2 levels normally change little over the short term, although high CO_2 levels can build up within dense canopies during the night period due to tissue respiration and soil to canopy fluxes (from respiration of soil organisms). These rapidly equilibrate with the bulk air CO_2 levels during early morning. Nevertheless, the intercellular CO_2 concentrations, C_i, can change: C_i values tend to increase in the late afternoon and during the night and decrease during the early morning. Additionally, in those species which exhibit midday closure, C_i values will change at this time. In CAM plants, also, C_i values vary greatly over each 24 h cycle. However, for most species and under most conditions, C_i remains fairly constant for much of the day.

In intact leaves of C3 plants in air of about 350 ppm CO_2, C_i usually remains between 200 and 230 ppm CO_2, although in *C. communis* Morison and Jarvis (1983) obtained C_i values somewhat higher at about

290 ppm CO_2. If C_i is artificially changed then a series of curves are obtained relating stomatal conductance to C_i at different light levels (Fig. 6.12). At the higher light levels stomatal conductance remains steady until C_i values near to ambient CO_2 are reached and then stomata begin to close. At lower light levels stomata respond to all C_i values, closing as C_i increases. The broken line in Fig. 6.12 intersects the curves at C_i values that would be obtained in normal air of about 320 ppm CO_2 at the various light levels. It is evident that C_i values remain fairly constant for all light levels except the lowest ones.

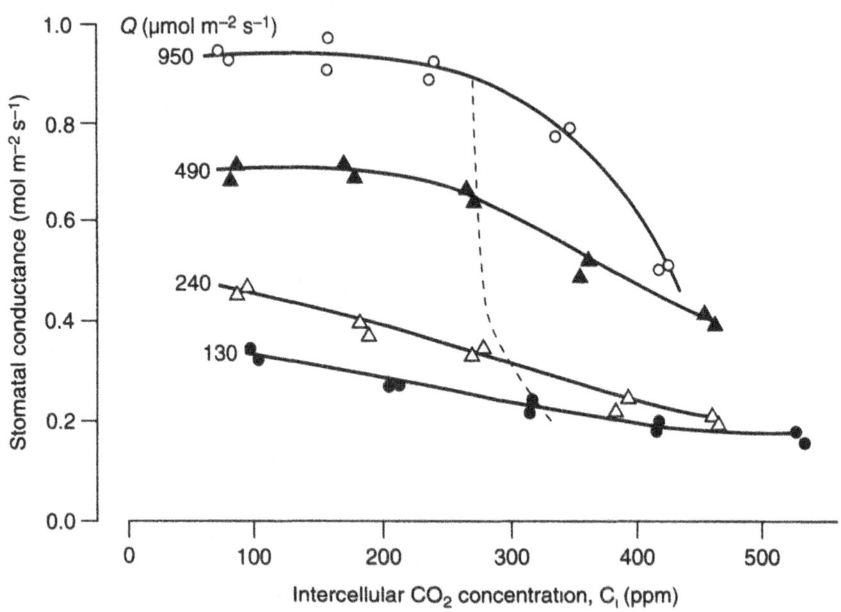

Figure 6.12 Effect of intercellular CO_2 concentration (C_i) on stomatal conductance in *Commelina communis* leaves at different photon flux densities of 950 (○), 490 (▲), 240 (△) and 130 (●) µmol m^{-2} s^{-1} The dashed line intersects the curves at C_i values equivalent to an ambient CO_2 concentration of 320 ppm CO_2 Leaf to air vapour pressure difference of 0 33 kPa was maintained at a leaf temperature of 20°C From Morison and Jarvis (1983)

Certain species appear insensitive to changing CO_2 concentrations under some conditions (usually of saturating light intensities for stomatal opening). The lack of response to CO_2 may occur when leaves have a high water potential and endogenous ABA is very low, but the lack of response has also been seen in a number of species with less than maximum leaf water potential. Species which are reported to be relatively insensitive to changing CO_2 concentrations include *Pinus sylvestris*, *Picea sitchensis* (see Morison and Jarvis, 1981) and *Gunnera tinctoria* (Osborne *et al.*, 1993)

Stomata of C4 species also appear to be more sensitive to CO_2 than those of C3 species and closure in C4 species occurs at much lower CO_2 levels than observed in C3 species. CO_2 levels below about 100 ppm may not cause further stomatal opening, at least in C3 species, while in C4 species stomata are apparently sensitive down to zero levels of CO_2.

The atmospheric CO_2 level in 1994 was about 355 μmol mol^{-1} (ppm) and is increasing at approximately 1.8 μmol mol^{-1} per annum (Houghton *et al.*, 1992). The responses of stomata to elevated CO_2 over the long-term are therefore of interest in order to determine if they acclimate to such an environment. In many cases stomatal frequency has been observed to decrease as CO_2 concentrations increase (see Chapter 2) but little is known about stomatal responses to prolonged elevated CO_2 levels. Some early work showed that even if plants of *Xanthium pennsylvanicum* and lettuce are grown in high CO_2 concentrations for long periods of time the stomatal response to CO_2 is the same as in plants grown at 0 ppm CO_2 (Mansfield and Jones, 1970). Stomata of these species, therefore, did not acclimatize to an elevated CO_2 level. More current information, however, presents a confusing picture. For example, Radoglou *et al.* (1992) also found that sensitivity of stomata to CO_2 concentration was not affected by the concentration of CO_2 that plants were grown at. However, Hollinger (1987) found that stomata of *Pinus radiata* and *P. menziesii*, but not *Nothofagus fusca*, were less sensitive to dry air when grown under normal CO_2 concentrations than when grown under elevated CO_2 concentrations. Furthermore, Berryman *et al.* (1994) observed that stomata in seedlings of *Eucalyptus tetrodonta* grown in CO_2-enriched air were not more sensitive to leaf-air vapour pressure deficits or C_i compared to stomata in plants grown at normal CO_2 concentrations, but their responses to temperature decreased and the equilibrium conductance at saturating light decreased in plants grown at elevated CO_2 levels. It appears, therefore, that elevated atmospheric CO_2 concentrations result in variable stomatal responses to environmental factors depending on the species and the environmental parameter being examined.

6.4.1 Stomatal movements in CAM plants

Stomata in CAM plants usually have substantial opening periods during the night and may close during the day. The exact pattern of stomatal movements, however, depends on the species and upon the pretreatment of the plant, particularly regarding diurnal temperature changes and the lengths of the photoperiod. In general, three patterns of stomatal movements have been observed (Neales, 1975):

1. Stomata which open towards the end of the dark period and may

remain open for the entire light period. These plants exhibit weak CAM.

2. Stomata which open for most of the dark period followed by increased opening immediately after the light period begins and then closure. Further opening occurs, again, at the end of the same light period. Such stomatal behaviour is found in plants exhibiting 'full' CAM.
3. Stomata which open during the dark period and close during the entire light period. These plants exhibit 'very strong' CAM.

Plants which can only function in 'full' or 'very strong' CAM mode are known as obligate while those plants which can exhibit C3-type or CAM stomatal behaviour are known as facultative.

The behaviour of stomata in CAM plants is part of an adaptive mechanism evolved by plants which grow in hot and arid environments to conserve water. At night, when temperatures fall, stomata open to allow CO_2 to enter the plant. The CO_2 is 'fixed' via PEP carboxylase and stored as malic acid. During the day when temperatures may be very high stomata close, preventing excessive water loss from the plant. The malic acid produced at night is essentially a CO_2 store and during the day it is decarboxylated, releasing CO_2 which is refixed via the Calvin cycle.

Originally it was not clear whether stomata of CAM plants functioned relatively independently of the changing CO_2 concentrations within the leaves which occur due to the carboxylation and decarboxylation phases of CAM. However, after removal of the epidermis, CO_2 exchange remains much the same as when the epidermis is present. Moreover, a reasonably good positive correlation exists between intercellular CO_2 levels within the tissues and stomatal aperture (Fig. 6.13). It appears, therefore, that stomatal responses of CAM plants are primarily controlled by the changing intercellular CO_2 concentration in the following manner. When malate is being decarboxvlated during the light period CO_2 levels within leaves will be high and stomata will close; when malate is being formed during the dark period CO_2 will be rapidly and efficiently fixed by PEP carboxylase, thereby lowering leaf CO_2 levels, and stomata will open.

6.5 Temperature

As the leaf temperature is raised, the metabolic activity within the guard cells and the leaf as a whole will increase, reach an optimum, and then decrease as more and more cell damage occurs. The effect of the increased metabolic activity within guard cells will tend to stimulate opening.

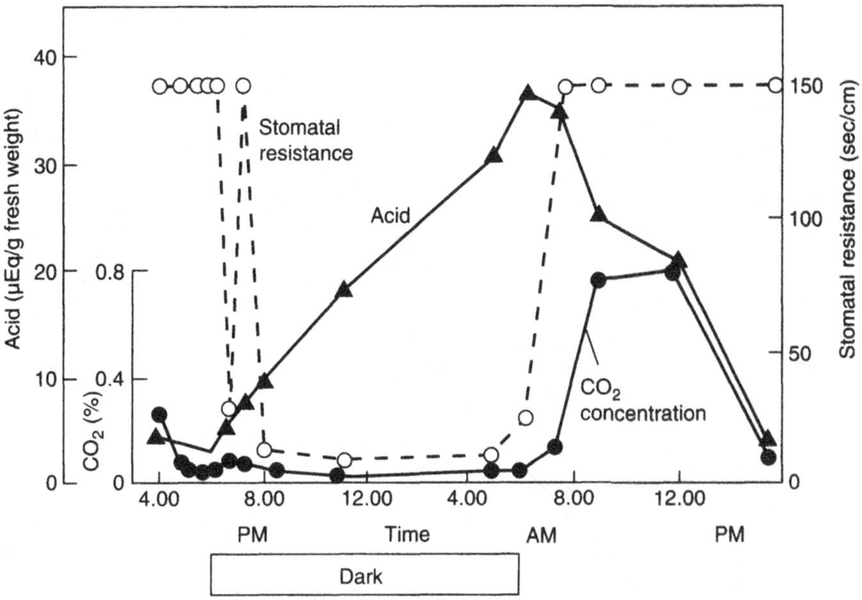

Figure 6.13 Diurnal changes in internal gas phase CO_2 concentration, stomatal resistance to diffusion of water vapour and acid content in *Agave deserti*. From Cockburn *et al.* (1979)

There are also indirect affects of temperature on stomatal behaviour. It is possible that the temperature increases will affect internal CO_2 concentrations which, in turn, will affect stomatal movements. If respiration and photorespiration outpaces photosynthesis as the temperature increases, CO_2 levels will increase within the leaf which will tend to bring about stomatal closure. Also, an increase in leaf temperature will result in an increase of the water vapour pressure gradient between the leaf and the surrounding air which may ultimately cause stomatal closure either through direct or indirect effects (see Sections 6.2 and 6.6).

It is not surprising, therefore, that there is considerable variation in the temperature values obtained for optimal stomatal opening in intact leaves (see Fig. 6.14b). For example, maximum opening in *V. faba* occurred at between 35 and 40°C (Stålfelt, 1962) or even greater than 40°C for other species (Schulze *et al.*, 1973) under some circumstances. Even in epidermal strips incubated in media of constant water (osmotic) potential, considerable variation of temperature optima for stomatal opening is observed. Thus, maximum opening in epidermal strips of the fern *Polypodium vulgare* varied at between 20 and 28°C depending on the pretreatment of the plants (Lösch, 1977), while for stomata in epidermal strips of *C. communis*, maximum apertures were achieved at about 35°C (Willmer, 1980). Usually, however, maximum apertures are achieved at values of around 30°C (see Fig. 6.14a).

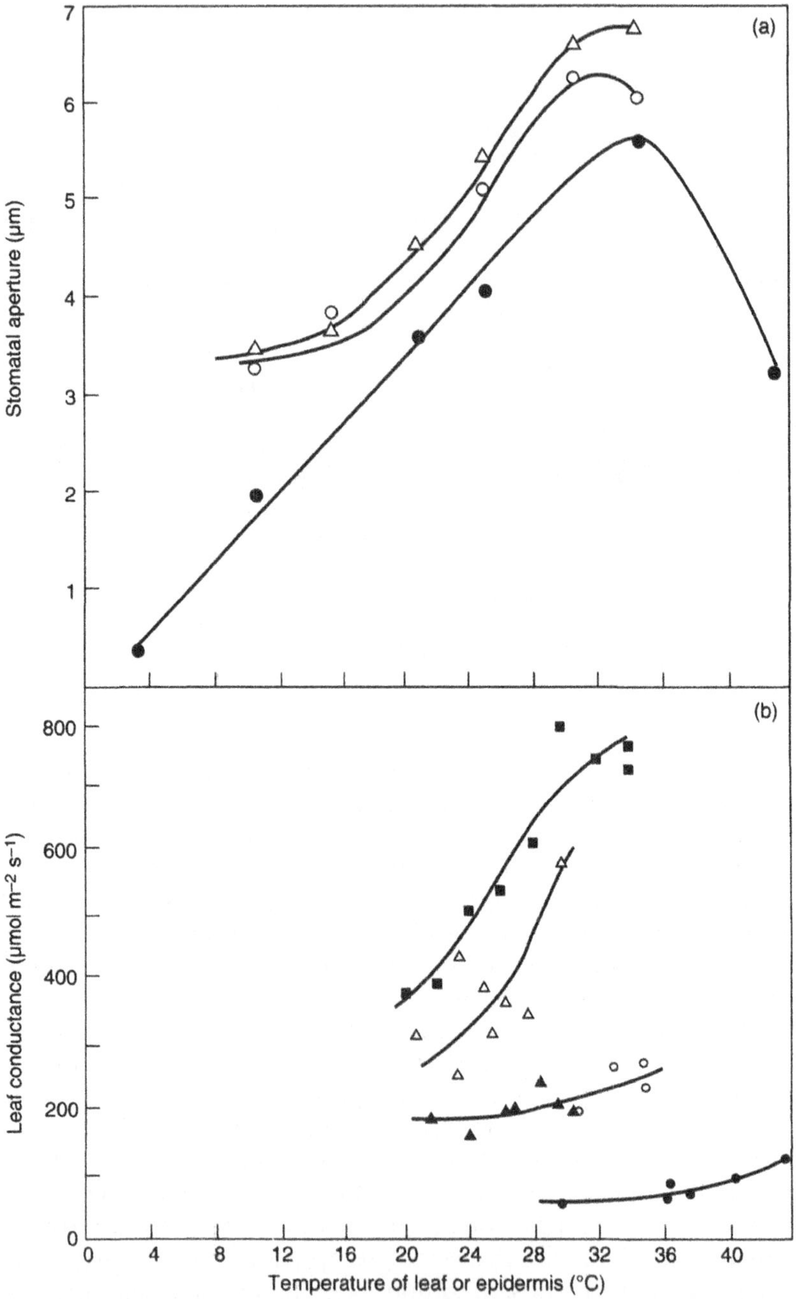

Figure 6.14 The effects of temperature on stomatal responses in epidermal strips (a) and in leaves (b). Data in (a) for *Bellis perennis* (△) and *Cardamine pratensis* (○) is from Honour *et al.* (1994) and for *Commelina communis* (●) is from Willmer (1984). Data in (b) for *Sesamum indicum* with a constant vapour pressure difference of 7 mbar (■) is from Hall and Kaufmann (1975); for *Citrus sinensis* with a constant vapour pressure difference of 5 (△) and 15 mbar (▲) is from Hall *et al.* (1975); for *Prunus armeniaca* with a constant vapour pressure difference of 15 mb (○) and 40 mb (●) is from Schulze *et al.* (1974).

Increased temperature can overcome the closing effects of increased CO_2 levels and darkness For example, increasing the temperature from 27 to 36°C resulted in stomatal opening in *Xanthium* leaves in the dark without removal of CO_2 (Mansfield, 1965) Darwin (1898) also observed that closure occurred more slowly in the dark at high rather than low temperatures Additionally, at high temperatures stomata of some species remain open in the light if leaves do not suffer from water stress (e g Stålfelt, 1962) even if CO_2 is being released from the leaves (Raschke, 1970) In part this is due to the CO_2 feedback system being uncoupled from the stomatal responses at the higher temperature

In ivy leaves a 30 min period of high temperature (44–48°C) causes stomata to be sluggish in response to light/dark transitions, apertures remaining smaller than in control leaves and, additionally, the sluggish responses persist for many days (Bauer, 1979) The finding by Bauer (1979) that flushing the leaves with low CO_2 air resulted in more normal stomatal responses suggested that photosynthesis had been impaired raising C_i which influenced stomatal behaviour Equally, guard cell functioning may be impaired directly by increased temperatures, although guard cells may also be able to tolerate higher temperatures than mesophyll cells

Cold treatment of plants can also subsequently effect stomatal behaviour A number of studies have shown that low night temperatures result in decreased rates of stomatal opening coupled with low rates of photosynthesis in a variety of species (e g Pasternak and Wilson, 1972, Drake and Raschke, 1974, Drew and Bazzaz, 1982)

In the natural environment plants do experience extremes of temperatures and a number of interesting studies have been made which conclude that, in some species, the responses of stomata to night temperature are of adaptive importance For example, Drew and Bazzaz (1979) showed that transpiration and photosynthesis of *Populus deltoides* leaves from a Southern US location were reduced to a greater extent by cool nights than ones from a Northern US location Furthermore, Drew and Bazzaz (1982) found that species which germinate or grow in early spring when temperatures are low (but not freezing) are insensitive to cool nights, i e their stomatal conductances in the following days are unaffected by low night temperatures In later spring and summer, however, stomatal conductances of emerging species were inhibited by low temperatures Drew and Bazzaz (1982), however, were careful to point out that adaptive differences in responses to night temperatures were also related to both phenology of germination and growth and habitat type

If plants are not chilling resistant species, such as *Phaseolus vulgaris*, then stomata appear to be locked open during chilling periods despite severe wilting, chill-hardened plants or chill-resistant plants, in contrast, are able to reduce stomatal apertures and maintain positive leaf turgor during chilling (e.g. Eamus *et al.*, 1983).

As already indicated there will be direct effects of temperature on guard cell metabolism. Mawson and Cummins (1991), for example, found that guard cell chloroplasts of *Saxifraga cernua* acclimated to low temperatures (10°C relative to 20°C) in terms of their photosynthetic electron transport functions. Likewise, an interesting correlation between the temperature sensitivity of guard cell respiration and stomatal conductances in Pima cotton was observed by Lu *et al.* (1993b). They used an F_2 population of cotton from a cross between a heat-resistant, high yielding line, and a heat sensitive primitive cotton and found that conductances were 3- to 4-fold higher at 40°C than at 25°C and the range of conductances was much greater at the higher temperature. Furthermore, they found that in mechanically isolated, enzymatically cleaned epidermis, guard cell respiration was highest from those plants that had highest stomatal conductances. Additionally, the slope of guard cell respiration as a function of temperature increased linearly with stomatal conductance. The authors concluded that the co-segregation of rates of guard cell respiration and stomatal conductances indicate that both properties are under genetic control and that guard cell respiration is a component of the sensory transduction of the stomatal response to temperature.

6.6 Plant water status and atmospheric humidity

Tissue water potential has an important influence on stomatal behaviour. Bulk leaf water potential is the resultant of water uptake and storage by the plant and water loss from the plant via transpiration. Therefore any factor which affects these processes such as soil water availability, temperature, atmospheric humidity and wind will tend to perturb leaf water potentials.

Figure 6.15(a and b) shows the relationship of leaf water potential and leaf turgor against stomatal conductance for a number of plant species. Although there are species differences in stomatal sensitivity to water stress, with the C4 species generally being less sensitive, normally stomata begin to close only after certain levels of leaf water potential have been reached and then tend to close rapidly as the potential continues to decrease. For some species, such as Sorghum growing under field conditions, leaf water potentials may drop to very low values before the stomata close. Figure 6.15(b) also shows that when the stomatal response to water stress is related to leaf turgor rather than leaf water potential the differences in apparent sensitivity between species is decreased.

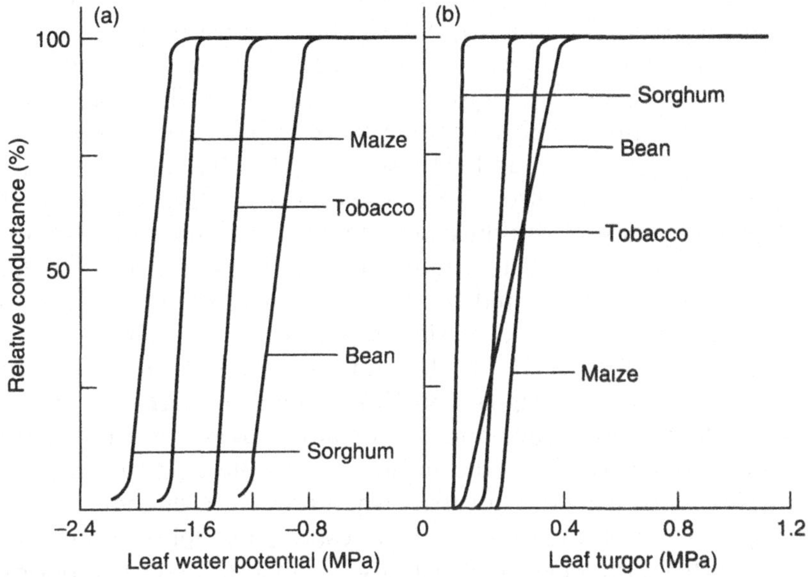

Figure 6.15 Leaf conductances of several species, at similar irradiances, as a function of (a) leaf water potential and (b) turgor pressure in leaf tissue 'Bean' = *Phaseolus vulgaris L* Data from Davies *et al* (1980) adapted from Turner (1974)

Stomatal responses to humidity can also be very rapid, and can occur when the bulk leaf water potential does not alter, suggesting the turgor relations of the epidermis are relatively independent of the mesophyll tissue (e.g. Edwards and Meidner, 1978). Meidner (1975) found that the epidermis had a high hydraulic conductance and could act as a major route for water movement. In addition, high rates of water evaporation from the inner epidermal cell walls would combine to affect the turgor relations of the cells in the epidermis, including the guard cells. Under such conditions, changes in VPD would modulate evaporation through the pore and hence stomatal conductances through a negative feedback loop localized primarily in the epidermis.

However, not all stomatal responses to humidity can be explained in terms of negative feedback loops, where the maximum response to an increase in VPD would only lead to pore closure and subsequent reduction in transpiration back to the initial level. Cowan (1977) and Farquhar (1978) have described situations when the imposed change in the rate of transpiration causes a change in the conductance that was independent of the resulting change in transpiration rate. They termed this a feedforward response and it becomes most obvious under conditions where the decrease in conductance actually reduces transpiration under conditions of increased evaporative demand and leaf water content may even increase. Thus, feedforward responses cannot be explained by mechanisms that involve water evaporation from within the leaf passing through the pore. The only hypothesis that is consistent with feedforward responses is that direct cuticular water loss from externally facing surfaces of the guard and subsidiary cells influences the stomata combined with an appreciable resistance to water flow to the guard cells (Farquhar, 1978; Jarvis and Morison, 1981). Based on work by Schönherr (1982), Schulze (1986) presented models indicating that non-linear changes in conductance through porous cuticular membranes with changing relative humidity closely paralleled a range of stomatal responses to humidity. The cuticle may therefore be instrumental in the mechanism underlying the control phenomena.

For example Lange *et al.* (1971) found that guard cells in epidermal strips of *Polypodium vulgare* and *Valerianella locusta* responded to repeated changes in relative humidity of air jets blown over the external face of the epidermis, whilst the air on the mesophyll side was saturated with water. As stomata did not react when they were directly in contact with liquid water on their inner side, a restricted supply of water to the guard cells from the epidermal cells also seemed to be required for the response. Lange *et al.* (1971) concluded that stomata reacted to changes of humidity in the ambient air and that evaporation through the cuticle of the guard cells and neighbouring cells (termed peri-stomatal transpiration) was the most likely mechanism leading to localized changes in the water potential around the guard cells that affected their turgor. The guard cells were postulated to act as humidity sensors responding to the difference in water potential of the air inside and outside the leaf. Lösch and Schenk (1978) subsequently demonstrated that guard cells of *Valerianella locusta* showed rapid decreases in aperture in response to an increase in VPD, whilst the K^+ content of the guard cells declined over a much longer time period. Equally, when stomata were induced to open by a decrease in VPD, the increase in K^+ content of the guard cells followed with a lag of about 20 min. Thus the initial humidity-dependent stomatal movements were hydropassive in this species, with metabolic adjustment of guard cell turgor through changes in guard cell K^+ levels occurring as secondary processes. In contrast, there was a direct correlation between K^+ content in the guard cells and stomatal aperture in response to light and dark stimuli.

The stimulus perceived by the guard cells that initiates hydroactive responses to humidity has not been well characterized. Possible sensors include a turgor or volume measuring mechanism in the guard cells, either of which might be triggered by the altered mechanical interactions between the guard cells and neighbouring cells. For example, Shackel and Brinkmann (1985) directly measured rapid changes in the turgor of epidermal cells from *T. virginiana* in response to a change in

VPD, which subsequently triggered changes in stomatal aperture after a short lag period. Stomatal movements continued for about 1 h before reaching a new steady state conductance. However, Grantz and Schwartz (1988) found that solute content of guard cells did not alter in epidermal strips of *C. communis* subjected to osmotic stress to simulate a change in external water potential, i.e. under these conditions there was no hydroactive adjustment of guard cell solute content. In leaf discs, however, where the epidermis remained in contact with the mesophyll tissue, they reported a marked change in guard cell solute levels in response to the same osmotic stresses. Grantz and Schwartz (1988) concluded that hydroactive guard cell responses in *Commelina* were not regulated by epidermal water status directly, but required signals originating in the mesophyll tissue, although Grantz (1990) also suggested that cuticular transpiration could play a role in the delivery and redistribution of such signal metabolites. These conclusions differ from work by Losch (1977, 1979) on *Polypodium*, who showed that humidity responses were intrinsic to stomata in epidermal peels.

In some species, there appear to be specialized regions of cell wall with reduced cuticle that may assist in feedforward humidity responses. For example, Appleby and Davies (1983) found that the cuticle on the outer face of guard cells from *Quercus* was much thinner than other regions of the epidermis and tracer applied to the transpiration stream accumulated in the wall underlying this region. They suggested that this resulted from increased transpiration from these regions and could account for the feedforward humidity response in this species. Canny (1990) does however caution against interpretation of tracer accumulation as a reliable means to trace water fluxes and presents evidence that the bulk of the water movement in leaves is symplastic on leaving the xylem vessels, with tracer movement in the apoplast being via diffusion. Nevertheless, in *Pinus* similar regions of reduced cuticle exist in the anticlinal walls between the guard cells and the subsidiary cells and within the stomatal pore in the ventral wall of the guard cells (Appleby and Davies, 1983). Jarvis and Morison (1981) suggest that the latter site would be masked during closure, and could not therefore sense humidity in closed stomata, although Appleby and Davies (1983) found that these sites may also be exposed externally during stomatal movements, as the guard cells in *Pinus* rotate relative to the plane of the epidermis. In contrast, the thinner cuticular regions in *Populus* lie within the stomatal throat and are not exposed during stomatal movements. As predicted, this species does not exhibit as strong a response to humidity at high VPD as *Pinus*. Appleby and Davies conclude that the stomatal response to humidity in both angiosperms and gymnosperms depends upon the presence of an evaporation site in the guard cell wall combined with the critical orientation of the guard cell with respect to the plane of the epidermal cells. However, Sheriff (1977a) has shown that while species that exhibit direct responses to humidity may have larger substomatal cavities than those species which do not respond to humidity, there was no correlation with plant habitat, leaf venation pattern, stomatal pore depth or the presence of specialised cuticular regions.

In at least some species, the first stomatal movement in response to a change in humidity is in the opposite direction from the subsequent stomatal movement (e.g. Kappen *et al.*, 1987). These passive responses are caused by a reduction in the back pressure of the neighbouring cells as epidermal or subsidiary cells lose more water than the guard cells. The response is reversed as the guard cells also begin to lose turgor or hydroactive changes in guard cell osmotic potential are triggered. Meidner (1987) also observed two-phase kinetics of stomatal movements in response to changes of VPD using a viscous flow porometer.

Thus, in *Tradescantia albiflora*, Kappen and Haeger (1991) directly observed changes in the morphology of cells in the epidermis during changes in relative humidity using a video system and found that a decrease in VPD led to shrivelling of the epidermal cells initially and transient stomatal opening. This was followed by stomatal closure, presumably from hydropassive or hydroactive adjustment of guard cell turgor. With the subsequent decline in transpiration the subsidiary and epidermal cells regained turgor and volume. Similarly, Maier-Maercker (1979) followed changes in subsidiary cell morphology in response to changing evaporative demand in several species, where the walls of the subsidiary cells were observed to wrinkle as the cells lost water and decreased in volume. With the ensuing stomatal closure, the cells regained volume and the walls became smooth again. Maier-Maercker (1979) suggested that the changes in subsidiary cell volume resulted from increased water flux through the guard cells rather than loss via the subsidiary cell cuticle, as the subsidiary cells rehydrated after the stomata closed despite the continued low external relative humidity.

This highlights the importance of the leaf morphology and degree of cuticularization in directing water fluxes and hydropassive movements in humidity responses, however, there is still considerable uncertainty on the precise hydraulic pathways for water fluxes within the leaf (Meidner, 1983; Boyer, 1985; Canny, 1990). With equal availability of water on surfaces of epidermal, mesophyll and vascular cells, evaporation will be greatest where the surrounding air is least saturated, i.e. near the stomatal pores (Meidner, 1975; Tyree and Yianoulis, 1980). However, evaporation has a substantial cooling effect and the epidermal tissue may well be cooled by external air currents and absorb less radiant energy than the mesophyll so that liquid water on the walls of green mesophyll tissue may be at a slightly higher temperature and thus be subject to more vigorous evaporation (Meidner, 1983). Sheriff (1977b) has shown that water droplets can condense on the internal epidermal cell walls under some circumstances in *Nicotiana glauca* In addition, the epidermal cell walls facing the substomatal cavity show varying levels of cuticularization between different species which will reduce water evaporation from these sites (Boyer, 1985) Indeed, Nonami and Schulze (1989) found using simultaneous measurements of cell turgor pressure and osmotic pressures in *T virginiana* that the epidermal cells had a higher water potential than the mesophyll cells, contradicting expectations if it is assumed that the major site of water loss is via evaporation from cells in the epidermis Nonami *et al* (1990) further suggested that competition between water flux to the guard cells and evaporation from the mesophyll could explain the concurrent decrease in transpiration rate with an increase in VPD

Thus so far experiments have highlighted the number of factors that may contribute to the humidity response in different species including, possible localized evaporation sites for water, the potentially non-linear dependence of cuticular transpiration on relative humidity and varying levels of water supply to the guard cells depending on the leaf morphology, hydraulic conductivity and degree of cuticularization However there is no general concensus as to the relative importance of each of these factors, even in species, such as *Tradescantia*, where several sets of measurements have been made

Debate has also focused on the appropriate humidity parameter that may actually be sensed by the guard cells (Assmann, 1993) Experimentally, stomatal conductance may correlate well with a range of humidity parameters, including, the relative humidity at the leaf surface (Ball *et al* , 1987), the absolute humidity difference from the leaf to the air (Kaufmann, 1982), the leaf to air vapour pressure difference (Assmann and Grantz, 1990, Aphalo and Jarvis, 1991) or the rate of evaporation (Mott and Parkhurst, 1991) The work of Mott and Parkhurst (1991) using *Hedera helix* is particularly innovative in this respect as they used a combination of helium and oxygen (Helox) to allow separation of the rate of evaporation from the vapour pressure difference, as the diffusion rate in response to the same VPD is 2 33 times faster in helox than in air It will be of considerable interest in the future to apply this strategy to measurements on other species, but it is evident there is still much confusion about the interpretation of humidity responses

Humidity responses do not occur in isolation *in vivo* but interact with a number of other factors (reviewed in Losch and Tenhunen, 1981) For example, Losch (1977, 1978) found that humidity responses in *Polypodium* were not markedly affected by temperature in unstressed plants, but a significant interaction between humidity, temperature and water stress developed once a threshold of water potential was exceeded Assmann and Grantz (1990) found that stomatal responses to blue light in several species depended on the VPD, with faster responses and overshoot phenomena observed at high but not low VPD, whilst Kappen and Haeger (1991) found that the humidity response in *Tradescantia* could be overridden by the CO_2 response.

6.6.1 Midday closure

Stomata of diverse species such as coffee, *Arbutus unedo* and onion show a distinct 'midday' closure particularly when growing under hot, sunny conditions (Fig. 6.16). It was suggested that the closure was due to water stress brought about by the high temperatures which occurred about the middle of the day (Loftfield, 1921; Stocker, 1956). With mild water stress the decreased stomatal conductance at midday will tend to compensate for the increased evaporative demand due to the higher temperatures, such that transpiration rates may be relatively constant. However, under such conditions the pattern of CO_2 exchange may show midday stomatal closure with two peaks of net photosynthesis. With extreme water stress two peaks of transpiration can occur with a depression at midday (e.g. Tenhunen *et al.*, 1981). Essentially the midday closure was considered to be a feedback response to water loss preventing further tissue dehydration. However, according to feedback control mechanisms proposed by Farquhar (1978) a *reduction* of transpiration with increasing evaporative demands cannot occur by a feedback loop since stomatal closure will immediately improve tissue water status. It was therefore suggested that stomata may act as 'humidity sensors' (Lange, 1969; Schulze *et al.*, 1972) immediately changing their conductance in response to changing atmospheric humidity. Cowan and Farquhar (1977) called this a feedforward controlling system (see Section 6.2) enabling plants to optimize water use efficiency.

An extensive field study of the diurnal course of stomatal resistances and transpiration of some Mediterranean perennials was made by Lösch *et al.* (1982). They observed that midday closure was most evident in the dry summer season and disappeared during the wet season of autumn and in the later, less stressful, winter season. Hence, they concluded that midday closure was an adaptive response to the environment, assisting plants to survive drought periods of the summer. In *Quercus suber*, a sclerophyllous species of Mediterranean regions, the midday depression of CO_2 uptake is considered to be a result of a decrease in the carboxylation efficiency coupled with an increase in the CO_2 compensation point, caused by elevated temperatures and decreased VPD, and a decrease in the CO_2 saturated photosynthetic capacity (Tenhunen *et al.*, 1984). As a consequence of the changes in photosynthesis, C_i remained essentially constant despite stomatal closure.

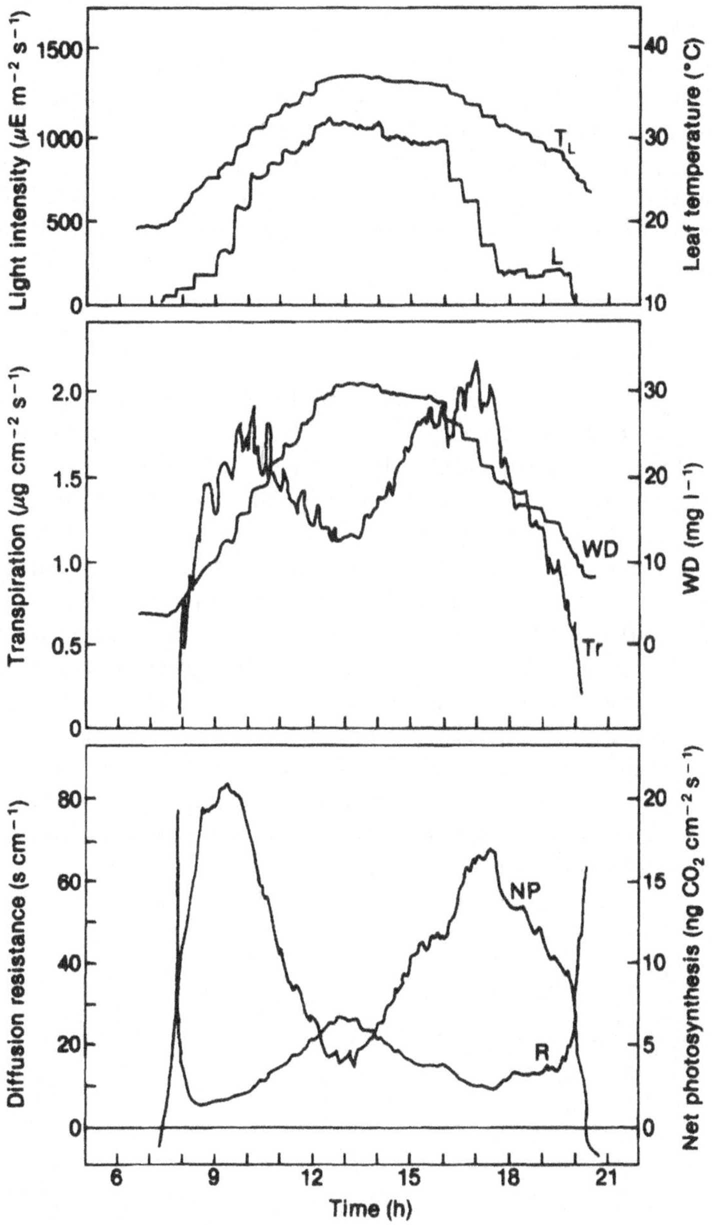

Figure 6.16 Daily time course of light intensity (L) incident on leaves in the gas exchange cuvette, leaf temperature (T_L), transpiration rate (Tr), vapour pressure difference (WD), leaf diffusion resistance (R) and net photosynthetic rate (NP) measured for leaves of well watered potted *Arbutus unedo* plants under simulated natural conditions in an environmental chamber From Tenhunen *et al* (1980)

However, earlier workers considered midday closure was essentially a CO_2 response. As temperatures increase, respiration and photorespiration outpaces photosynthesis and high CO_2 levels accumulate in the leaves which close stomata. Evidence for this view came from Meidner and Heath (1959) who found that, using onion leaves, stomatal closure normally occurs between 25 and 35°C but this was prevented if the central leaf cavity was flushed with CO_2-free air. However, as indicated earlier, it is difficult to maintain constant vapour pressure deficits with changing temperatures and humidity factors may have still influenced the stomatal activity to some extent.

6.6.2 Iwanoff effect

When a leaf is excised in air it has long been observed that there is a temporary increase in stomatal opening before the stomata close (e.g. Darwin, 1897; Iwanoff, 1928) which has been termed the 'Iwanoff effect'. Later, other investigators occasionally found that there was a momentary stomatal closing effect before opening occurred followed by the closure (Fig. 6.17) (Meidner, 1965). Although there was some initial controversy over interpretation of the observations (e.g. Allerup, 1961; Heath, 1963; Brun, 1965) it is now generally explained as follows. When the leaf is excised the tension of the water columns in the xylem vessels is released. This results initially in a transitory pressure on the guard cells with concomitant slight closure as water first hydrates the surrounding epidermal and subsidiary cells, but then the guard cells successfully compete for the water with resulting stomatal opening. Ultimately, due to transpirational water loss, water becomes limiting to the whole leaf and the stomata close.

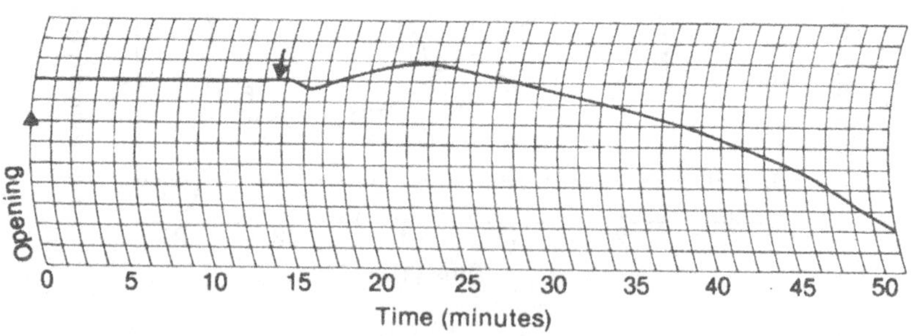

Figure 6.17 A viscous flow porometer trace of stomatal responses following leaf excision The stomata were open in the light and the leaf was excised at the time indicated by the arrow Note the initial small closure followed by a period of opening before the final closing response occurs From Mansfield (1971)

If a leaf is excised under water usually there is an initial slight stomatal closure often followed by small oscillatory movements (Heber *et al* , 1986) Thus, the sudden uptake of water by the leaf swells the epidermal cells leading to hydropassive stomatal closing followed by a series of oscillations of aperture until a new steady-state is established

6.7 *Mineral nutrition*

In general, anything that affects photosynthesis or leaf water potential will indirectly effect stomatal response to some degree Thus, if a mineral deficiency lowers the photosynthetic capacity of a leaf then stomata will tend to close as CO_2 increase Hence investigators have found that nitrogen, phosphorous or potassium deficiency causes at least slight decreases of stomatal aperture (e g Peaslee and Moss, 1968, Wallace and Frolich, 1965, Desai, 1937) Figure 6 18 (a and b) shows that when plants are grown deficient in nitrogen or potassium, stomatal conductances are lower than in control plants grown sufficient in the minerals at all levels of leaf water potential (except at the higher values in the potassium treatment) However, a number of reports indicate that a high K^+ fertilization reduces transpiration rates (e g Brag, 1972, Jensen, 1982) and stomatal conductance (e g Bradbury and Malcolm, 1977, Lindhauer, 1985) relative to plants grown at lower K^+ levels A high K^+ fertilization rate also improved water use efficiency in cereals (e g Anderson *et al* , 1992) which was assumed to be due to better control of transpirational water loss Later work (Losch *et al* , 1992), however, suggested that the increased water use efficiency at the high K^+ fertilization might be due to decreased stomatal density and altered stomatal shape (guard cells in the flag leaves of barley were shorter but wider than in leaves from the low K^+ treatment)

It is also possible, of course, that the metabolism of guard cells themselves is directly affected by mineral deficiency Indeed, since K^+ is the major osmoticum accumulated by guard cells during stomatal opening then it is possible that if K is severely limiting there will be a direct effect of potassium deficiency on stomatal functioning with predictable decreased opening

6.8 *Circadian rhythms and shorter-term oscillations in stomatal movements*

Plants and animals exhibit rhythmic behaviour in their biochemical and physiological activity, usually as a result of the diurnal pattern of day and

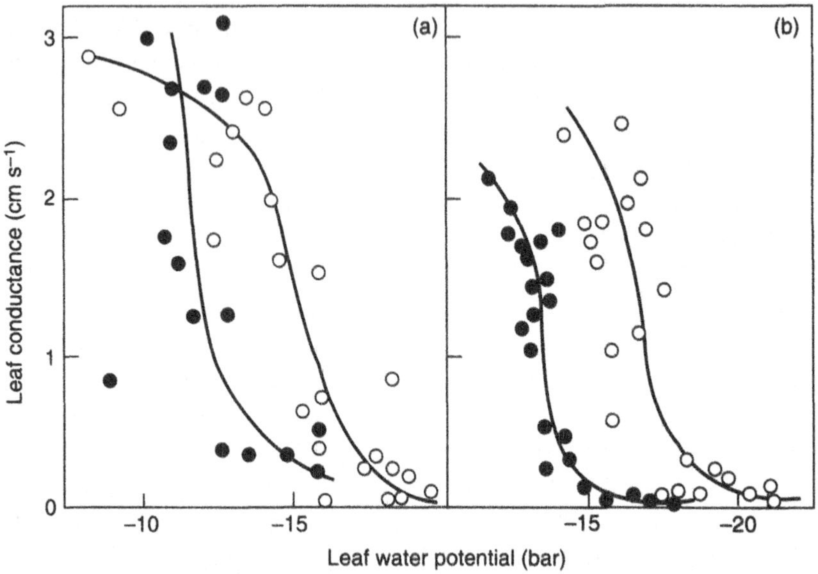

Figure 6.18 The effect of nitrogen (a) or phosphorous (b) deficiency in cotton plants on leaf conductance (normally equivalent to stomatal conductance) at different leaf water potentials. ●, Phosphorous and nitrogen deficient plants; ○, phosphorous and nitrogen sufficient plants From Radin (1984) and Radin *et al.* (1982).

night which is 'imprinted' into the organism. A diurnal rhythm is one having a periodicity of precisely 24 h. However, almost all free-running rhythms in animals and plants have a periodicity of between 21 and 28 h, but not exactly 24 h, and such rhythms are known as circadian (*circa* = about; *diem* = day). Circadian rhythms and many other longer- or shorter-term rhythms are endogenous, i.e. a rhythm programmed from within the organism as opposed to one directly and immediately brought about by exogenous stimuli.

To demonstrate that a rhythm is circadian and endogenous it has been considered that at least three features should be observed:

1. The period of the rhythm should be about 24 h but not necessarily match environmental cyclical variations.
2. The rhythm should persist under constant environmental conditions.
3. It must be possible to phase-shift the rhythm by a suitable stimulus.

Such endogenous rhythms are also arrested by anoxia. Additionally, the period of the oscillation should show a high degree of temperature compensation so that it changes little at different ambient temperatures. The temperature compensation in the period of the circadian oscillator has resulted in it being called a biological clock. A number of reviews on the subject have been written (e g Wilkins, 1992, Hopmans, 1971)

Circadian rhythms in stomatal movements were observed long ago (Darwin 1898, Lloyd 1908), but even today our understanding of them is not clear It is not known to what extent they control stomata and nor is their value to the plant understood

When a plant is kept in continuous light the stomata of many species, and probably all species, open and close with a period of about 24 h (e g Mansfield and Heath, 1963, Martin and Meidner, 1971, Hennessey *et al* , 1993, see Fig 6 19a) There is a damping or gradual decrease in the amplitude of the free-running rhythm with time Although this is highly suggestive of an endogenous rhythm it is difficult to establish whether the rhythm is a result of processes residing in the guard cells (a direct effect) or due to processes occurring in other cell types of the leaf which, in turn, effect stomatal behaviour (indirect effects) For example, if there is an intracellular circadian rhythm of CO_2 production by the mesophyll, as in CAM plants (Wilkins, 1959), this alone could effect stomatal movements, giving an appearance that the stomata themselves were controlling this rhythm However, if the guard cells produced the CO_2 then the stomatal movements would be a direct circadian rhythm Recently, Wilkins (1993) has shown that removal of the epidermis from leaves of *Bryophyllum fedtschenkoi* abolishes the circadian rhythm of CO_2 assimilation in continuous light Thus stomata appear to either interact with the CO_2 fixing capacity of the mesophyll cells to generate the rhythm or the stomata themselves open and close to create the rhythm

The stomatal rhythm in continuous light can be phase shifted so that a 6-h dark period given during the natural 'light' phase causes an inversion of the original pattern

Stomatal circadian rhythms in continuous dark have also been observed in many species although they rapidly dampen out after one or two cycles (e g Martin and Meidner, 1971, Stålfelt 1963, Hennessey *et al* , 1993, see Fig 6 19b)

The opening cycle in *T virginiana* in continuous dark can be phase-shifted when a period of light occurs during the natural 'dark' phase, but there is little or no phase shift when a period of light occurs during the natural 'light' period (Martin and Meidner, 1972) Also a low intensity light-treatment (10–1500 lux) of *Xanthium pennsylvanicum* plants at the beginning of the dark treatment causes a phase shift, or, effectively, delayed opening in the dark (Mansfield and Heath, 1963) Additionally, the timing of both light and dark treatments affects the timing and duration of opening in darkness Furthermore, the low light level effect which delays opening in the dark is temperature dependent, no delay occurs in light of 10 lux at 15°C whereas, at higher temperatures, opening in the dark is delayed (Mansfield and Heath, 1964)

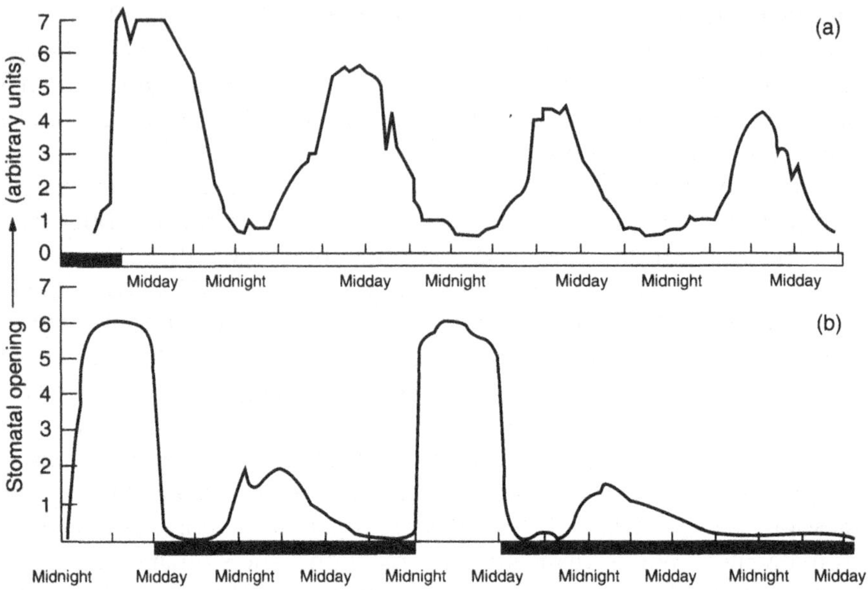

Figure 6.19 Stomatal behaviour in leaves of *Tradescantia virginiana* exposed to continuous light (1500 lux) (a) or continuous darkness after a 12 h photoperiod (b). From Martin and Meidner (1971).

Hennessey *et al.* (1993) also found that the circadian rhythm in *P. vulgaris* leaves rapidly damped out in constant light and CO_2-free air and under conditions conducive to high photosynthetic rates. They concluded that at least part of the stomatal circadian rhythm, at least in *P. vulgaris*, is not a direct guard cell effect but 'active' photosynthesis (not just the presence of light) is necessary for sustained expression of the rhythm.

Using epidermal strips of *C. communis* (Meidner and Willmer, 1993) and *V. faba* (Gorton *et al.*, 1989) showed that the 'clock' regulating the circadian rhythm resides in the epidermis, possibly in the guard cells.

Occasionally in the literature the term 'night opening' of stomata is referred to. This description normally refers to stomatal opening which occurs before sunrise; stomata may also remain open after sunset. It represents part of an endogenous rhythm which overlaps with other opening stimuli during daylight and night hours. Stomatal opening also occurs at night in CAM plants, but this appears to be largely controlled by intercellular CO_2 concentrations. It is likely that there is also some endogenous control of apertures by the guard cells in CAM plants.

The basic mechanism which times the endogenous rhythm (biological clock) of stomatal movements or any other endogenous rhythmic process has not been identified. Rhythmic change in the levels of some metabolite or energy source may be key controls but in turn, some basic oscillating process must control the cycling of the metabolite levels. It has been suggested that the cellular energy supply is concerned in the control of endogenous rhythms since anoxia, in some cases, inhibits the rhythm. The lack of certain metabolites or energy source may account for the unsustained rhythm of stomatal opening when leaves are kept in continuous darkness.

The function of circadian rhythms in the control of stomatal aperture is not understood. Presumably they contribute to a general synchronization of stomatal responses in a plant (a coarse tuning of stomatal behaviour). Also the rhythms result in anticipatory behaviour in which stomata can open at dawn before light is sufficient for photosynthesis (so-called night opening) and presumably this is of some advantage to the plant. The rhythms may be of some other adaptive significance; indeed, in some organisms they have been shown to be of survival value.

Stomata also exhibit a variety of shorter-term oscillations of aperture. In general these are brought about by sudden changes of environmental factors which effect the water balance of the plant.

In some species oscillations in stomatal aperture with a period of about 20–40 min (they may be up to 2.5 h in the dark) are caused by time lags in the rehydration of leaves. The time lags, in turn, are caused by a combination of the hydraulic resistances and hydraulic capacitances in the plant. The sequence of events can be summarized as follows: If the transpiration rate is suddenly increased (e.g. by increased irradiance) the rate of water uptake can lag behind the transpiration rate. This results in the leaf becoming water stressed with resultant hydropassive stomatal closure (in some species there may be a transient stomatal opening; see the Iwanoff effect, Section 6.6.3). The leaf rehydrates and the stomata reopen. This sequence of events is then repeated until, finally, an equilibrium situation is reached and the stomatal oscillations damp out.

Fluctuations of transpiration (with a period of about 1–10 min) may occur without concomitant changes in stomatal aperture. This has been explained in much the same way as described above for stomatal oscillations except that the rehydration and dehydration cycles of the symplast are affected much less than the apoplast. Thus the turgor relations of the cells are not greatly affected and therefore stomatal movements do not occur.

6.9 Stomatal activity during the life cycle of a leaf

In general as a leaf matures the stomatal responses are more rapid and apertures achieved are larger until an optimum level of activity occurs. Thereafter, as the leaf continues to age, stomatal responses become more sluggish and maximum apertures reached are smaller (e.g. Ludlow and Wilson, 1971; Vaclavik, 1973; Davis and McCree, 1978; Willmer *et al.*,

1988; see Fig. 6.20). There are a number of reviews discussing the behaviour of stomata during the ontogeny and growth of a plant (e.g. Solárová and Pospisilová, 1983; Field, 1987). Such patterns of stomatal behaviour in response to CO_2 and to ABA are also observed in epidermal strips taken from leaves of different ages (e.g. Willmer *et al.*, 1988) (Fig. 6.21).

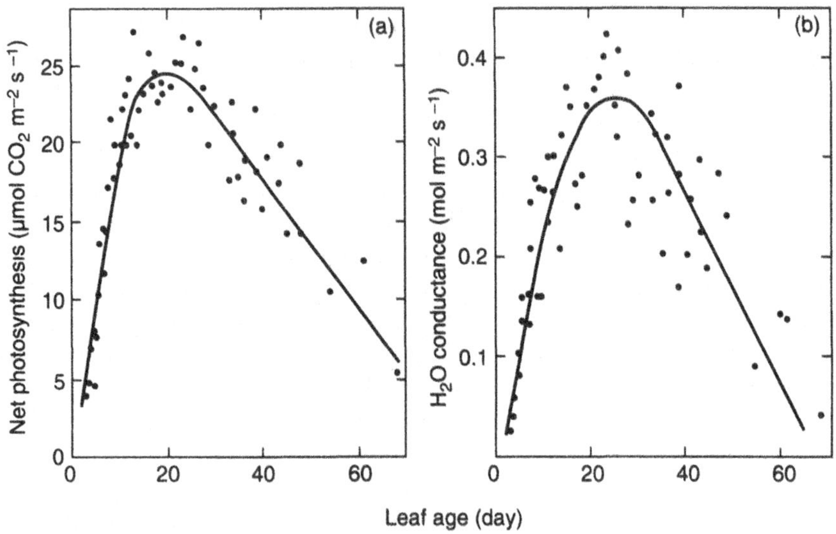

Figure 6.20 Photosynthesis (a) and stomatal conductance (b) measured at saturating light intensity, a CO_2 partial pressure of 33 Pa, a leaf temperature of 27 8°C and a vapour concentration gradient of 21 mmol mol^{-1} in leaves of cotton varying in age Photosynthesis and conductance in these experiments were not influenced by leaf insertion level Redrawn from Constable and Rawson (1980)

Usually, stomata of senescing leaves remain operable well after the mesophyll cells of the leaf have turned yellow. Thus, stomata of senescing leaves of *V. faba* can respond to changing CO_2 concentrations and to kinetin (Wardle and Short, 1983) and guard cells in epidermal strips from senescing leaves of *Gingko biloba* (Zeiger and Schwartz, 1982), *V. faba* (Wardle and Short, 1983) and *C. communis* (Willmer, unpublished) can develop turgor and the stomata open when illuminated in the presence of KCl. Furthermore, chloroplasts in guard cells of senescing leaves of *G. biloba* are green and exhibit typical fluorescence transitions associated with electron transport and photophosphorylation at a time when chloroplasts in mesophyll cells show no signs of such activity. Similarly, guard cells of *C. communis* have some functionality and maintain green chloroplasts well after the mesophyll cells of the leaf are chlorophyll-free (Willmer, unpublished). Ozuna *et al.* (1985), found that

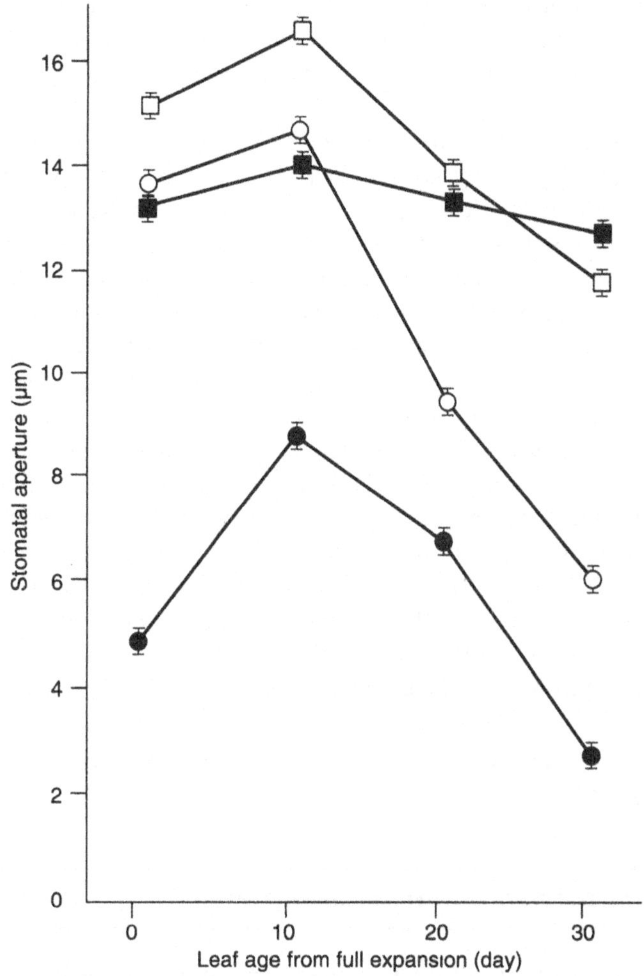

Figure 6.21 Changing stomatal response to ABA and CO_2 in epidermal strips as a function of leaf age □, CO_2-free air, minus ABA, ○, normal air, minus ABA, ■, CO_2-free air, plus ABA, ●, normal air, plus ABA The bars labelled with the symbols are standard errors of the means From Willmer *et al* (1988)

in senescing leaves of *Nicotiana glauca*, although conductances were only 5–10% of those in non-senescing leaves, guard cells in epidermal strips from the senescing leaves could reach full turgor in light and accumulate K^+ from the incubation medium. Thus the stomata have the potential to function fully but factors within the senescing leaf such as C_i or the mineral ion content determine the actual stomatal aperture. The significance of why the stomata remain functional when mesophyll photosynthesis has stopped is not understood. Senescence in guard cells and mesophyll cells do not proceed together and it appears that signals which trigger senescence in mesophyll cells do not do so in guard cells. Thus, the senescence processes in guard and mesophyll cells do not seem to be closely coupled, the guard cells being 'isolated' from the rest of the leaf.

It was concluded by Thimann and Satler (1979a,b) that leaf senescence was influenced by stomatal aperture such that opening caused delayed senescence. However, this view was modified in later work when it was considered that senescence was not under the 'direct control of changes in stomatal aperture' (Park and Thimann, 1990). Such conclusions were reached using leaves or leaf sections which were floated on test solutions in the light or dark but unfortunately interpretation of the cause of the resultant stomatal apertures is notoriously difficult. This is basically because gas diffusion at the leaf /liquid interface is very slow and in light CO_2 levels drop due to photosynthesis stimulating stomatal opening while in the dark CO_2 levels build up due to respiration promoting closure and so the apertures are not necessarily due to substances in the incubating medium. Park and Thimann (1990) and Thimann and Tan (1988) also concluded that the 'synthesis of one or more proteins controls both the opening and closing of the stomata.' Numerous other reports indicate that inhibitors of protein synthesis also inhibit opening (e.g. Pallaghy and Fischer, 1974). Perhaps this is not too surprising considering that opening and perhaps closing are active processes and guard cells contain an abundance of the metabolic machinery for making proteins.

6.10 Stomatal behaviour in canopies

Stomatal behaviour is often monitored in individual leaves of a plant grown as an isolated experimental specimen. In reality plants rarely grow as isolated individuals but as part of a community, in and amongst other plants. The canopy type can vary greatly and may contain many different species, often with definite regions within the canopy, as is found in a rain forest, or essentially one species, as in a field of barley or a grass sward. Such varied canopies can result in greatly different microclimates around leaves which, in turn, will affect stomatal behaviour. Thus stomatal responses in the leaves of an isolated plant are likely to be quite different to those in leaves of plants growing in a community and often the relationship between photosynthesis and stomatal conductivity, which normally exists when individual plants or leaves are investigated, breaks down in canopies. A major reason for this is that the canopy is uncoupled from the environment above and sensitivity to environmental signals (photon flux density, temperature, etc.) is reduced. Thus, as the scale of size increases from an individual leaf, to an individual plant and to a large community, transpiration is less dependent on stomatal control. Such aspects have been especially considered by Jarvis and co-workers (e.g. Jarvis, 1993; Jarvis and McNaughton, 1986) and is now an area of research receiving major attention.

6.11 Gaseous environmental pollutants and stomata

Major gaseous environmental pollutants which may effect plants are SO_2, oxides of nitrogen, ozone and certain hydrocarbons. These substances are produced by industry, often in very localized areas, by combustion engines and domestic heating systems in their exhaust gases, and some pollutants may be formed from photochemical reactions in the atmosphere. The latter air pollution is known as 'photochemical smog'.

Two areas of major interest are: (1) the effect of the pollutant on stomatal behaviour and the method of entry of the pollutant into leaves, and (2) the effect of the pollutant on the physiology and growth of the plant. The latter aspect will be neglected in the discourse here but further information on this aspect can be obtained from a variety of sources (e.g. Ziegler, 1975; Rao *et al.*, 1983; Shimazaki *et al.*, 1984; Schulte-Hostede *et al.*, 1988).

Sulphur dioxide has received most attention, but reports on the effects of this pollutant on stomatal behaviour are conflicting. Many researchers find that relatively low concentrations of SO_2 (50–2860 $\mu g\ m^{-3}$, equivalent to 0.02–1 $\mu l\ l^{-1}$ or ppm) increase stomatal opening and transpiration rates while reducing photosynthesis. Other investigators observe either no effect of low SO_2 concentrations on transpiration or that stomatal closure is brought about by relatively high SO_2 concentrations (1–2 ppm). To account for the differences in stomatal behaviour in some species to high and low SO_2 concentrations Black and Black (1979) suggested the following. Associated with wider stomatal opening in *V. faba* leaves exposed to SO_2 (from 50 to 200 $\mu g\ m^{-3}$) was increased epidermal cell damage which could result in hydropassive opening, while at higher SO_2 concentrations (500 $\mu g\ m^{-3}$ and above) guard cells were also damaged which would result in stomatal closure.

Curiously, irreversible and reversible stomatal responses have been observed on an attached sunflower leaf exposed to 1.5 ppm SO_2 and it appears some stomata and patches of mesophyll are more susceptible to the pollutant than others (Omasa *et al.*, 1981). Environmental factors also appear to have an important control in the sensitivity of stomata to SO_2. For example, the effects of SO_2 on stomata in some species are dependent on atmospheric humidity (Mansfield and Majernick, 1970; Black and Black, 1979), while growth of the grass, *Phleum pratense*, is reduced to a greater extent to SO_2 pollution in conditions of low irradiance and short days (Davies, 1980). Figure 6.22 shows the effects of SO_2 on stomatal behaviour in *V. faba* leaves as a function of VPD; 35 ppb SO_2 in the atmosphere resulted in larger conductances than in control leaves except at higher VPD values when a sudden drop in conductance occurred.

The extent of leaf damage appears to be related to stomatal aperture. Generally, symptoms of leaf injury are not observed if stomata are closed. Leaf injury may also be related to the initial degree of stomatal opening and the subsequent rate of closure upon contact with SO_2. For

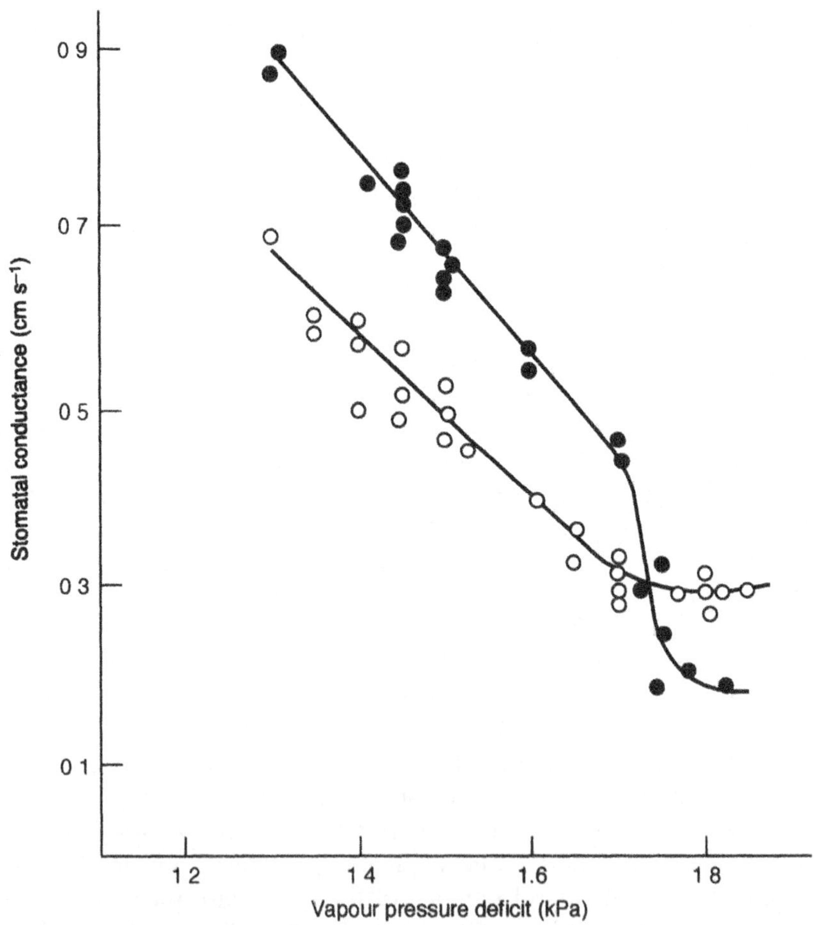

Figure 6.22 The effect of water vapour pressure deficit on stomatal conductance in *Vicia faba* leaves exposed to charcoal-filtered air (○) or air containing 35 ppb SO_2 (●) From Unsworth and Black (1981)

example, Bonte and Longuet (1975) found that increased atmospheric humidity slowed down stomatal closure in *Pelargonium zonale (= Pelargonium hortorum)* leaves which, in the presence of SO_2, resulted in increased leaf necrosis compared to leaves exposed to SO_2, at lower atmospheric humidities. Patchiness in stomatal responses to SO_2 and extent of tissue damage within a leaf have also been observed (e.g. Omasa and Onoe, 1984).

Some species are much more sensitive to SO_2 damage than others for reasons not understood. Peanut and tomato plants, for example, appear more resistant to the effects of SO_2 than radish or spinach.

Photochemical smogs occur over large cities wherever high levels of oxides of nitrogen and hydrocarbons accumulate in the atmosphere. Photochemical reactions occur which result in further products such as ozone and peroxyacetyl nitrate being formed. Photochemical smog induces stomatal closure and, in general, increased stomatal aperture and damage to plants are positively correlated.

The effects of individual components of photochemical smogs on stomatal behaviour and leaf damage have been studied. Ozone is a widespread and a very damaging air pollutant to plants in all major industrialised regions of the world. It originates from two sources, i.e. from photochemical reactions as indicated above and from stratospheric/troposphoric exchange. Ozone generally closes stomata or increases stomatal resistance (e.g. Lehnherr *et al.*, 1987; Pearson and Mansfield, 1993), although some studies, find no effects or even decreased stomatal resistance (e.g. Eamus *et al.*, 1990). Ozone also directly affects photosynthesis (e.g. Reich and Amundson, 1985) and it appears that decreases in stomatal conductance are a result of both its direct affect on stomata and an increase in C_i due to the inhibition of photosynthesis. Exposure of foliage to a few hundred parts per billion (0.01–0.1 ppm) of ozone for a few hours can result in damage to some species. One symptom of ozone damage is decreased chlorophyll levels within leaves.

The effects of ozone on stomata and gas exchange are complex and may vary depending on many factors including the environmental conditions, the exposure period and concentration of ozone, the age of the plants and general growing conditions. Pearson and Mansfield (1993) for example, found that ozone increased stomatal resistance in leaves of well-watered, young, beech trees, but in unwatered trees ozone reduced the increase in resistance as water-stress developed. The concensus of opinion is that if stomata are closed there is much less damage to foliage. Also, some plants are more resistant to the effects of ozone than others. This increased resistance has been correlated with a lower stomatal frequency which can result in slower penetration of the pollutant (e.g. MacDowall, 1965; Engle and Gabelman, 1966), although not all studies have confirmed this (e.g. Ting and Dugger, 1968).

Less is known about the effects of nitrogen oxides (NO_x) on stomata, although they are commonly occurring pollutants. Most investigators observe decreased transpiration rates or stomatal apertures (e.g. Omasa *et al.*, 1981) and photosynthesis (Capron and Mansfield, 1976) in the presence of nitrogen dioxide (NO_2). Nitric oxide (NO) probably has similar effects and is also reported to reduce photosynthesis (Capron and Mansfield, 1976). Nitrous oxide (N_2O), however, has no effect on stomata and has even been utilised with a diffusion porometer.

To summarize, some plants seem more resistant to any given pollutant than others and, generally, there is less damage to a plant if the stomata are closed. Of course, closed stomata may not be a total barrier to the pollutants which may enter leaves by penetrating the cuticle. Thus, the rate of entry of a pollutant and, therefore, the external concentration of the pollutant and the exposure period are important in determining the extent of damage to a plant.

The long-term effects of exposure to a pollutant on stomatal behaviour and plant damage are also likely to be considerably different to shorter-term exposures. Environmental factors may also play a large role in determining the extent of the damage caused by the pollutants. Exposure of plants to different combinations of pollutants also result in markedly different and possibly more drastic effects than exposure of plants to individual pollutants.

There is extensive literature dealing with urban and rural pollution and its effects on plant growth and specifically on stomata. Reviews of this topic include those by Zeigler (1975), Black (1981) and Mansfield and Freer-Smith (1981).

6.12 Plant pathogens and stomata

Although mycorrhizal fungi and probably soil microflora generally are beneficial to the healthy growth of plants, certain fungi and bacteria have deleterious effects. Those that affect plants by influencing stomatal behaviour and plant water relations are considered here.

6.12.1 Entry of pathogens into leaves via stomata

A number of plant pathogens, including some of economic importance, gain entry into foliage either solely via stomata or as an alternative to direct penetration of epidermal cells and tissues. Such pathogens include parasitic bacteria, certain downy mildews, aecidiospores and uredospores of *Puccinia* spp (rusts) and several Fungi Imperfectii.

Infection of plants via stomata involves two vital processes, i.e. the location of stomata by the pathogen and the subsequent formation of infection structures which occur at the stomata

The nature of the stimulus which attracts zoospores and germ tubes to stomata has received considerable attention A number of explanations have been forwarded including attraction of the pathogens along chemical or water vapour gradients or that the pathogen is directed by the surface topography of the leaf (a thigmotropic response)

Royle (1976) has suggested that zoospores of *Pseudoperonospora humuli*, which tend to select open stomata, are attracted by two stimuli One stimulus is surface topography since zoospores on perspex replicas of leaf surfaces tend to settle on open stomata rather than closed ones The other stimulus was a chemical one and it was suggested to 'involve photosynthesis It is enticing, therefore, to suggest that a CO_2 gradient attracts zoospores to stomata However, the evidence indicating that CO_2 concentrations regulate entry into stomata is not strong although high, non-physiological, levels of CO_2 have been observed to suppress penetration by some pathogens including *Puccinia graminis*

Another intriguing question is how surface topography can direct germ tubes to stomata (see Burnett, 1976) A partial answer to this question is that the germ tubes are directed to grow between rows of wax crystals which form a regular lattice on some leaf surfaces (Lewis and Day, 1972) This may also explain why a spore from an obligate parasite which germinates on a plant other than its normal host may not develop a proper infection structure In other words, the germ tube is unable to adapt to different dimensions of the regular lattice of wax crystals of the host leaf surface and becomes disorientated

Orientation of epidermal cells (Staples and Macko, 1980) and cuticular ridges (Pring, 1980) have also been suggested to assist in the direction-finding of germ tubes

More recently Edwards and Bowling (1986) concluded that uredospore germ tubes of *Uromyces viciae-fabae* were attracted to stomata via pH gradients on the leaf surface They observed a marked pH gradient increasing away from the guard cells of closed stomata in *C communis* and *T virginiana*, although not of open ones and germ-tube attraction occurred in both non-host and host (*V faba*) species of the fungus

The nature of the stimulus which causes entry of the pathogen into stomata and encystment or appressorium formation has also received considerable attention and also may result from physical and/or chemical signals Entry through stomata by fungal pathogens may be conveniently grouped into three types (see Wheeler, 1968) One type is that observed when cereal leaves are infected by uredospores of *Puccinia* spp Upon germination of a uredospore the germ tube shows directional growth towards a stoma Upon reaching a stoma the cytoplasm of the germ tube accumulates at the tip which swells to form an appressorium within the stomatal pore. A cross-wall forms, cutting off the empty germ tube, and from the appressorium a wedge-shaped hypha grows through the pore. Inside the substomatal cavity the hypha swells, receiving the contents of the appressorium.

Another type of pathogen entry is found in *Fulvia* (formerly *Cladosporium*) *fulvum* in which spores germinate on the leaf surface in a water film or in moist air. Growth of the germ tube appears to be haphazard and they often grow over stomata. When penetration of a stoma occurs a lateral hypha forms which grows through the pore into the substomatal cavity.

A third type of stomatal penetration by a fungal pathogen is exemplified by that found in *Plasmopara viticola*. Motile zoospores swim in a water film on the leaf surface and, upon nearing a stoma, are attracted to them by some stimulus. The zoospores then settle over the stomata and encyst. A hypha emerges from the cyst and grows through the stomatal pore into the substomatal cavity.

Attempts have also been made to find a relationship between stomatal frequency and their arrangement and structure and resistance to pathogen entry. Although there are conflicting reports it is likely that stomatal frequency and structure will influence the rate and numbers of pathogen infection. Some pathogens enter the leaf through open stomata while others can force an entry even through closed stomata. In *Citrus nobilis*, a species of orange, the stomatal anatomy regulates pathogen entry. The upper cuticular lips of the guard cells are very pronounced and these, apparently, prevent entry of *Pseudomonas citri* (a bacterial canker) and *C. nobilis* is, therefore, resistant. In *C. grandis*, however, the cuticular lips are not so prominent and the bacterial pathogens can more readily enter the stomatal pores, making this species of orange much more susceptible.

However, probably most usually the resistance to infection is located at stages later than entry into stomatal pores. Indeed, in both susceptible and resistant cultivars of coffee, development of appressoria and infection structures in the substomatal cavity by *Hemileia vastatrix* occurs in an identical way (Coutinho *et al.*, 1993).

6.12.2 Effects of pathogens on stomatal behaviour

When pathogens infect plants, stomata may be affected in a variety of ways which may be reflected in different patterns of transpiration and photosynthesis. The host–pathogen interaction is often complex and there may be direct and indirect effects of the pathogen on stomata. The direct effects may be due to substances produced by the host (phytoalexins), or substances produced by the pathogen (toxins, enzymes) which affect the stomata in some way An example of how stomata are affected in an indirect way by a pathogen is found in the vascular wilts (e g *Fusarium oxysporum Lycopersici)* which can cause rapid wilting of leaves and eventual death of the host The major cause of these symptoms is that the vessels become plugged with fungal hyphae, cell debris from damaged walls or tyloses and the water supply to the leaves becomes restricted (There is some controversy about whether the penetration of hyphae into vessels causes cavitation, although it is known that cavitation stimulates tylose formation which would block xylem vessels) As a result, stomatal apertures may fluctuate even before visual symptoms of infection appear, although ultimately stomata will close through lack of water supply This is the situation observed in banana plants infected with *Pseudomonas solanacearum* in which initially large, short-term oscillations of stomatal aperture occur before leaf wilting begins due to blockage of the water transporting system (Beckman *et al* , 1962) In only a very few cases is there evidence that membrane damaging toxins are involved in the vascular wilt syndrome

Often leaves infected by fungal pathogens exhibit lower stomatal conductances and movements may be more lethargic than in non-infected leaves These are some of the symptoms observed, for example, in potato leaves infected with *Verticillium dahliae* (Bowden and Rouse, 1991)

Infection of barley and bean plants with rust fungi generally results in the plants being more susceptible to water stress The decline of leaf water potential in barley has two main causes, i e a decrease of leaf diffusive resistance due to cuticular rupture from fungal sporulation and a decline in stomatal responsiveness to leaf water status (Berryman *et al* , 1991a) Additionally there is a decrease in root hydraulic conductivity (Berryman *et al* , 1991b) so that there is an increased water loss from the leaf and a decreased water supply from the roots The turgor relations of epidermal cells of rust infected barley leaves are complex and depend on the distance from the centre of the infection and the stage of infection, but cell turgor is always lower in infected areas than in control leaves (Berryman *et al* , 1991c)

In general, when pathogens infect leaves, stomata lose their control of gas exchange between the leaf and the environment In some cases stomata in infected leaves close or eventually fail to open widely in light which will not only reduce transpiration but also photosynthesis and growth of the plant In other cases stomata remain wide open even in the dark and there will be uncontrollable loss of water Both opening and closing responses may be brought about by production of phytoalexins by the host or phytotoxins by the pathogen which affect guard cell processes directly (see Sections 7 5 and 7 6) Wide stomatal opening and uncontrollable loss of water may also be due to destruction and digestion of the cuticle and cell walls of the epidermal layer by enzymes secreted by the pathogen. Figure 6.23 shows an extremely wide open stoma in a leaf of *V. faba* infected with *Botrytis fabae*. The state of the stoma is believed to be due to enzymes liberated by the pathogen which have partially digested the walls of the guard cells and surrounding epidermal cells thereby allowing an unnatural extension of the guard cells. Phytotoxins, extracellular enzymes and phytoalexins may also affect membrane permeability (see Chapter 7 for further details).

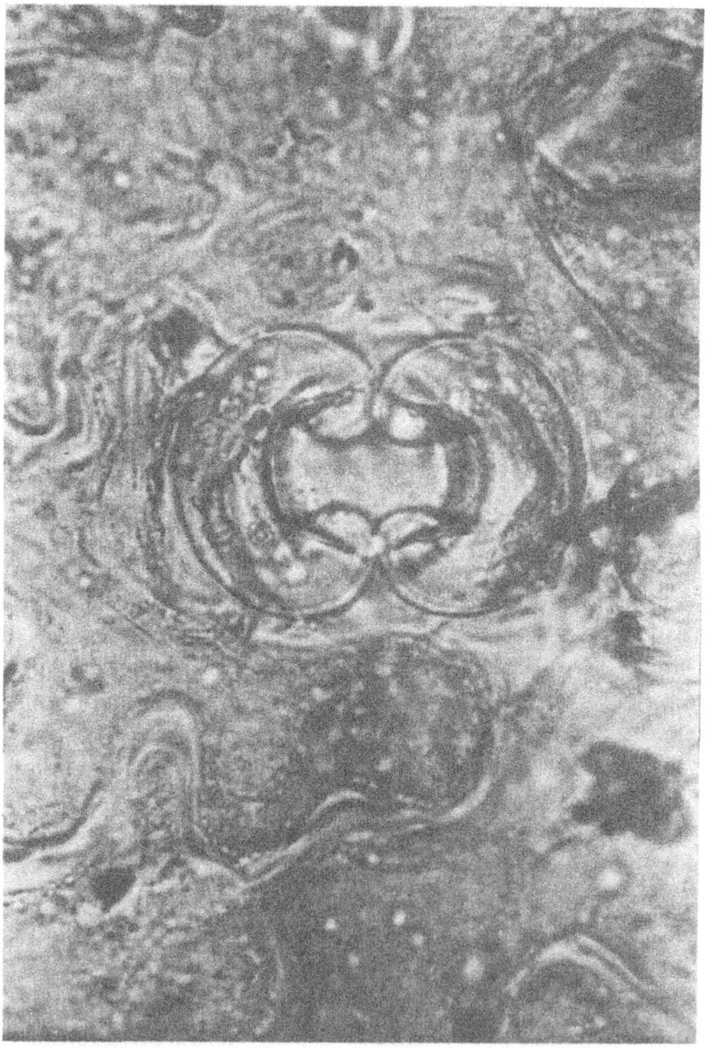

Figure 6.23 Unnaturally wide stomatal opening in *Vicia faba* as a result of infection by *Botrytis fabae*. Unpublished work of Mansfield and Willmer, from Willmer (1983)

More detailed information on the effects of pathogens on stomatal behaviour may be found in a report by Duniway (1976)

6.12.3 Symbiotic fungal associations with plants and stomatal behaviour

Of particular note are the symbiotic fungal associations with grass species such as tall fescue (*Festuca arundinacea* Scheb) and perennial ryegrass (*Lolium perenne* L) The fungi (*Acremonium* spp Morgan-Jones & Gams, section *Alba-lanosa*) complete their entire life cycles within the plants and are non pathogenic although the grass–fungus interaction is responsible for toxicities to many grazing animals Some studies have suggested that the infected plants may be more drought resistant than non-infected plants as a result of greater stomatal closure of the former plants However, Richardson *et al* (1993) found that certain genotypes of tall fescue containing the fungal endophyte, *A coenophialum*, had higher stomatal conductance values under water stress than those grass genotypes not containing the endophyte The higher conductance values in the infected leaves were attributed to the maintenance of higher turgor in these leaves relative to non-infected leaves although how higher turgor resulted was not clear

References

Allerup, S (1961) Stem cutting and water movement in young barley plants *Physiol Plant* , **14**, 632–637

Andersen, M N , Jensen, C R and Losch, R (1992) The interaction effects of potassium application and drought in field grown barley II Nutrient relations, tissue water content and morphological development *Acta Agric Scand* , **42**, 45–56

Aphalo, PJ and Jarvis, PG (1991) Do stomata respond to relative humidity? *Plant Cell Environ* , **14**, 127–132

Appleby, R F and Davies, WJ (1983) The structure and orientation of guard cells in plants showing stomatal responses to changing vapour pressure difference *Ann Bot* , **52**, 459–468

Assmann, S M (1988) Enhancement of the stomatal response to blue light by red light, reduced intercellular concentrations of CO_2, and low vapour pressure differences *Plant Physiol* , **87**, 226–231

Assmann, S M (1993) Signal transduction in guard cells *Ann Rev Cell Biol* , **9**, 345–375

Assmann, S M and Grantz, D A (1990) The magnitude of the stomatal response to blue light *Plant Physiol* , **93**, 701–709

Bauer, H (1979) Photosynthesis of ivy leaves (*Hedera helix* L) after heat stresss III Stomatal behaviour *Z Pflanz* , **92**, 277–284

Beckman, C H , Brun, WH and Buddenhagen, I W (1962) Water relations in banana plants infected with *Pseudomonas solanacearum* *Phytopathology*, **52**, 1144–1148

Berryman, C A , Eamus, D and Farrar, J F (1991a) Water relations of leaves of barley infected with brown rust *Physiol Plant Pathol* , **38**, 393–405

Berryman, C A , Eamus, D and Farrar, J F (1991b) The hydraulic conductivity of roots of rust-infected barley seedlings *Physiol Plant Pathol* , **38**, 407–415

Berryman, C A , Eamus, D and Farrar, J F (1991c) Variation in epidermal cell turgor of rust-infected barley seedlings *New Phytol* , **119**, 535–540

Berryman, C A , Eamus, D and Duff, G A (1994) Stomatal responses to a range of variables in two tropical tree species grown with CO_2 enrichment *J Exp Bot* , **45**, 539–546

Black, C R and Black, VJ (1979) The effects of low concentrations of sulphur dioxide on stomatal conductance and epidermal cell survival in field bean (*Vicia faba* L) *J Exp Bot* , **30**, 291–298

Bonte J and Longuet, P (1975) Interrelations entre la pollution par le dioxyde de soufre et le mouvement des stomates chez le *Pelargonium hortorum* effets de l'humidite relative et de la teneur en gaz carbonique de l'air *Physiol Veg* , **13**, 527–537

Bowden, R L and Rouse, D I (1991) Effects of *Verticillium dahliae* on gas exchange of potato *Phytopathology*, **81**, 293–301

Boyer, J S (1976) Photosynthesis at low water potentials *Phil Trans Roy Soc Lond* , *Ser B*, **273**, 501–512

Boyer, J S (1985) Water transport *Ann Rev Plant Physiol* , **36**, 473–516

Bradbury, I K and Malcolm, D C (1977) The effect of phosphorus and potassium on transpiration, leaf diffusive resistance and water use efficiency in Sitka spruce (*Picea sitchensis*) seedlings *J Appl Ecol* , **14**, 631–642

Brag, H (1972) The influence of potassium on the transpiration rate and stomatal opening in *Triticum aestivum* and *P sativum* *Physiol Plant* , **26**, 250–257

Brun, WA (1965) Rapid changes in transpiration in banana leaves *Plant Physiol* , **40**, 797–802

Burnett, J H (1976) In *Fundamentals of Mycology*, (ed J H Burnett), Edward Arnold and Crane Russak, London

Canny, M J (1990) What becomes of the transpiration stream? *New Phytol* , **114**, 341–368

Capron, T M and Mansfield, T A (1976) Inhibition of net photosynthesis in tomato in air polluted with NO and NO_2 *J Exp Bot* , **101**, 1181–1186

Cockburn, W, Ting, I P and Sternberg, L O (1979) Relationships between stomatal behaviour and internal carbon dioxide concentration in crassulacean acid metabolism plants *Plant Physiol*, **63**, 1029–1032

Constable, G A and Rawson, H M (1980) Effect of leaf position, expansion, and age on photosynthesis, transpiration and water use efficiency of cotton *Aust J Plant Physiol*, **7**, 89–100

Contour-Ansel, D and Louguet, P (1985) Short-term effect of light on phenolic compounds in isolated leaf epidermis of *Pelargonium hortorum* *J Plant Physiol*, **120**, 223–231

Coutinho, T A, Rijkenberg, F H J and van Asch, M A J (1993) Development of infection structures by *Hemileia vastatrix* in resistant and susceptible selections of *Coffea* and in *Phaseolus vulgaris* *Can J Bot*, **71**, 1001–1008

Cowan, I R (1977) Stomatal behaviour and environment *Adv Bot Res*, **4**, 117–229

Cowan, I R and Farqhuar, G D (1977) Stomatal function in relation to leaf metabolism and environment *Soc Exp Biol Symp*, **31**, 471–505

Darwin, F (1897) Observations on stomata by a new method *Proc Cambridge Phil Soc*, **9**, 303–308

Darwin, F (1898) Observations on stomata *Phil Trans Roy Soc Lond*, *Ser B*, **190**, 531–621

Davies, T (1980) Grasses more sensitive to SO_2 pollution in conditions of low irradiance and short days *Nature*, **284**, 483–485

Davies, W J, Wilson, J A, Sharpe, R E and Osonubi, O (1981) Control of stomatal behaviour in water-stressed plants, in *Stomatal Physiology*, (eds P G Jarvis and T A Mansfield), Cambridge University Press, Cambridge, pp 163–185

Davis, S D and McCree, K J (1978) Photosynthetic rate and diffusion conductance as a function of age in leaves of bush bean (*Phaseolus vulgaris* L) *Crop Sci*, **18**, 280–282

Desai, M C (1937) Effects of certain nutrient deficiencies on stomatal behaviour *Plant Physiol*, **12**, 253–281

Dietzer, G F and Frosch, S H (1990) Multiple action of far-red light in photoperiodic induction and circadian rhythmicity *Photochem Photobiol*, **52**, 173–179

Drake, B and Raschke, K (1974) Pre-chilling of *Xanthium strumarium* L reduces net photosynthesis and independently, stomatal conductance, while sensitizing the stomata to CO_2 *Plant Physiol*, **53**, 808–812

Drew, A P and Bazzaz, F A (1979) Response of stomatal resistance and photosynthesis to night temperature in *Populus deltoides* *Oecologia*, **41**, 89–98

Drew, A P and Bazzaz, F A (1982) Effect of night temperature on daytime stomatal conductance in early and late successional plants *Oecologia*, **54**, 76–79

Dunıway, J M (1976) Water status and imbalance, in *Encyclopedia of Plant Physiology, Vol 4, Physiological Plant Pathology*, (eds R Heitefuss and P H Williams), Springer, Berlin

Eamus, D and Murray, M (1991) Photosynthetic and stomatal conductance responses of Norway spruce and beech to ozone, acid mist and frost – a conceptual model *Environ Pollut*, **72**, 23–44

Eamus, D, Fenton, R and Wilson, J M (1983) Stomatal behaviour and water relations of chilled *Phaseolus vulgaris* L and *Pisum sativum* L *J Exp Bot*, **34**, 434–441

Eamus, D, Barnes, J D, Mortensen, L *et al* (1990) Persistent stimulation of CO_2 assimilation and stomatal conductance by summer ozone fumigation in Norway spruce *Environ Pollut*, **63**, 365–379

Edwards, M C and Bowling, D J F (1986) The growth of rust germ tubes towards stomata in relation to pH gradients *Physiol Mol Plant Pathol*, **29**, 185–196

Edwards, M and Meidner, H (1978) Stomatal responses to humidity and the water potentials of epidermal and mesophyll tissue *J Exp Bot*, **29**, 771–780

Englee, R L and Gabelman, WH (1966) Inheritance and mechanism for resistance to ozone damage in onion *Allium cepa* L *Proc Am Soc Hort Sci*, **89**, 423–430

Evans, L T and Allaway, WG (1972) Action spectrum for the opening of *Albizzia julibrissin* pinnules and the role of phytochrome in the closing movements of pinnules and of stomata of *Vicia faba* *Aust J Biol Sci*, **25**, 885–93

Farquhar, G D (1978) Feedforward responses of stomata to humidity *Aust J Plant Physiol*, **5**, 787–800

Field, C B (1987) Leaf-age effects on stomatal conductance, in *Stomatal Function*, (eds E Zeiger, G D Farquhar and I R Cowan), Stanford University Press, Stanford, CA, pp 365–384

Fitzsimons, PJ and Weyers, J D B (1983) Separation and purification of protoplast types from *Commelina communis* L leaf epidermis *J Exp Bot*, **34**, 55–66

Fitzsimons, PJ and Weyers, J D B (1986) Volume changes of *Commelina communis* L guard cell protoplasts in response to K^+, light and CO_2 *Physiol Plant*, **66**, 463–468

Gorton, H L, Williams, WE, Binns, M E *et al* (1989) Circadian stomatal rhythms in epidermal peels from *Vicia faba* *Plant Physiol*, **90**, 1329–1334

Gotow, K, Kondo, N and Syono, K (1982) Effects of CO_2 on volume changes of guard cell protoplasts from *Vicia faba* L *Plant Cell Physiol*, **23**, 1063–1070

Grantz D A (1990) Plant responses to atmospheric humidity *Plant Cell Environ*, **13**, 667–679

Grantz, D A and Schwartz, A (1988) Guard cells of *Commelina communis* L do not respond metabolically to osmotic stress in isolated epidermis implications for stomatal responses to drought and humidity *Planta*, **174**, 166–173

Habermann, H M (1973) Evidence for two photoreactions and possible involvement of phytochrome in light-dependent stomatal opening *Plant Physiol* , **51**, 543–548

Hall, A E and Kaufmann, M R (1975) Stomatal responses to environment with *Sesamum indicum* L *Plant Physiol* , **55**, 455–459

Hall, A E , Camacho, B and Kaufmann, M R (1975) Regulation of water loss by citrus leaves *Physiol Plant* , **33**, 62–65

Heath, O VS (1963) Rapid changes in transpiration in plants *Nature*, **200**, 190–191

Heath, O VS and Russell, J (1954) Studies in stomatal behaviour VI An investigation of the light responses of wheat stomata with the attempted elimination of control by the mesophyll Part II *J Exp Bot* , **5**, 269–292

Heber, U , Neimanis, S and Lange, O L (1986) Stomatal aperture, photosynthesis and water fluxes in mesophyll cells as affected by the abscission of leaves Simultaneous measurements of gas exchange, light scattering and chlorophyll fluorescence *Planta*, **167**, 554–562

Heller, F-O , Kausch, W and Trapp, L (1971) UV-mikroskopischer Nachweis von Strukturanderungen in Schliesszellen von *Vicia faba* L *Naturwissenschaften*, **58**, 419

Hennessey, T L , Freeden, A L and Field, C B (1993) Environmental effects on circadian rhythms in photosynthesis and stomatal opening *Planta*, **189**, 369–376

Hollinger, D Y (1987) Gas exchange and dry matter allocation responses to elevation of atmospheric CO_2 concentration in seedlings of three tree species *Tree Physiol* , **3**, 192–202

Holmes, M G and Klein, WH (1985) Evidence for phytochrome involvement in light mediated stomatal movements in *Phaseolus vulgaris* L *Planta*, **166**, 348–353

Honour, S J , Webb, A A R and Mansfield, T A (1994) The responses of stomata to abscisic acid and temperature are interrelated *Proc Roy Soc Lond* , *Ser B*, in press

Hopmans, PA M (1971) Rhythms in stomatal opening of bean leaves *Meded Landbouwhogeschool Wageningen*, **71**, 1–86

Hsiao, T C , Allaway, WG and Evans, L Y (1973) Action spectra for guard cell Rb^+ uptake and stomatal opening in *Vicia faba* *Plant Physiol* , **51**, 82–88

Iino, M , Ogawa, T and Zeiger, E (1985) Kinetic properties of the blue light response of stomata *Proc Natl Acad Sci USA*, **82**, 8019–8023

Iwanoff, L (1928) Zur Methodik der Transpirations-bestimmung am Standort *Ber Deutsch Bot Ges* , **46**, 306–310

Jarvis, P.G. (1993) Water losses of crowns, canopies and communities, in *Water Deficits*, (eds J.A.C. Smith and H. Griffiths), BIOS Scientific, Oxford, pp. 285–315.

Jarvis, P.G. and Morison, J.I.L. (1981) Stomatal control of transpiration and photosynthesis, in *Stomatal Physiology*, (eds P.G. Jarvis and T.A. Mansfield), Cambridge University Press, Cambridge, pp. 247–280.

Jarvis, P.G. and McNaughton, K.G. (1986) Stomatal control of transpiration: scaling up from leaf to region. *Adv. Ecol. Res.*, **15**, 1–49.

Jensen, C.R. (1982) Effect of soil water osmotic potential on growth and water relations in barley during soil water depletion. *Irrig. Sci.*, **3**, 111–121.

Johnsson, M., Issaias, S., Brogardh, T. and Johnsson, A. (1976) Rapid, blue-light-induced transpiration response restricted to plants with grass-like stomata. *Physiol. Plant.*, **36**, 229–232.

Kappen, L. and Haeger, S. (1991) Stomatal responses of *Tradescantia albiflora* to changing air humidity in light and in darkness. *J. Exp. Bot.*, **42**, 979–986.

Kappen, L., Andresen, G. and Lösch, R. (1987) *In situ* observations of stomatal movements. *J. Exp. Bot.*, **38**, 126–141.

Karlsson, P.E. (1986) Blue light regulation of stomata in wheat seedlings. II. Action spectrum and search for action dichroism. *Physiol. Plant.*, **66**, 207–210.

Karlsson, P.E. (1988) Phytochrome is not involved in the red-light-enhancement of the stomatal blue-light-response in wheat seedlings. *Physiol. Plant.*, **74**, 544–548.

Karlsson, P.E. and Assmann, S.M. (1990) Rapid and specific modulation of stomatal conductance by blue light in ivy (*Hedera helix*). *Plant Physiol.*, **94**, 440–447.

Karlsson, P.E., Hoglund, H.-O. and Klockare, R. (1983) Blue light induces stomatal transpiration in wheat seedlings with chlorophyll deficiency caused by SAN 9789. *Physiol. Plant.*, **57**, 417–421.

Karlsson, P.E., Bogomolni, R.A. and Zeiger, E. (1992) HPLC of pigments from guard cell protoplasts and mesophyll tissue of *Vicia faba* L. *Photochem. Photobiol.*, **55**, 605–610.

Kaufmann, M.R. (1982) Leaf conductance as a function of photosynthetic photon flux density and absolute humidity difference from leaf to air. *Plant Physiol.*, **69**, 1018–1022.

Keck, R.W. and Boyer, J.S. (1974) Chloroplast response to low leaf water potentials. III. Differing inhibition of electron transport and photophosphorylation. *Plant Physiol.*, **53**, 474–479.

Kirschbaum, M.U.F., Gross, L.J. and Pearcy, R.W. (1988) Observed and modelled stomatal responses to dynamic light environments in the shade plant, *Alocasia macrorrhiza*. *Plant Cell Environ.*, **11**, 111–121.

Kuiper, PJ C (1964) Dependence upon wavelength of stomatal movement in epidermal tissue of *Senecio odoris* *Plant Physiol*, **39**, 952–955

Lange, O L (1969) Wasserumsatz und Bewegungen *Fortschritte Bot*, **31**, 76–86

Lange, O L, Losch, R, Shulze, E -D and Kappen, L (1971) Responses of stomata to changes in humidity *Planta*, **100**, 76–86

Lasceve, G, Gautier, H, Jappe, J and Vavasseur, A (1993) Modulation of the blue light response of stomata of *Commelina communis* by CO_2 *Physiol Plant*, **88**, 453–459

Lee, J and Bowling, D J F (1992) Effect of the mesophyll on stomatal opening in *Commelina communis* *J Exp Bot*, **43**, 951–957

Lehnherr, B, Grandjean, A, Machler, F and Fuhrer, J (1987) The effect of ozone in ambient air on ribulosebisphosphate carboxylase/oxygenase activity decreases photosynthesis and grain yield in wheat *J Plant Physiol*, **130**, 189–200

Lewis, B G and Day, J R (1972) Behaviour of uredospore germ-tubes of *Puccinia graminis tritici* in relation to the fine structure of wheat leaf surfaces *Trans Br Mycol Soc*, **58**, 139–145

Lindhauer, M G (1985) Influence of K^+ nutrition and drought on water relations and growth of sunflower (*Helianthus annuus* L) *Z Pflanz Bodenkd*, **148**, 654–669

Lloyd, F E (1908) The physiology of stomata *Publ Carnegie Inst Wash 82*

Loftfield, J V G (1921) The behaviour of stomata *Publ Carnegie Inst Wash 314*

Losch, R (1977) Responses of stomata to environmental factors – experiments with isolated epidermal strips of *Polypodium vulgare* I Temperature and humidity *Oecologia*, **29**, 85–97

Losch, R (1979) Responses of stomata to environmental factors in experiments with isolated epidermal strips of *Polypodium vulgare* II Leaf bulk water potential, air humidity, and temperature *Oecologia*, **39**, 229–238

Losch, R and Schenk, B (1978) Humidity responses of stomata and the potassium content of guard cells *J Exp Bot*, **29**, 781–787

Losch, R and Tenhunen, J D (1981) Stomatal responses to humidity – phenomenon and mechanism, in *Stomatal Physiology*, (eds PG Jarvis and TA Mansfield), Cambridge University Press, Cambridge, pp 137–162

Losch, R, Tenhunen, J D, Pereira, J S and Lange, O L (1978) Diurnal courses of stomatal resistance and transpiration of wild and cultivated Mediterranean perennials at the end of the summer dry season in Portugal *Flora*, **172** 138–160

Losch, R , Jensen, C R and Anderson, M N (1992) Diurnal courses and factorial dependencies of leaf conductance and transpiration of differently potassium fertilized and watered field grown barley plants *Plant and Soil*, **140**, 205–224

Lu, Z , Quinones, M A and Zeiger, E (1993a) Abaxial and adaxial stomata from Pima cotton (*Gossypium barbadense* L) differ in their pigment content and sensitivity to light quality *Plant Cell Environ* , **16**, 851–858

Lu, Z , Quinones, M A and Zeiger, E (1993b) The temperature sensitivity of guard cell respiration co-segregates with stomatal conductance in a F2 population of Pima cotton *Plant Physiol Suppl* , **102**, 138

Ludlow, M M and Wilson, G L (1971) Photosynthesis of tropical pasture plants III leaf age *Aust J Biol Sci* , **24**, 1077–1087

MacDowall, F D H (1965) Predisposition of tobacco to ozone damage *Can J Plant Sci* , **45**, 1–12

Maier-Maecker, U (1979) Peristomatal transpiration and stomatal movement a controversial view III Visible effects of peristomatal transpiration on the epidermis *Z Pflanz* , **91**, 225–238

Mansfield, T A (1965) Stomatal opening in high temperature in darkness *J Exp Bot* , **16**, 721–731

Mansfield, T A (1971) Stomata versatile sensory devices but difficult experimental subjects *J Biol Educ* , **5**, 115–123

Mansfield, T A and Heath, O V S (1963) Studies in stomatal behaviour IX Photoperiodic effects on rhythmic phenomena in *Xanthium pennsylvanicum* *J Exp Bot* , **15**, 334–352

Mansfield, T A and Heath, O V S (1964) Studies in stomatal behaviour X An investigation of responses to low intensity illumination and temperature in *Xanthium pennsylvanicum* *J Exp Bot* , **15**, 114–124

Mansfield, T A and Meidner, H (1966) Stomatal opening in light of different wavelengths effects of blue light independent of carbon dioxide concentration *J Exp Bot* , **17**, 510–521

Mansfield, T A and Jones, R J (1970) Increases in the diffusion resistances of leaves in a carbon dioxide-enriched atmosphere *J Exp Bot* , **21**, 951–958

Mansfield, T A and Majernick, O (1970) Can stomata play a part in protecting plants against air pollutants? *Environ Pollut* , **1**, 149–154

Mansfield, T A and Freer-Smith, P H (1981) Effects of urban air pollution on plant growth *Biol Rev* , **56**, 343–368

Martin, E S and Meidner, H (1971) Endogenous stomatal movements in *Tradescantia virginiana* *New Phytol* , **70**, 923–928

Martin, E S and Meidner, H (1972) The phase response of the dark stomatal opening in *Tradescantia virginiana* to light and dark treatments *New Phytol* , **71**, 1045–1054

Martin, E S and Stevens, R A (1979) Circadian rhythms in stomatal movements, ın *Structure, Functıon and Ecology of Stomata*, (ed D N Sen), Bıshen Sıngh and Mahendra Pal Sıngh, Dehra Dun, Indıa, pp 251–268

Mawson, B T and Cummıns, WR (1991) Low temperature acclımatıon of guard cell chloroplasts by the arctıc plant *Saxıfraga cernua* L *Plant Cell Envıron* , **14**, 569–576

Meıdner, H (1965) Stomatal control of transpıratıonal water loss *Symp Soc Exp Bıol* , **XIX**, 185–204

Meıdner, H (1975) Water supply, evaporatıon, and vapour dıffusıon ın leaves *J Exp Bot* , **26**, 666–673

Meıdner, H (1983) Our understandıng of plant water relatıons *J Exp Bot* , **34**, 1606–1618

Meıdner, H (1987) The humıdıty response of stomata and ıts measurement *J Exp Bot* , **38**, 877–882

Meıdner, H and Heath, O VS (1959) Stomatal responses to temperature and carbon dıoxıde concentratıon ın *Allıum cepa L* and theır relevance to mıdday closure *J Exp Bot* , **10**, 206–219

Meıdner, H and Wıllmer, C M (1993) Cırcadıan rhythms of stomatal movements ın epıdermal strıps *J Exp Bot* , **44**, 1649–1652

Morıson, J I L and Jarvıs, PG (1981) The control of transpıratıon and photosynthesıs by stomata, ın Stomatal Physıology, (eds PG Jarvıs and TA Mansfıeld), Cambrıdge Unıversıty Press, Cambrıdge, pp 247–279

Morıson, J I L and Jarvıs, PG (1983) Dırect and ındırect effects of lıght on stomata II In *Commelına communıs* L *Plant Cell Envıron* , **6**, 103–109

Mott, K A and Parkhurst, D F (1991) Stomatal responses to humıdıty ın aır and helox *Plant Cell Envıron* , **14**, 509–515

Muralı, N S and Saxe, H (1984) Effects of UV-C radıatıon on net photosynthesıs, transpıratıon and dark respıratıon of *Spathyphyllum wallısıı Physıol Plant* , **60**, 192–196

Neales, T F (1975) The gas exchange patterns of CAM plants, ın *The Envıronmental and Bıologıcal Control of Photosynthesıs*, (ed R Marcelle), Junk, The Hague

Negash, L and Bjorn, L O (1986) Stomatal closure by ultravıolet radıatıon *Physıol Plant* , **66**, 360–364

Negash, L , Jenson, P and Bjorn, L O (1987) Effect of ultravıolet radıatıon on accumulatıon and leakage of $^{86}Rb^+$ ın guard cells of *Vıcıa faba Physıol Plant* , **69**, 200–204

Nelson, S D and Mayo, J M (1975) The occurrence of functıonal nonchlorophyllous guard cells ın *Paphıopedılum* spp *Can J Bot* , **53**, 1–7

Nonamı, H and Schulze, E -D (1989) Cell water potentıal, osmotıc potentıal, and turgor ın the epıdermıs and mesophyll of transpırıng leaves *Planta*, **177**, 35–46

Nonani, H., Schulze, E.-D. and Ziegler, H. (1990) Mechanisms of stomatal movement in response to air humidity, irradiance and xylem water potential. *Planta*, **183**, 57–64.

Ogawa, T., Ishikawa, H., Shimada, K. and Shibata, K. (1978) Synergistic action of red and blue light and action spectra for malate formation in guard cells of *Vicia faba* L. *Planta*, **142**, 61–65.

Omasa, K., Hashimoto, Y. and Aiga, I. (1981) A quantitative analysis of the relationship between SO_2 or NO_2 sorption and their acute effects on plant leaves using image instrumentation. *Environ. Control Biol.*, **19**, 59–67.

Omasa, K. and Onoe, M. (1984) Measurement of stomatal aperture by digital image processing. *Plant Cell Physiol.*, **25**, 1379–1388.

Osborne, B.A., O'Connell, C., Campbell, G.C. and Weyers, J.D.B. (1994) Stomata of *Gunnera tinctoria* do not respond to light, CO_2 or ABA. *J. Exp. Bot. Suppl.*, **44**, 4.

Osman A.M. and Milthorpe, F.L. (1971) Photosynthesis of wheat leaves in relation to age, illumination, and nutrient supply. II. Results. *Photosynthetica*, **5**, 61–70.

Ozuna, R., Yera, R., Ortega, K. and Tallman, G. (1985) Analysis of guard cell viability and action in senescing leaves of *Nicotiana glauca* (Graham), tree tobacco. *Plant Physiol.*, **79**, 7–10.

Pallaghy, C.K. and Fischer, R.A. (1974) Metabolic aspects of stomatal opening and ion accumulation by guard cells of *Vicia faba*. *Z. Pflanz.*, **71**, 332–344.

Pallas, J.E. (1980) An apparent anomaly in peanut leaf conductance. *Plant Physiol.*, **65**, 848–851.

Park, J. and Thimann, K.V. (1990) Senescence and stomatal aperture as affected by antibiotics in darkness and light. *Plant Physiol.*, **92**, 696–702.

Pasternak, D. and Wilson, G.L. (1972) After-effects of night temperatures on stomatal behaviour and photosynthesis of *Sorghum*. *New Phytol.*, **71**, 683–689.

Pearson, M. and Mansfield, T.A. (1993) Interacting effects of ozone and water-stress on the stomatal resistance of beech (*Fagus sylvatica* L). *New Phytol.*, **123**, 351–358.

Peaslee, D.E. and Moss, D.N. (1968) Stomatal conductivities in K-deficient leaves of maize (*Zea mays*, L.) *Crop Sci.*, **8**, 427–430.

Pemadasa, M.A. (1982) Abaxial and adaxial stomatal responses to light of different wavelengths and to phenylacetic acid on isolated epidermis of *Commelina communis* L. *J. Exp. Bot.*, **33**, 92–99.

Pring, R.J. (1980) A fine structural study of the infection of leaves of *Phaseolus vulgaris* by uredospores of *Uromyces phaseoli*. *Physiol Plant Pathol.*, **17**, 269–276.

Quinones, M.A., Lu, Z. and Zeiger, E. (1993) Zeaxanthin concentrations co-segregate with the magnitude of the blue light response of adaxial guard cells and leaf stomatal conductances in an F2 population of Pima cotton. *Plant Physiol. Suppl.*, **102**, 15.

Radin, J.W. (1984) Stomatal responses to water stress and to abscisic acid in phosphorus-deficient cotton plants. *Plant Physiol.*, **76**, 392–394.

Radin, J.W., Parker, L.L. and Guinn, G. (1982) Water relations of cotton plants under nitrogen deficiency. V. Environmental control of abscisic acid accumulation and stomatal sensitivity to abscisic acid. *Plant Physiol.*, **70**, 1066–1077.

Radoglou, K.M., Aphalo, P. and Jarvis, P.G. (1992) Response of photosynthesis, stomatal conductance and water use efficiency to elevated CO_2 and nutrient supply in acclimated seedlings of *Phaseolus vulgaris* L. *Ann Bot.*, **70**, 257–264.

Rao, I.M., Amundson R.G., Alscher-Herman, R. and Anderson, L.E. (1983) Effects of SO_2 on stomatal metabolism in *Pisum sativum* L. *Plant Physiol.*, **72**, 573–577.

Raschke, K. (1970) Temperature dependence of CO_2 assimilation and stomatal aperture in leaf sections of *Zea mays*. *Planta*, **91**, 336–363.

Raschke, K. (1975) Stomatal action. *Ann. Rev. Plant Physiol.*, **26**, 309–40.

Raschke, K. (1979) Movements of stomata, in *Encyclopedia of Plant Physiology, Vol. 7, Physiology of Movements*, (eds W. Haupt and M.E. Feinlieb), Springer, Berlin, pp. 381–441.

Reich, P.B. and Amundson, R.G. (1985) Ambient levels of ozone reduce net photosynthesis in tree and crop species. *Science*, **230**, 566–570.

Richardson, M.D., Hoveland, C.S. and Bacon, C.W. (1993) photosynthesis and stomatal conductance of symbiotic and nonsymbiotic tall fescue. *Crop Sci.*, **33**, 145–149.

Roth-Bejerano, N. and Itai, C. (1981) Involvement of phytochrome in stomatal movement. Effect of blue and red light. *Physiol. Plant.*, **52**, 201–206.

Roth-Bejerano, N. and Nejidat, A. (1987) Phytochrome effects on K^+ fluxes in guard cells of *Commelina communis*. *Physiol. Plant.*, **71**, 345–351.

Roth-Bejerano, N., Nejidat, A and Itai, C. (1985) Further support for the involvement of phytochrome in stomatal movements. *Physiol Plant.*, **64**, 501–506.

Royle, D. J. (1976) Structural features of resistance to plant-diseases, in *Biochemical Aspects of Plant–Parasite Relationships (Phytochem. Soc. Symp. Ser. 13)*, (eds J. Friend and D.R. Threlfall), Academic Press, London.

Saunders, M.J., Cordonnnier, M.-M., Palevitz, B.A. and Pratt, L.H. (1983) Immunofluorescence visualization of phytochrome in *Pisum sativum* L. epicotyls using monoclonal antibodies. *Planta*, **159**, 545–553.

Schnabl, H., Weissenbock, G., Sachs, G. and Scharf, H. (1989) Cellular distribution of UV-absorbing compounds in guard and subsidiary cells of *Zea mays* L. *J. Plant Physiol.*, **135**, 249–252.

Schonherr, J (1982) Resistance of plant surfaces to water loss transport properties of cutin, suberin and associated lipids, in *Encyclopedia of Plant Physiology, Vol 12B, Physiological Plant Ecology II Water Relations and Carbon Assimilation*, (eds O L Lange, PS Nobel, C B Osmond and H Ziegler), Springer, Heidleberg, pp 154–179

Schulte-Hostede, S , Darrall, N M , Blank, L W and Wellburn, A R (1988) *Air Pollution and Plant Metabolism*, Elsevier, Amsterdam

Schulze, E -D (1986) Carbon dioxide and water vapour exchange in response to drought in the atmosphere and in the soil *Ann Rev Plant Physiol* , **37**, 247–274

Schulze, E D , Lange, O L , Kappen, L *et al* (1973) Stomatal responses to changes in temperature at increasing water-stress *Planta*, **110**, 29–42

Schulze, E -D , Lange, O L , Evanari, M *et al* (1974) The role of air humidity and leaf temperature in controlling stomatal resistance of *Prunus armeniaca* L under desert conditions I A simulation of the daily course of stomatal resistance *Oecologia*, **17**, 159–170

Shackel, K A and Brinckmann, E (1985) *In situ* measurement of epidermal cell turgor, leaf water potential, and gas exchange in *Tradescantia virginiana* L *Plant Physiol* , **78**, 66–70

Sheriff, D W (1977a) The effect of humidity on water uptake by, and viscous flow resistance of, excised leaves of a number of species physiological and anatomical observations *J Exp Bot* , **28**, 1399–1407

Sheriff, D W (1977b) Evaporation sites and distillation in leaves *Ann Bot* , **41**, 1081–1082

Shimazaki, K -I , Ito, K , Kondo, N and Sugahara, K (1984) Reversible inhibition of the photosynthetic water-splitting enzyme system by SO_2 fumigation assayed by chlorophyll fluorescence and EPR signal *in vivo* *Plant Cell Physiol* , **25**, 795–803

Shimazaki, K -I , Iino, M and Zeiger, E (1986) Blue light-dependent proton extrusion by guard cell protoplasts of *Vicia faba* *Nature*, **319**, 324–326

Shimazaki, K , Igarashi, T and Kondo, N (1988) Protection by the epidermis of photosynthesis against UV-C radiation estimated by chlorophyll a fluorescence *Physiol Plant* , **74**, 34–38

Solarova, J and Pospisilova, J (1983) Photosynthetic characteristics during ontogenesis of leaves, VIII Stomatal diffusive conductance and stomatal reactivity *Photosynthesis*, **14**, 523–531

Stålfelt, M G (1955) The stomata as a hydrophotic regulator of the water deficit of the plant *Physiol Plant* , **8**, 572–593

Stålfelt, M G (1962) The effect of temperature on opening of stomatal cells *Physiol Plant* , **10**, 752–793

Stålfelt, M G (1963) Diurnal dark reactions in the stomatal movements *Physiol Plant* , **16**, 756–766

Staples, R C and Macko, V (1980) Formation of infection structures as a recognition response in fungi *Exp Mycol*, **4**, 2–16

Stocker, O (1956) Die Abhangigkeit der Transpiration von den Umweltfaktoren, in *Encyclopedia of Plant Physiology, Vol III*, (ed W Ruhland), Springer, Berlin, pp 436–488

Takahama, U (1988) Hydrogen peroxide-dependent oxidation of flavonoids and hydroxycinnamic acid derivatives in epidermal and guard cells of *Tradescantia virginiana* L *Plant Cell Physiol*, **29**, 475–481

Tenhunen, J D, Lange, O L, Braun, M *et al* (1980) Midday stomatal closure in *Arbutus unedo* leaves in a natural macchia and under simulated habitat conditions in an environmental chamber *Oecologia*, **47**, 365–7

Tenhunen, J D, Lange, O L, Gebel, J *et al* (1984) Changes in photosynthetic capacity, carboxylation efficiency, and CO_2 compensation point associated with midday stomatal closure and midday depression of net CO_2 exchange of leaves of *Quercus suber Planta*, **162**, 193–203

Teramura, A H, Tevini, M and Iwanzik, W (1983) Effects of ultraviolet-B irradiation on plants during mild water stress I Effects on diurnal stomatal resistance *Physiol Plant*, **57**, 175–180

Thimann, K V and Satler, S O (1979a) Relations between leaf senescence and stomatal closure senescence in light *Proc Natl Acad Sci USA*, **76**, 2295–2298

Thimann, K V and Satler, S O (1979b) Relation between senescence and stomatal opening senescence in darkness *Proc Natl Acad Sci USA*, **76**, 2770–2773

Thimann, K V and Tan, Z-Y (1988) The dependence of stomatal closure on protein synthesis *Plant Physiol*, **86**, 341–343

Ting, I P and Dugger, WM (1968) Factors affecting ozone sensitivity and susceptibility of cotton plants *J Air Pollut Control Ass*, **18**, 810–813

Travis, A J and Mansfield, T A (1981) Light saturation of stomatal opening on the adaxial and abaxial epidermis of *Commelina communis J Exp Bot*, **32**, 1169–1179

Turner, N (1974) Stomatal responses to light and water under field conditions, in *Mechanisms of Regulation of Plant Growth*, (ed R Bielski), Royal Society of New Zealand, Bull 12, pp 423–432

Tyree, M T and Yianoulis, P (1980) The site of evapouration from substomatal cavities, liquid path resistances and hydroactive stomatal closure *Ann Bot*, **46**, 175–193

Unsworth, M H and Black, VJ (1981) Stomatal responses to pollutants, in *Stomatal Physiology*, (eds PG Jarvis and T A Mansfield), Cambridge University Press, Cambridge, pp 187–203

Vaclavik, J (1973) Effect of different leaf ages on the relationship between the CO_2 uptake and water vapour efflux in tobacco plants *Biol Plant*, **15**, 233–236

Vierstra, R D , John, T R and Poff, K L (1982) Kaempferol 3-*O*-galactoside 7-*O*-rhamnoside is the major green fluorescing compound in the epidermis of *Vicia faba* *Plant Physiol* , **69**, 522–525

Wallace, A and Frolich, E (1965) Phosphorus-deficiency symptoms in tobacco and transpirational water loss *Nature*, **208**, 1131

Wardle, K and Short, K C (1983) Stomatal responses and the senescence of leaves *Ann Bot* , **52**, 411–412

Warrit, B , Landsberg, J J and Thorpe, M R (1980) Respones of apple leaf stomata to environmental factors *Plant Cell Environ* , **3**, 13–22

Watts, WR and Neilson, R E (1978) Photosynthesis in Sitka spruce (*Picea sitchensis* [Bong] Carr) VIII Measurements of stomatal conductance and $^{14}CO_2$ uptake in controlled environments, *J Appl Ecol* , **15**, 245–255

Weissenbock, G , Schnabl, H , Sachs, G *et al* (1984) Flavonol content of guard cell and mesophyll cell protoplasts isolated from *Vicia faba* leaves *Physiol Plant* , **62**, 356–362

Weissenbock, G , Hedrich, R and Sachs, G (1986) Secondary phenolic products in isolated guard cell, epidermal cell and mesophyll cell protoplasts from pea (*Pisum sativum* L) leaves distribution and determination *Protoplasma*, **134**, 141–148

Weissenbock, G , Schnabl, H , Scharf, H and Sachs, G (1987) On the properties of fluorescing compounds in guard and epidermal cells of *Allium cepa* L *Planta*, **171**, 88–95

Wheeler, B E J (1968) Fungal parasites of plants, in *The Fungi*, *An Advanced Treatise*, *Vol III*, *The Fungal Population*, (eds G C Ainsworth and A S Sussman), Academic Press, New York, pp 179–204

Wilkins, M B (1959) An endogenous rhythm in the rate of carbon dioxide output of *Bryophyllum* 1 Some preliminary experiments *J Exp Bot* , **10**, 377–90

Wilkins, M B (1969) Circadian rhythms in plants, in *Physiology of Plant Growth and Development*, (ed M B Wilkins), McGraw-Hill, London, pp 179–204

Wilkins, M B (1992) Circadian rhythms their origin and control *New Phytol* , **121**, 347–375

Wilkins, M B (1993) The role of stomata in generation of circadian rhythms in plant tissues *J Exp Bot Suppl* , **44**, 2

Willmer, C M (1984) Some characteristics of phosphoenolpyruvate car boxylase activity from leaf epidermal tissue in relation to stomatal functioning *New Phytol* , **84**, 593–602

Willmer, C M (1988) Stomatal sensing of the environment *Biol J Linn Soc* , **34**, 204–217

Willmer, C M , Wilson, A B and Jones, H G (1988) Changing sensitivities of stomata to abscisic acid and CO_2 as leaves and plants age *J Exp Bot* , **39**, 401–410

Willmer, C.M. and Pantoja, O. (1992) The plasma membrane and tonoplast of guard cells, in *Plant Membranes: A Biophysical Approach*, (ed. Y.Y. Lesham), Kluwer, Dordrecht, pp. 220–238.

Wong, S.C., Cowan, I.R. and Farquhar, G.D. (1979) Stomatal conductance correlates with photosynthetic capacity. *Nature*, **282**, 424–426.

Wong, S.C., Cowan, I.R. and Farquhar, G.D. (1978) Leaf conductance in relation to assimilation in *Eucalyptus pauciflora* Sieb. ex Spreng. Influence of irradiance and partial pressure of carbon dioxide. *Plant Physiol.*, **62**, 670–674.

Zeiger, E. and Hepler, P.K. (1979) Blue light-induced intrinsic vacuolar fluorescence in onion guard cells. *J. Cell Sci.*, **37**, 1–10.

Zeiger, E. and Schwartz, A. (1982) Longevity of guard cell chloroplasts in falling leaves: implication for stomatal functioning and cellular aging. *Science*, **218**, 680–682.

Zeiger, E., Field, C. and Mooney, H.A. (1981) Stomatal opening at dawn: possible roles of the blue light response in nature, in *Plants and the Daylight Spectrum*, (ed. H. Smith), Academic Press, New York, pp. 391–407.

Zeiger, E., Iino, M. Shimazaki, K.-I. and Ogawa, T. (1987) The blue-light response of stomata: mechanism and function, in *Stomatal Function*, (eds E. Zeiger, G.D. Farquhar and I.R. Cowan), Stanford University Press, Stanford, CA, pp. 207–227.

Ziegler, I. (1975) The effects of SO_2 pollution on plant metabolism. *Residue Rev.*, **56**, 79–105.

7 The influence of hormones and other naturally occurring compounds on stomatal behaviour

7.1 *Introduction*

Abscisic acid (ABA) is intimately involved in the regulation of stomatal behaviour; other groups of hormones, such as the cytokinins, may also be involved in the control of stomata. Additionally, many other naturally occurring compounds affect stomata including certain fatty acids, some phenolics, simpler metabolites, such as proline, and many phytotoxins, such as fusicoccin and victorin, which are produced by fungal pathogens.

The effects of certain environmental factors on stomatal behaviour may be mediated by hormones. For example, water stress, salt stress and chilling of plants can result in elevated ABA levels within leaves with subsequent stomatal closure. Developmental changes in a plant can also indirectly influence stomatal behaviour probably by affecting hormonal levels within leaves. For example, the panicle of pearl millet modified stomatal behaviour in the flag leaves possibly via influencing leaf ABA levels (Henson *et al.*, 1984; Henson and Mahlalakshmi, 1985). Similarly, debudding *Xanthium pennsylvanicum* plants or partial defoliation of *Phaseolus vulgaris* plants resulted in increased stomatal conductances relative to that in control plants (Krizec and Milthorpe, 1966; Meidner, 1970). Also, the development of fruit in some species is paralleled with wider stomatal openings in the leaves and with changes in the levels of ABA, cytokinins or gibberellins within the leaves. Furthermore, the history of the growing conditions of a plant may influence hormone levels in leaves which, in turn, may subsequently affect stomatal behaviour.

The 'after-effect of wilting may be an example of this

This chapter gives an account of the hormonal control of stomata at a physiological and a biochemical level and also discusses the effects of other naturally occurring products on stomatal behaviour The possible modes of action of ABA on stomata will be discussed in detail in Chapter 8 A section on the artificial control of stomata is also presented

7.2 Abscisic acid

Much attention has been paid to the biochemistry and physiology of ABA, particularly over the last decade and numerous articles have been written about its implication in stomatal behaviour (e g Walton, 1980, Zeevart and Creelman, 1988, Hartung and Slovic, 1991, Davies and Jones, 1992)

Abscisic acid is a terpene (terpenoid) compound Terpenes are made up of multiples of five carbon isoprene units The 15 carbon compounds, of which ABA is one, are known as sesquiterpenoids The structure of ABA is shown in below in Fig 7 5

7.2.1 Stomatal responses to ABA

In the late 1960s it was discovered that ABA inhibited transpiration (Little and Eidt, 1968, Mittelheuser and van Steveninck, 1969) and accumulated in leaf tissue under water stress (Wright, 1969, Wright and Hiron, 1969) (see Fig 7 1) Other types of stress also induce ABA biosynthesis Thus, ABA accumulates in leaf tissue of water logged plants (e g Jackson and Hall, 1987) and in tissues of cold-treated plants (e g Dale and Campbell, 1981) It is now established that ABA prevents stomatal opening and causes closure in virtually all plant species so far investigated Night opening of stomata in *Kalanchoe daigremontiana*, a CAM plant, is also prevented by ABA (Fig 7 2)

A few species of plants, however, are reported to be unresponsive to ABA These species include yellow lupin (*Lupinus luteus*) (Lancaster *et al* , 1977) and *Gunnera tinctoria* (Osborne *et al* , 1993) Further attention to these examples is needed

Abscisic acid will close stomata within minutes of application whether it is applied to the surface of leaves or via the transpiration stream of excised leaves, ABA will also prevent opening of stomata when supplied to leaves by either method Additionally, ABA, under most circumstances, closes stomata and prevents opening in epidermal strips (Fig 7 3) However, if the concentration of KCl in the incubation medium is increased the closing effects of ABA decrease (Willmer *et al* , 1978)

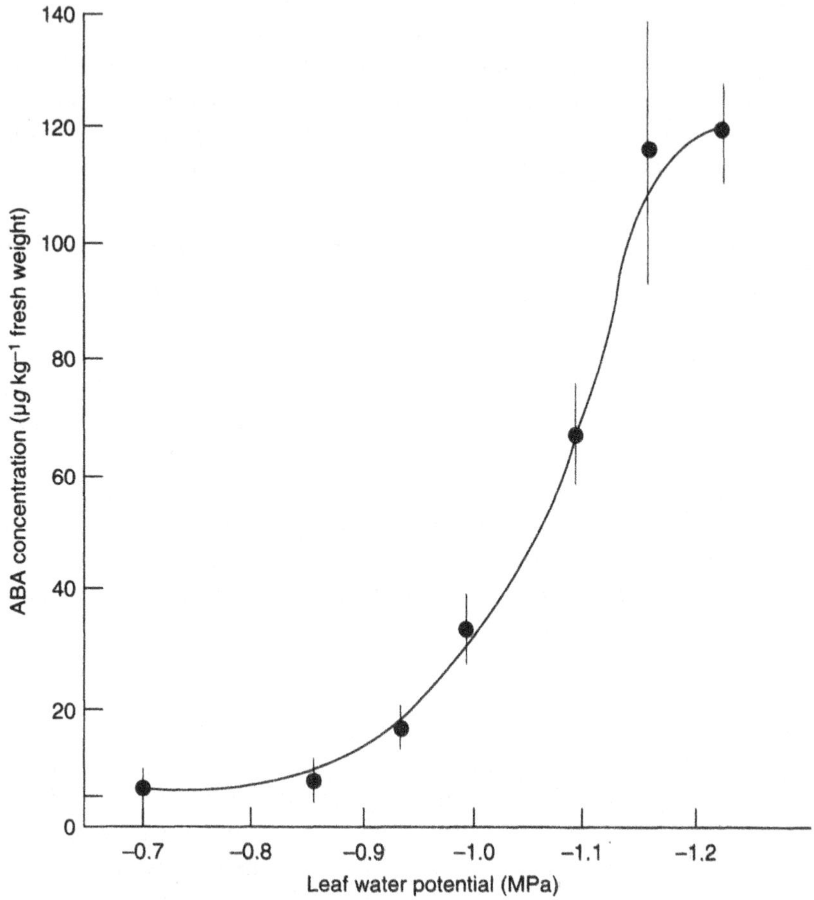

Figure 7.1 The effect of leaf water potential on abscisic acid levels in excised wheat leaves From Wright (1977)

Protoplasts are also sensitive to ABA. For example, Gotow *et al.* (1982) found that 10 μM ABA caused contraction of *Vicia faba* guard cell protoplasts (GCPs) in the light and dark; also Fitzsimons and Weyers (1987) found that GCPs of *Commelina communis* responded to ABA in a manner which was qualitatively and quantitatively similar to that of intact stomata, i.e. ABA prevented GCPs from swelling under low CO_2 conditions and caused swollen GCPs to shrink. According to Iverson *et al.* (1983), after 30 min exposure of GCPs of *V. faba* to 10 μM ABA diameters decreased significantly in white or green light while in red light no change of diameters was observed. They also found no effect of ABA on diameters of protoplasts from epidermal or mesophyll cells in green light.

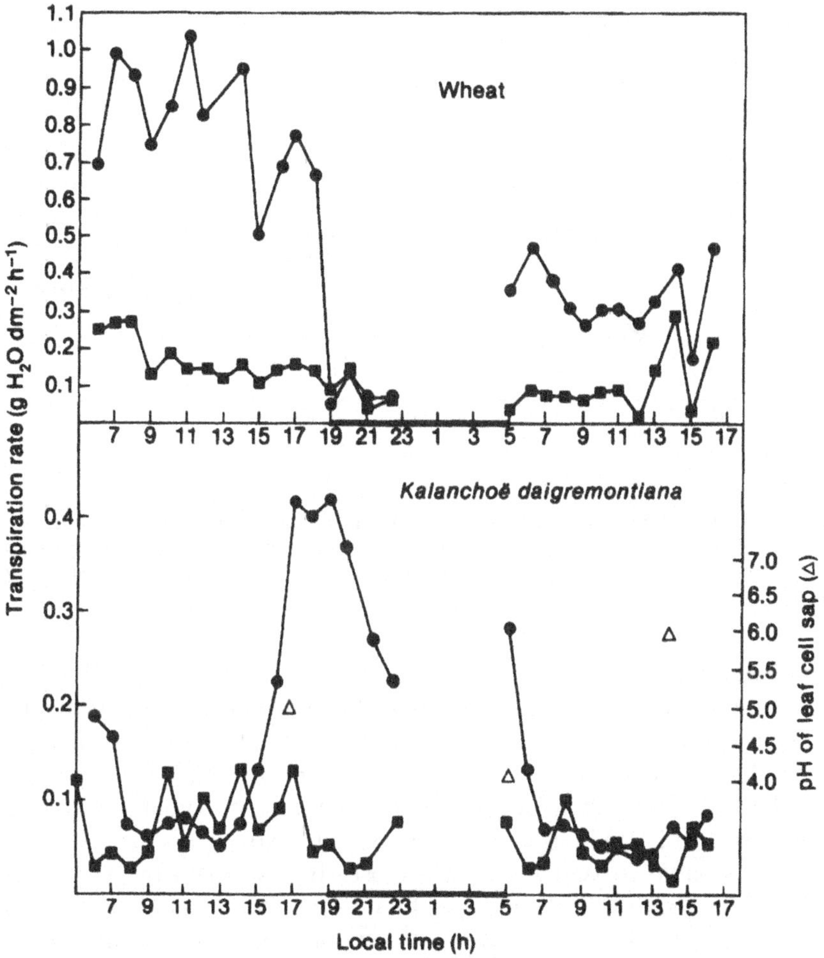

Figure 7.2 Transpiration rates of wheat and *Kalanchoe daigremontiana* with their petioles immersed in 10^{-4} M ABA (■) or water (●) The bar on the abscissa indicates the period of darkness From Scott-Murphy and Willmer (unpublished)

Interestingly some other types of cell protoplasts are also affected by ABA. For example, Chang *et al.* (1983) found that ABA increased the densities of maize root protoplasts possibly by losing water, which would result in smaller volumes, while Itai and Roth-Bejerano (1986) found that ABA (3.7 μM), in the presence of KCl, caused swelling of epidermal cell protoplasts of *C. communis*.

At one time the implication of ABA in the regulation of stomatal movements was questioned, but now there are compelling reasons for believing it has a major role in the hormonal control of stomatal functioning. It has been established that exceedingly low concentrations of

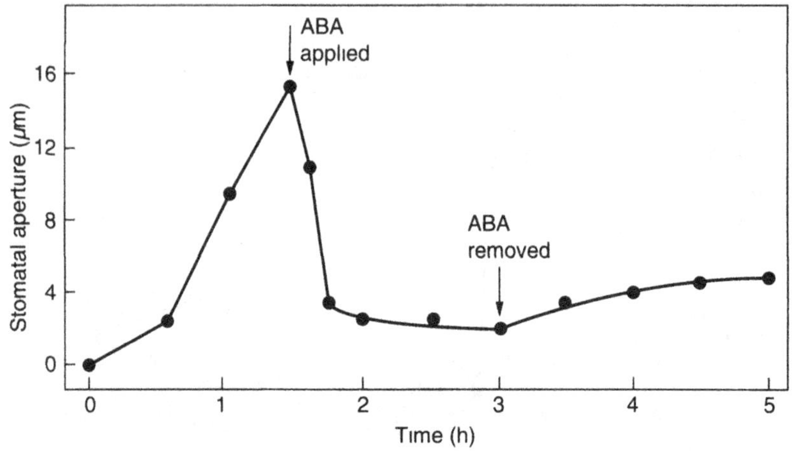

Figure 7.3 Effect of 10^{-4} M ABA on open stomata in epidermal strips of *Commelina communis* and their subsequent recovery in the absence of ABA. Strips were incubated in the light (500 Wm^{-2}) at 30°C with air bubbling through the medium. From Willmer *et al.* (1978).

ABA synthesized in tissues other than the guard cells close stomata and prevent them from opening, ABA produces its effects on stomata even in epidermal strips in the absence of the influence of leaf mesophyll, that ABA causes GCPs to contract in the manner of stomatal closing, and that ABA levels build up in leaves as water stress occurs. Also, wilty mutants unable to synthesize ABA close their stomata upon an exogenous supply of ABA. Indeed, now the mode of action of ABA on stomata at the molecular level is beginning to be unravelled (see Chapter 8).

In some instances, however, there is a poor correlation between bulk leaf water potential and ABA content of the leaves (e.g. Pierce and Raschke, 1980) or between stomatal aperture and ABA content of the leaves (e.g. Henson, 1981). There may be several reasons for the lack of correlation. The amount of ABA synthesized by a plant and the level of ABA in a leaf may depend on the type of stress (whether osmotic, water or cold stress), the extent and duration of the stress, the previous growing conditions of the plant, the developmental stage of the plant (e.g. whether flowering or fruiting) and on the inherent ability of a plant to manufacture and degrade ABA.

In some species it has been observed that the bulk leaf water potential is not the best indicator of the water stress experienced by the epidermis and, therefore, the stomata of that leaf (see Meidner, 1983; Harris and Outlaw, 1991). Hence, there may not be a good correlation between bulk leaf ABA levels and stomatal aperture. In addition, there may be a redistribution of ABA throughout the leaf (see Section 7 2 2) Thus, although total leaf ABA levels may not necessarily increase greatly during a stress response, the level in guard cells, for example, may do For example, Harris *et al* (1988) found that the ABA content of guard cells of *V faba* (including apoplastic space) was only 0 15% of the leaf ABA Thus any changes in ABA levels in guard cells would be masked by the overall leaf concentrations

Of further significance is the finding that leaf ABA levels appear more closely related to leaf *turgor* potentials rather than bulk leaf water potentials (Davies and Lakso, 1978), and as turgor approaches zero ABA levels rise sharply (Pierce and Raschke, 1980)

Besides ABA concentration changes in cells and tissues there may be differential sensitivity to the hormone under different plant conditions (see Trewavas, 1987, Weyers *et al* , 1987, Paterson and Weyers, 1988, Peng and Weyers, 1994)

Thus the complexity of the situation makes it difficult to interpret correlations between changes of bulk leaf ABA levels, leaf water potentials and stomatal conductances Nevertheless, under stress conditions bulk leaf ABA levels generally increase, especially in mesophytes but less so in xerophytes (Aquatic plants and particularly plants with submerged leaves and no stomata have a very limited ability to synthesize ABA under stress conditions) Moreover, it has been concluded that only a few percent over the initial bulk leaf ABA level is sufficient to initiate closure and that ABA can be synthesized rapidly enough to account for the observed rates of closure More specifically, abscisic acid levels also increase in guard cells as water stress proceeds, although the measured values have varied widely according to the investigation For example, Harris *et al* (1988) measured ABA concentrations in a guard cell of water-stressed *V faba* leaves of 8 8 fg per guard cell (equivalent to about 3 5 μM assuming a guard cell volume of 3 0 pl and ignoring compartmentation) In another study Harris and Outlaw (1991) measured maximum ABA concentrations of 15 μM in guard cells of water-stressed *V faba* leaves Additionally, they observed that over a 6 h period of water stress the ABA content of the mesophyll reached maximum levels after about 2 h of the stress, increasing by about 18-fold During that time the ABA content of guard cells increased by about 26-fold, upon removal of stress, there was an initial delay before guard cell ABA levels declined by 46 and 83% after 2 and 6 h, respectively A similar situation occurred in the mesophyll but with an immediate decline in ABA levels occurring upon removal of stress and within 6 h it reached prestress levels Some studies detect surprisingly high ABA levels in cells Thus, in guard cells of *Valerianella locusta*, Behl and Hartung (1986) measured 1 3 mM ABA in the cytoplasm and 0 45 mM in the vacuole (equivalent to an average guard cell concentration of 0 64 mM) and in the epidermal layer of leaves of

Tradescantia virginiana and *C. communis* a gradient of ABA concentration existed between the guard cells (mean, 2.49 mM), subsidiary cells (mean, 1.25 mM) and epidermal cells (mean, 0.86 mM) (Brinckmann *et al.*, 1990).

7.2.2 Effects of ABA on photosynthesis

It has been suggested that ABA may affect stomatal behaviour indirectly by modifying photosynthesis which, in turn, would change intracellular CO_2 concentrations.

Although ABA does not inhibit photosynthesis in isolated chloroplasts (e.g. Keck and Boyer, 1974; Kriedeman *et al.*, 1975), isolated cells (Mawson *et al.*, 1981; Sharkey and Raschke, 1980) or leaf sections (Willmer *et al.*, 1981), a number of investigators concluded that in intact leaves ABA was directly responsible for a decrease in photosynthetic rates, i.e. inhibition of photosynthesis was non-stomatal (e.g. Raschke, 1982; Cornic and Miginiac, 1983; Raschke and Hedrich, 1985; Bunce, 1987). Essentially, the conclusion that photosynthesis was directly affected by ABA was obtained from the observation that the photosynthetic rate at a given, *calculated* intercellular CO_2 concentration, C_i, was depressed relative to untreated leaves.

These conflicting reports remained an enigma until, at about the same time, various groups discovered that the apparent, non-stomatal, inhibition of photosynthesis was an artefact attributable to 'patchy' stomatal opening (also see Chapter 2). ('Patchy' stomatal opening has also been referred to as non-uniform or heterogeneous stomatal opening but these terms are misleading because, as all who have worked with stomata appreciate, there can be a wide distribution of stomatal apertures within a leaf and stomata are unlikely to be ever uniformly open; see Chapter 2). The 'patchy' opening resulted in overestimates of C_i and a spurious C_i/assimilation relationship (Downton *et al.*, 1988; Raschke and Patzke, 1988; Terashima *et al.*, 1988). Additionally, caution is needed in the interpretation of results of experiments in which cells, tissues and particularly whole leaves are treated with hormones or other substances. This is because there may be problems concerning entry of the substances into cells (determined in part by molecular size, lipid solubility, ionization state at a particular pH) and also a physiological response as a result of the applied substance does not necessarily imply that it is directly concerned in that response. For example, ABA promotes senescence in leaves with protein and chlorophyll levels decreasing and DNase activity increasing (e.g. Jung and Gossman, 1985). Thus, photosynthetic rates may rapidly decrease in ABA-treated leaves due to factors not directly concerned in photosynthesis. Moreover, ABA is known to influence ion fluxes which may affect membrane potentials and ATP levels and ultimately the consequences may impinge on enzyme activity and photosynthesis (see Seeman and Sharkey, 1987).

7.2.3 Distribution and site of action of ABA

Abscisic acid is ubiquitously located in all parts of higher plants and is also synthesized by certain algae (Tietz and Kasprik, 1986) and several phytopathological fungi (Dorffling *et al.*, 1984). Abscisic acid is not found in unicellular green algae where lunularic acid may be a replacement for it (Ullrich and Kuntz, 1984), but it is present in liverworts (Li *et al.*, 1994) and mosses. Curiously, ABA has also been located in the brain of some mammals (Le Page-Degivry *et al.*, 1986) although it may have originated from plants in the diets of animals.

Not only has ABA been detected in the various tissues of higher plants, it is also considered to be synthesized in virtually all cells containing amyloplasts or chloroplasts (see Taiz and Zeiger, 1991). Although a few reports indicated that ABA was synthesized (Cornish and Zeevart, 1986) and catabolized by guard cells (Grantz *et al.*, 1985) the current consensus of opinion is, however, that it is not (see Section 7.2.7) and therefore ABA must be synthesized elsewhere in a plant and translocated to guard cells to bring about a stomatal response.

The amount of ABA in tissues or a cell depends on rates of biosynthesis, catabolism and transport across membranes. Only the undissociated, neutral form of ABA (ABAH) is freely permeable across membranes, the ABA^- anion having almost zero conductance, and therefore ABA behaves ideally according to the anion trap mechanism for weak acids in the absence of specific transporters (see below) (Heilmann *et al.*, 1980; Slovik *et al.*, 1992). According to model analysis (Cowan *et al.*, 1982; Slovik *et al.*, 1992; Slovik and Hartung, 1992) the anion trap mechanism and resulting ABA redistribution plays a key role in stomatal regulation. Thus, the distribution of ABA within different compartments of cells depends on the pH, the more alkaline a compartment the more ABA being accumulated and trapped (Cowan *et al.*, 1982). Since pH values of compartments within a cell (e.g. apoplast, cytosol, vacuole and chloroplast stroma) are different ABA will accumulate to different extents in each compartment. Furthermore, the pH values of some compartments (apoplast, cytosol and stroma but not the vacuole) change under plant water stress, and the pH of at least some of the compartments will change under light/dark transitions, stimulating a redistribution of ABA (see Hartung and Slovik, 1991). Additionally, large pH changes occur in the vacuole and cell wall of guard cells during stomatal movements.

Within a mesophyll cell ABA accumulates mainly in the chloroplasts and apoplast which have the highest pH. However, under water stress conditions ABA synthesis is stimulated and, coupled with pH changes of the different compartments, redistribution of ABA occurs resulting in increased levels in the apoplast Abscisic acid is then considered to diffuse apoplastically, being swept along in the transpiration stream to the guard cell walls where it brings about the stomatal response (see Chapter 8 for the mode of action of ABA) Abscisic acid could also travel symplastically from the cytoplasm of mesophyll cells to the cytoplasm of cells neighbouring guard cells since there are plasmodesmatal connections between all cells in between and this may also be a major route for water movement Figure 7 4 summarizes the possible movements of ABA at the cell, tissue and whole plant levels

There is also evidence that there are a number of pools of ABA within a leaf presumably if ABA is not catabolized (this occurs in the cytoplasm) it may be stored at various sites within the cells or even accumulate in the epidermal layer where it may be released when the need arises (see Hartung *et al* , 1990) The latter situation would allow for a more rapid stomatal response since the distance the ABA would travel to the target site would be shorter

The mechanism by which ABA crosses membranes in leaf tissue is not certain In root tip cells and suspension cells, Rubery and Astle (1982) concluded that transport across the plasma membrane was facilitated by a highly specific ABA carrier In leaf cells results so far indicate that a saturatable carrier system may exists only in guard cells possibly at the plasma membrane and/or tonoplast (see Hartung and Slovik, 1991) There is also some uncertainty about the form of ABA which is transported but in the absence of a carrier mechanism only the uncharged form (ABAH) would be able to diffuse across the lipid part of the membrane The ABAH conductance decreases in the order, chloroplast envelope >> plasma membrane > tonoplast (within the same cell type) and probably epidermis (guard cells, epidermal cells) > mesophyll (chlorenchyma cells, phloem cells) (Hartung and Slovik, 1991)

7.2.4 Root to shoot communication

A plant survives as an integrated system and signals to a leaf come from all parts of the plant to produce a net response The roots are the obvious sensors of the soil environment and stressed roots produce substances which are translocated to the guard cells via the transpiration stream There is good evidence that ABA is a major hormonal factor which is transported from the roots (see, e g Davies and Zhang, 1991) although there may be other substances at least in some species (Munns and King, 1988) Zhang *et al* (1987) consider that it is the total amount of ABA, rather than the concentration, which is important in determining the extent of the stomatal response However, the rate of xylem flow from the roots to the shoots will be a major regulator of ABA levels in

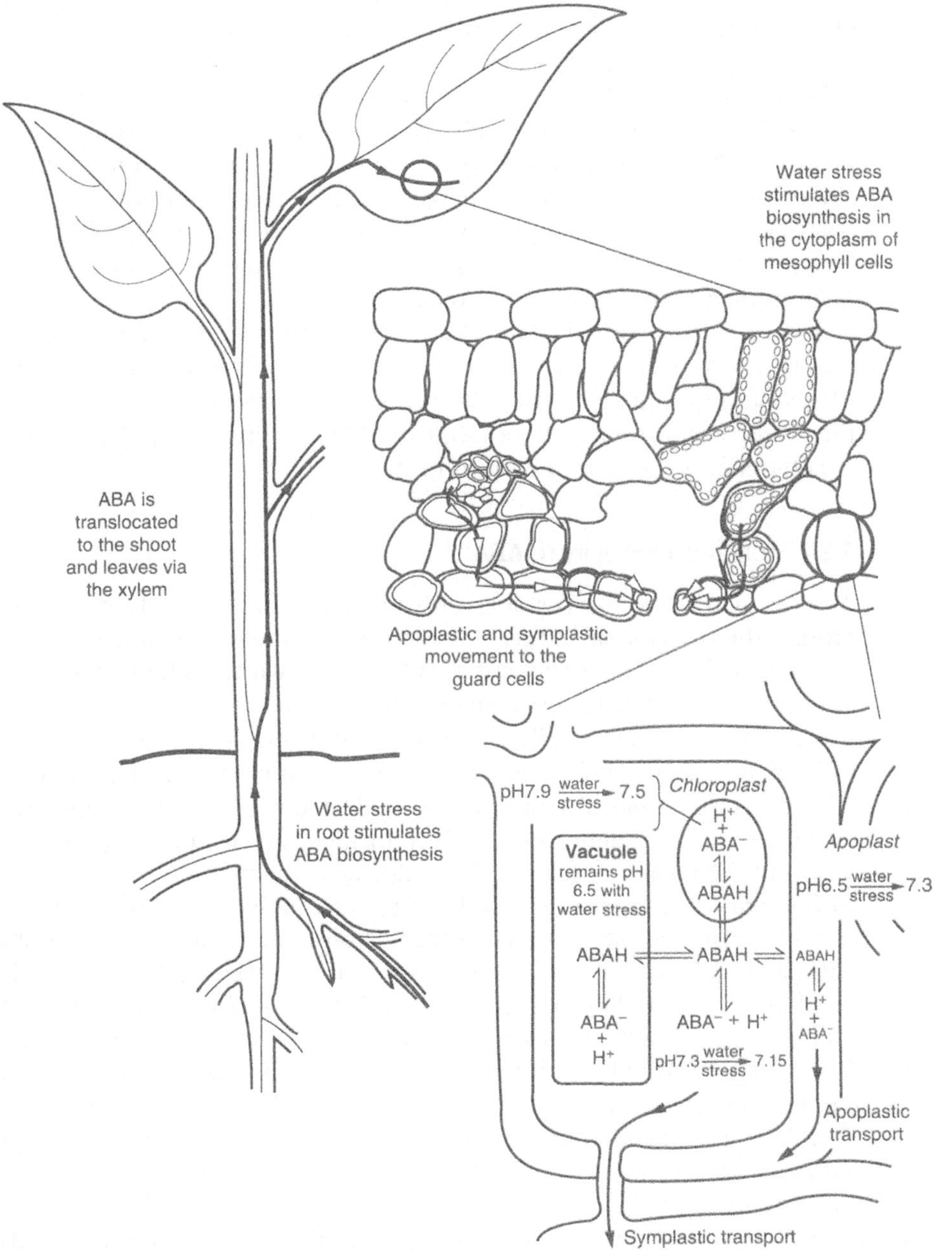

Figure 7.4 The possible pathways of ABA movement in a plant from the root or the leaf mesophyll to the guard cells An idealized mesophyll cell shows the compartmentation of ABA which is determined by pH (only the protonated form of ABA is membrane permeable) Water stress is considered to alter the pH values of the compartments as indicated Values from Hartung and Slovik (1991)

the shoot. The pH of the xylem sap will also modify the response of the stomata because, depending on the ionization state of ABA, transport into cells will be greater or smaller, lower pH values favouring greater uptake (e.g. Schurr *et al.*, 1992).

Possibly there is also a hydraulic signal from the roots which is transmitted to the leaves and which, in turn, stimulates stomatal closing processes. It is known, for example, that ABA can alter the water permeability of some tissues and cell types (e.g. Eamus and Tomos, 1983; Markhart, 1984). Thus, under conditions of low soil water availability the synthesis of ABA in the roots may alter their water conductivity ultimately affecting leaf water potentials.

Abscisic acid is much more abundant in the phloem than the xylem and is transported to the roots. Hence, it is conceivable that a signal, ABA, could be transmitted to the roots from stressed leaves to change the water conductivity of root cell membranes in an appropriate way which benefits the water relations of the plant.

7.2.5 Stomatal specificity to ABA

Abscisic acid has three possible sites of isomerization which are located around the two double bonds of the side chain and the asymmetric C1′ of the ring structure. The asymmetrical C1′ is known as a chiral centre and depending on whether an attached group protrudes forward or backwards, the molecule is called *S* or *R*, respectively. In plants ABA occurs only as the *S* enantiomer; the *R* enantiomer is unnatural. The asymmetric C1′ also confers optical activity to the molecule which results in its ability to rotate polarized light to the right (clockwise) or to the left (anticlockwise). In the former condition the molecule is called the (+) enantiomer, in the latter condition it is called the (–) enantiomer. The *S*-ABA synthesized by plants is the (+)-enantiomer. [Note that commercially available ABA is usually a 1:1 mixture of (+)- and (–)-ABA.] Additionally ABA exhibits geometrical isomerism as a result of the double bonds of the side chain. Depending on whether the various groups associated with carbons 2=3 and 4=5 are on opposite sides or on the same side the molecule can be denoted *cis* or *trans* for each double bond. Thus the naturally occurring ABA, as shown in Fig. 7.5 is (+)-*S*-ABA (with C2-*cis*, C4-*trans* configuration). Plants can interconvert the *cis* and *trans* isomers but not the (+) and (–) ones.

Although most investigators agree about the most active form of ABA and its related compounds which affect stomata, there are difficulties in establishing which structure is the most effective since applied compounds may be metabolized to more, or less, active forms within the tissues and diffusion to the site of action will depend on how polar a compound is, and the pH at which it is applied.

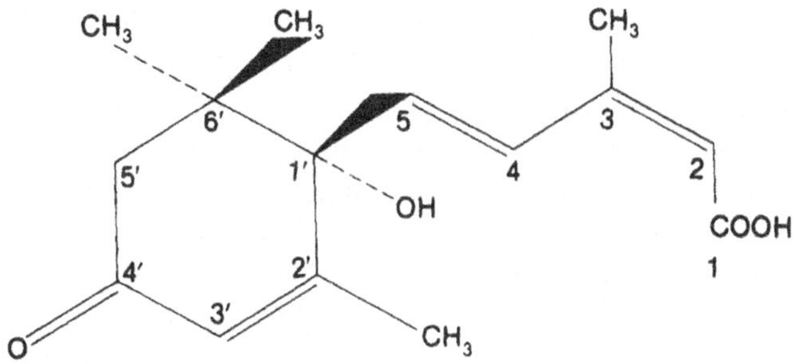

Figure 7.5 (+)-*S*-ABA (C2-*cis*, C4-*trans*). Note that C1′ is the only asymmetric carbon and confers optical activity to the molecule.

Stomata respond to the naturally occurring (+) enantiomer but very little to the unnatural (–) enantiomer. In some biological systems, however, both enantiomers are active. The C2-*cis*, C4-*trans* isomer is active while the C2-*trans*, C4-*trans* isomer is inactive. Because ABA is relatively unstable being deactivated due to photoisomerization and biological breakdown, it has some side effects, and it is expensive to isolate from plant tissue or synthesize, many analogues of ABA have been tested for their efficacy at reducing transpiration (e.g. Ogunkanmi *et al.*, 1974; Jung and Grossmann, 1985; Flores and Dorffling, 1990). The presence of an oxygen atom at the C1 position appears essential for high activity although the complete 2,4-pentadiene side chain or a ring carbonyl and hydroxyl group appear less essential for stomatal activity. However, Hite *et al.* (1994) found that if hydrazones were substituted for the 4′ carbonyl group then ABA was ineffective at inhibiting stomatal opening. According to Jung and Grossmann (1985) a cyclohexane unit and a six-carbon side chain appear necessary for biological activity. So far it appears that an ABA analogue which functions as an antitranspirant and which is commercially attractive has not been discovered.

7.2.6 Metabolism of ABA

Initially it was considered that ABA was synthesized in chloroplasts of leaf mesophyll cells but now it is generally accepted that it is both synthesized and degraded in the cytoplasm (e.g. Hartung *et al.*, 1980). However, there was much debate about whether it was also synthesized in the guard cells but currently the concensus is that ABA synthesis, degradation and turnover of conjugates does not occur, or is at insignificant levels, within guard cells (e.g. Behl and Hartung, 1986; Lahr and Raschke, 1988). There is also uncertainty about ABA metabolism in epidermal tissue. Singh *et al.* (1979) found rates of synthesis and metabolism of ABA in the epidermis to be similar to those in mesophyll cells, but Dorffling *et al.* (1980) concluded that isolated epidermal strips were unable to synthesize ABA.

Protoplasts in general are probably unable to synthesize ABA (Loveys and Robinson, 1987; Lahr and Raschke, 1988)

Although certain fungi such as *Cercospora rosicola* and *Botrytis cinerea* are able to synthesize ABA via ionylidene derivatives, there is no evidence for this pathway in higher plants. In higher plants the pathway of ABA synthesis remains to be elucidated but research has centred on two other pathways as indicated in Fig. 7.6 (see Zeevart and Creelman, 1988). In the direct pathway it is considered that ABA is formed from farnesyl pyrophosphate via unknown intermediate reactions which may involve xanthoxin. In the indirect pathway, which is currently favoured, it is considered that ABA is formed as a degradative product of some xanthophyll (a type of cartenoid), such as violaxanthin, via xanthoxin. Figure 7.6 also shows the derivation of other sesquiterpenoids, which affect stomata, such as farnesol.

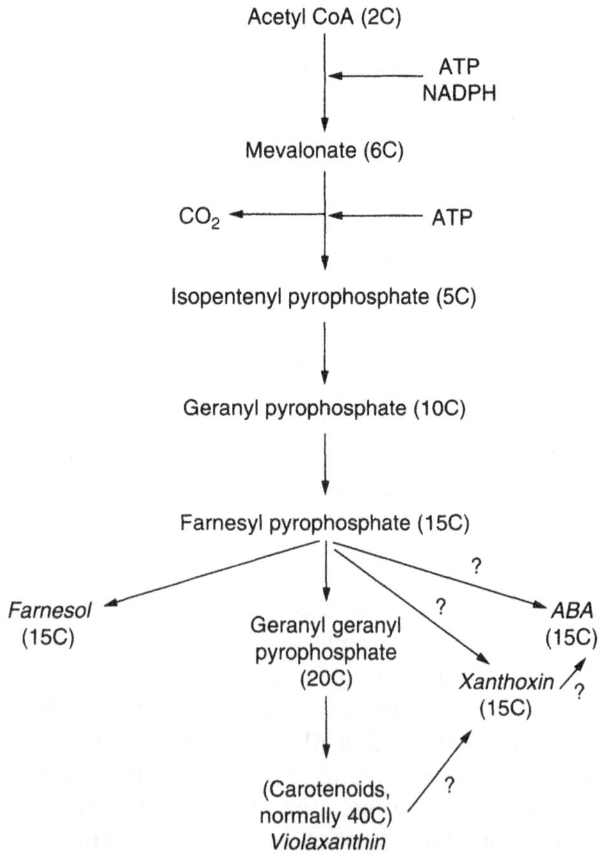

Figure 7.6 Possible pathways of ABA biosynthesis. The compounds in italics have all been found to affect stomatal behaviour. After Zeevart and Creelman (1988).

Although loss of cell turgor stimulates ABA synthesis it is not known how cell turgor regulates its synthesis There is also a possibility that ABA is synthesized by different pathways in turgid leaves and stressed leaves or the turnover of ABA is much more rapid in stressed leaves (Creelman *et al* , 1987)

Abscisic acid is degraded, via the unstable intermediate 6′hydroxyl-methyl-ABA, to phaseic acid and then to 4-dihydrophaseic acid Abscisic acid is also conjugated to form ABA-β-D-glucosyl ester (ABA-GE) In water-stressed leaves ABA and its catabolites, phaseic acid and dihydrophaseic acid accumulate ABA-GE, however, increases only slightly, its formation apparently being irreversible, and it is sequestered in the vacuole where it is unavailable as a source of stress-induced free ABA A detailed appraisal of ABA metabolism is reported in Zeevart and Creelman (1988)

7.2.7 After-effect of wilting

If a water-stress period is relieved, in some species there is a period of hours or even days before stomata open to their full potential (Fig 7 7) Stålfelt (1955) first described the phenomenon and assumed it to be a 'safety mechanism Although the cause of the effect is not certain it has been considered to be due to ABA slowly being removed from in and around the guard cells (via various means such as conjugation, export via phloem by degradation and redistribution), thereby allowing only slow stomatal recovery Dorffling *et al* (1977), for example, observed in a variety of mesophytic plants that the delay in stomatal opening was directly correlated to the leaf ABA content Curiously, however, in hygrophytes (*Tradescantia andersoniana*, *Menyanthes trifoliata* and *Mentha aquatica* were classified in this group) they found that stomatal opening was not delayed during recovery of leaf turgor and leaf ABA levels did not increase in response to water stress but remained constant (*Menyanthes*) or decreased (*Tradescantia* and *Mentha*)

Hartung and Slovik (1991) have also proposed that although there is a fast redistribution of ABA levels in the guard cell walls after cessation of stress, cytosolic ABA levels can remain high for many hours because of carrier mechanisms possibly present in the guard cell tonoplast This might explain why some species exhibit this after-effect while others do not

Some authorities have disagreed with the involvement of ABA in the after-effect since they found that leaf ABA levels and stomatal recovery did not correlate well (e g Beardsell and Cohen, 1975) For example, prestress ABA levels have been recorded in leaves before stomatal recovery and, in other cases, stomata have been found to open after application of ABA but before appreciable breakdown of ABA has

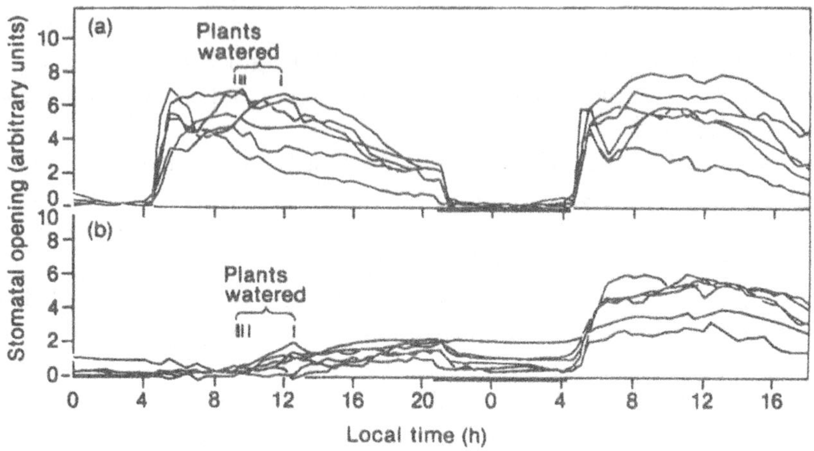

Figure 7.7 The after-effect of water stress on the stomatal behaviour in leaves of *Vicia faba* (a) Unstressed control plants (b) Plants were not watered for 3–6 days until the fourth expanded leaf from the apex had wilted for 2 days At the points indicated by the short bars individual plants were watered In each treatment the stomatal behaviour in six different leaves was recorded with a viscous flow porometer Black bars on the abscissa indicate the dark periods After Allaway and Mansfield (1970)

occurred. Furthermore, Hartung and Slovik (1991) indicate that total ABA content per unit leaf area declines only within 1–2 weeks of the stress period. Nevertheless, the lack of correlation between bulk leaf ABA levels and stomatal aperture may be due to one or more of the many reasons cited above.

There is a suggestion that ABA may be involved in short-term after-effects (over the first few hours) but not longer-term effects since Allaway and Mansfield (1970) found stomata in *V. faba* did not fully recover from water-stress until nearly 24 h after the stress was relieved, while Harris and Outlaw (1991) found that ABA levels in guard cells of *V. faba* dropped to pre-stressed values within 6 h after relief of a water-stress period.

Presumably, if ABA is a factor in the after-effect of wilting, stress effects could modulate ABA activity in some way such as altering gene expression for ABA or even changing the sensitivity of receptor sites to ABA (e.g. Peng and Weyers, 1994).

7.2.8 CO_2/ABA interactions

Raschke (1975) suggested that there is an interdependence, at least in some species, between CO_2 and ABA in bringing about stomatal closure since he found that stomata of *Xanthium strumarium* did not close in the light in response to increased CO_2 concentrations unless the leaves had been fed ABA. Conversely, he found, stomata would not close if leaves were fed ABA unless CO_2 was present in the air. This interaction of stomatal responses to CO_2 and ABA was also found in *Gossypium hirsutum* and *C. communis* though it was much weaker than in *X. strumarium*. Mansfield (1976), however, using *X. strumarium*, found no statistical interaction of ABA and CO_2 in their stomatal responses. It is difficult to reconcile the two reports, although differences in plant pretreatments might explain the differing conclusions; the study of Raschke, for example, was carried out at a much higher photon flux densities than that of Mansfield, and at high light levels stomata in some species may become less responsive to stimuli. Raschke *et al.* (1976) also found that stomata of *X. strumarium* were sensitized to CO_2 by a cold pretreatment of plants, i.e. the stomata closed upon exposure to high CO_2 concentrations. They related the sensitization of the stomata to increased leaf ABA levels caused by the cold stress.

Undoubtedly ABA levels around guard cells and leaf CO_2 levels will play some interrelated role in controlling stomatal aperture since, individually, these both bring about closure as their concentrations increase. Moreover, under normal conditions ABA is normally always present at low levels in leaf tissue and this level is elevated under stress conditions. It is pertinent to note, however, that stomata in epidermal strips of *C. communis* respond to ABA in the absence of CO_2 by closing or by not opening and respond to changes of CO_2 concentration in the absence of exogenously supplied ABA.

7.2.9 Wilty mutants

Wilty tomato mutants (*flacca*, *sitiens* and *notabilis*), a pepper mutant (*Capsicum scabrous diminutive*) and a potato mutant have been found in which stomata remain open even under water stress resulting in rapid wilting of the plants. There is also an increased root resistance to water flow in the mutants compared with normal plants which contributes to the wilting of the plants. The reason for the stomata remaining open under plant water stress is that a single gene mutation results in decreased ABA synthesis. The concentration of ABA in flacca leaves is only 20–40% of that found in wild-type leaves (Neill and Horgan, 1985). Tal and Imber (1970) reported the presence of high levels of auxin and kinetin-like activity in the flacca mutant which they considered might stimulate stomatal opening. Additionally Tal *et al.* (1979) found higher rates of ethylene evolution in flacca which they linked, in part, to an increased auxin content. However, Neill *et al.* (1986) could find no difference in indole-3-acetic acid (IAA) levels or of ethylene evolution (and of levels of ACC, the immediate precursor to ethylene biosynthesis) between Flacca and the wild-type Exogenously supplied ABA brings about stomatal closure, thereby preventing wilting (Imber and Tal, 1970), although there is a report that short-term applications of ABA are ineffective at closing the stomata

A number of ABA-insensitive mutants have been identified (in terms of general growth and development) though little has been done to investigate their stomatal responses However, the transpiration rate of a barley mutant, cool , was unaffected by application of 1 mM ABA and the evaporative cooling of leaves was always higher than in the wild-type (Raskin and Ladyman, 1988) Additionally, they found that the stomata were insensitive to exogenously applied ABA, darkness and drought stress although the stomatal density, ABA levels and metabolism of plants and the guard cell morphology were unaltered from the wild-type

*7.3 Other sesquiterpenoids (xanthoxin, all-*trans *farnesol, vomifoliol, phaseic acid)*

A number of sesquiterpenoids which are related to ABA have been studied for their effects on stomatal behaviour, i e xanthoxin, all-*trans* farnesol, vomifoliol and phaseic acid (see Fig 7 8a–d for their structures)

Xanthoxin is about 50% as active as ABA at causing stomatal closure when applied to leaves via the transpiration stream, but is inactive when applied to epidermal strips of *C communis* or *V faba* (Raschke *et al* , 1975) Water stress does not increase xanthoxin levels within leaves Xanthoxin may be a precursor of ABA (see Fig 7 6) and the above observations would support this contention presumably, when xanthoxin is applied to leaves via the transpiration stream, it is first converted to ABA in the mesophyll before it affects the stomata There is no effect of xanthoxin on stomata in epidermal strips possibly because the epidermis cannot convert xanthoxin to ABA

All-*trans* farnesol levels increase in water-stressed leaves of *Sorghum sudanense* (Wellburn *et al* , 1974) Farnesol also causes stomatal closure, which can be reversed, when applied to Sorghum leaves, it also causes closure and prevents opening of stomata in epidermal strips of *C communis* However, Farnesol causes structural changes to cells damaging membranes, including chloroplast envelopes, at non-physiological concentrations (10 M and above)

Phaseic acid was considered at one time to be involved in the control of stomata since water stress and certain photoperiodic changes led to increased phaseic acid levels in leaves which were correlated with stomatal closure (e g Loveys and Kriedemann, 1974) The effectiveness of

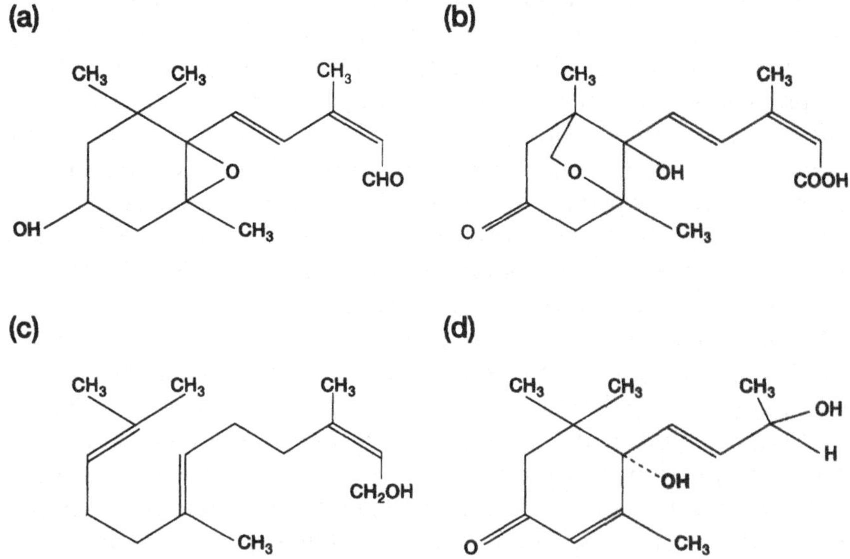

Figure 7.8 The structure of some sesquiterpenoids related to ABA. (a) Xanthoxin; (b) phaseic acid; (c) farnesol; (d) vomifoliol.

phaseic acid at closing stomata in epidermal strips and intact leaves varies from species to species. For example, it closes stomata in epidermal strips of *C. communis* but not of *V. faba*. Some investigators found that the acid inhibited photosynthesis (e.g. Kriedemann *et al.*, 1975) thereby elevating leaf CO_2 levels which would close stomata, but other investigators found no inhibition of photosynthesis. In general, the effects of phaseic acid on stomatal behaviour are not as marked as the effects of ABA, the former compound being a degradative product of the latter though retaining some activity.

(+)-Vomifoliol is also reported to be as active as ABA in causing stomatal closure in epidermal strips of *Eichhornia crassipes* (Stuart and Coke, 1975). (+)-Vomifoliol, the natural isomer is related structurally to ABA but the side-chain has two fewer carbon atoms than that of ABA (Fig. 7.8d).

7.4 Other hormones

7.4.1 Cytokinins

Although the effects of cytokinins on stomata are still not clear there is now a general view that they stimulate stomatal opening within monocot and dicot species (see Incoll and Jewer, 1987). However, in the past it was considered that kinetins increased rates of transpiration in grass species but not other monocotyledons or dicotyledons It had been suggested that the increased transpiration observed in grasses was due to kinetin delaying senescence of the excised leaves relative to rates of senescence in water controls In turn, the delayed senescence would extend the CO_2 fixing ability of the leaves resulting in lower leaf CO_2 levels As a consequence of the lowered CO_2 levels, stomata will open more widely in the kinetin treated leaves Pallas and Box (1970), however, suggested that the cytokinins acted on stomata indirectly by affecting water potentials elsewhere in the plant

A direct effect of cytokinin on guard cells was indicated when it was found that a range of synthetic and natural cytokinins enhanced stomatal opening in epidermal strips of the grass, *Anthephora pubescens* (Incoll and Whitelam 1977) In general, however, reports on the effects of kinetins on stomata are contradictory In intact leaves of a variety of monocot and dicot species cytokinins increase transpiration rates (e g Livne and Vaadia, 1965, Meidner, 1967, Biddington and Thomas, 1978, Horton, 1991) In epidermal strips, results are conflicting In some studies kinetin had no significant effect on stomatal behaviour in epidermal strips of *C communis*, *V faba* (Tucker and Mansfield, 1971, Ogunkanmi *et al* , 1973, Wardle and Short, 1981), *Pisum sativum* (Jewer *et al* , 1982) and *Tridax procumbens* (Das and Raghavendra, 1976) but in other studies kinetin and synthetic cytokinins induced opening in epidermal strips of *A pubescens* and *Kalanchoe daigremontiana* (Jewer and Incoll, 1980, 1981) and in one study kinetin closed stomata in epidermis of *C communis* (Blackman and Davies, 1983)

There may also be some interaction between cytokinins, ABA and CO_2 concentrations (e g Das and Raghavendra, 1976, Blackman and Davies, 1984)

7.4.2 Auxins

As with the cytokinins, reports on the action of auxins are rather contradictory and difficult to evaluate Ferri and Rachid (1949) and Mansfield (1967), for example, found that synthetic auxins induced stomatal closure in leaves from a variety of monocot and dicot species while Livne and Vaadia (1965) found no effect of IAA on stomata in intact leaves of wheat Zelitch (1961) also found that when tobacco leaf discs were floated on solutions of the synthetic auxins, 2,4-dichlorophenoxyacetic acid (2,4-D) or naphthalene-1-acetic acid (1-NAA), stomata closed Contrasting results are reported for the effects of IAA on stomata in epidermal strips of *C communis*, Tucker and Mansfield (1971) and Ogunkanmi *et al* (1973) found no effects on stomatal opening while Pemadasa (1982) generally observed a slight stimulation of opening in abaxial stomata and a large stimulation of opening in adaxial ones. Levitt *et al.* (1987) also observed that IAA (0.01–1.0 mM) promoted stomatal opening in epidermal strips of *V. faba* although the KCl concentration used in the incubation medium (50 mM) is much higher than normally used (1–10mM) for this species. Cox *et al.* (1985) concluded that although IAA stimulates stomatal opening, its synthetic analogues are inhibitory and that some of the earlier misleading results with epidermal strips were due to experimental conditions that obscured IAA activity (Davies and Mansfield, 1987).

Nevertheless, Lohse and Hedrich (1992) found that IAA and its synthetic analogues stimulated opening in epidermal strips of *V. faba* incubated in buffered 50 mM KCl; in darkness IAA, 1-NAA and 2,4-D increased apertures reaching maximum values after 4–5 h. Both NAA and 2,4-D gave bell-shaped dose–response curves with a maximum response at a concentration of 5 μM.

Indole-3-acetic acid has also been found to reduce the stomatal closing effects of ABA and also there appears to be an interaction between IAA, ABA and CO_2 concentration in the regulation of stomatal responses (e.g. Snaith and Mansfield, 1982; Eamus, 1986).

7.4.3 Gibberellins

Gibberellins have been reported to increase transpiration rates in excised barley leaves (Livne and Vaadia, 1965) but most investigators find no effects in intact leaves (e.g. Horton, 1991). In epidermal strips gibberellins are also reported to have no effects on stomatal opening (e.g. Tucker and Mansfield, 1971; Horton, 1971).

7.4.4 Ethylene

Carbon dioxide is a competitive inhibitor of ethylene action in many plant responses; the synthesis of ethylene has also been observed to increase in plants under water stress. Stomatal responses are also sensitive to changing CO_2 concentrations and water stress, and, hence, investigators have attempted to connect stomatal functioning and ethylene responses. Many reports indicate that ethylene tends to close stomata. Kays and Pallas (1980) found that in certain species, notably sunflower and peanut, exposure of the plants to 1.0 $\mu l\ l^{-1}$ ethylene for 2 h resulted in a substantial reduction of photosynthesis and in a later report (Pallas and Kays, 1982) concluded that at least in peanut leaves it was due to decreased stomatal conductance. Reports by Vitagliano and Hoad (1978), Bradford and Hsiao (1982) and Madhavan *et al.* (1983) also indicate increased stomatal resistance, in a variety of species, in response to ethylene. Whether ethylene directly brought about stomatal closure thereby reducing photosynthesis, or whether the ethylene effect on stomata was an indirect one due to the gas inhibiting photosynthesis which, in turn, would increase leaf CO_2 levels to close stomata, is not clear in many of the reports

However, Pallaghy and Raschke (1972) could find no effect of ethylene within the concentration range of 1 to 100 000 ppm on stomatal behaviour in corn or pea Nor have ethylene effects been observed on stomatal behaviour in leaves of pinto bean or *Sedum* (Madhavan *et al* , 1983) or in excised barley leaves (at 100 $\mu l\ l^{-1}$ ethylene) (Horton, 1991) Contrary to all these reports, Levitt *et al* (1987) conclude from their studies with ethephon, which is converted to ethylene in cells, that ethylene opens stomata in epidermal strips of *V faba*

7.5 Fatty acids

Short-chain saturated fatty acids (of carbon chain length C6 to C11) accumulate in leaves of barley and french bean under water stress (Willmer *et al* , 1978) The C9 (nonanoic), Cl0 (decanoic) and C11 (undecanoic) acids also prevent stomatal opening and cause closure of stomata in epidermal strips of *C communis* The C9 and C11 fatty acids are most effective but even they are not as effective as ABA at inhibiting stomatal opening and causing closure (see Fig 7 9) There is a relatively narrow concentration range (between 1 and 100 μM) over which they assert their action on stomata and stomatal closure is slow in their presence Like farnesol, high concentrations (in the order of l00 μM) cause cell and particularly membrane damage (Willmer *et al* , 1978) Indeed, the C9 and C10 fatty acids are now used commercially as active ingredients of a broad spectrum contact herbicide (Bestman, 1993)

The longer chain unsaturated fatty acids, linolenic (C18, three double bonds) and linoleic (C18 two double bonds), also prevent stomatal opening and bring about some closure (Fig 7 9)

Further investigation is needed into the effects of fatty acids to ascertain their significance in stomatal control

7.6 Phytotoxins

A number of phytotoxins produced by plant pathogens affect stomata either by promoting or inhibiting opening Little is known about most of the compounds, but fusicoccin has been investigated in some depth

Fusicoccin is a phytotoxin (the structure is depicted in Fig 7 10) produced by submerged cultures of the fungal pathogen, *Fusicoccum*

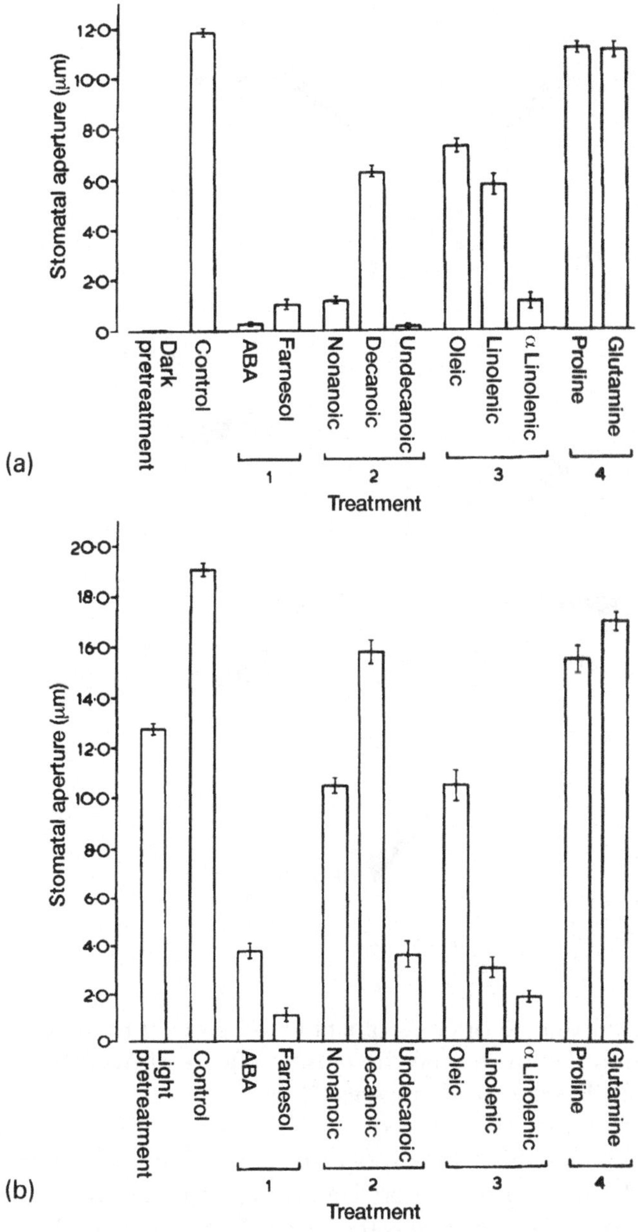

Figure 7.9 The effects of some sesquiterpenoids, fatty acids or amino acids on (a) stomatal opening and (b) open stomata in epidermal strips of *Commelina communis*. All compounds were at a concentration of 0.1 mM suspended or dissolved in a buffered (20 mM MES, pH 6.0) medium containing 75 mM KCl. CO_2-free air was bubbled through the medium and the epidermal strips illuminated (195 μmol m^{-2} s^{-1}) at a constant 27°C. Stomatal apertures were measured after 2 h incubation period. 1, Sesquiterpenoids; 2, saturated short-chain fatty acids; 3, unsaturated longer chain fatty acids; 4, amino acids. From Plumbe and Willmer (1986b).

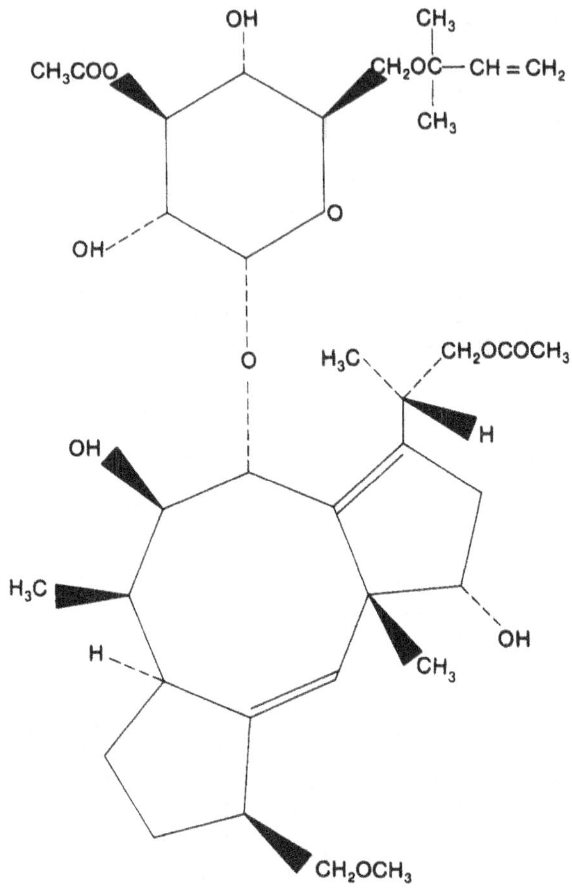

Figure 7.10 The structure of fusicoccin I

amygdali. The substance opens stomata when applied to leaves or epidermal strips and can overcome the closing effects of ABA (Squire and Mansfield, 1972). Fusicoccin also swells GCPs.

Fusicoccin increases H^+ extrusion and K^+ uptake in many types of cells and tissues and also hyperpolarizes membrane potentials (e.g. Marre, 1979). Fusicoccin may, therefore, bring about stomatal opening by influencing ion transport across the plasmalemma of guard cells (see Chapter 8 for further details).

Victorin (a pentapeptide joined to a secondary amine) is a phytotoxin produced by the fungus *Helminthosporium victoriae* which causes stomatal closure. Little more is known about its effects on stomata.

A toxin produced by *Helminthosporium maydis*, race T, causing southern corn leaf blight, also causes solute leakage from roots of susceptible plants and induces rapid stomatal closure possibly by inhibiting K^+ uptake (Arntzen *et al*, 1973)

Syringomycin (a 11 224 Da peptide) is a phytotoxin produced by the bacterium *Pseudomonas syringae* pv *syringae* which also reduces stomatal apertures in epidermal strips of *V faba* and in detached leaves of *Xanthium strumarium* (Mott and Takemoto, 1989) These authors found that syringomycin was as effective as ABA at closing stomata and neither compound affected the photosynthetic capacity of the mesophyll

7.7 Phytoalexins

Phytoalexins are antimicrobial substances produced by certain host plants normally in response to pathogen attack Many of these compounds have been found to affect stomatal behaviour For example, wyerone acid, a phytoalexin produced by *V faba*, inhibits stomatal opening in *V faba* and *C communis* epidermal strips although not as effectively as ABA Other phytoalexins, such as pisatin and phaseollin, have similar effects on stomata in epidermal strips of *V faba* and *C communis* (Plumbe and Willmer, 1986a) Pisatin, a phytoalexin produced by peas, has been observed to inhibit stomatal opening in pea leaves due to the inability of the guard cells to accumulate solutes (Ayres, 1980)

Many phytoalexins damage membranes Their effects on stomata may also be due, mainly, to changing membrane permeability of guard cells

Investigations so far indicate that phytoalexins are not produced by plants grown under water stress in the absence of pathogens unless, in some cases, tissue death occurs However, certain compounds produced under water stress such as short chain fatty acids and farnesol, inhibit spore germination of *Cladosporium herbarum*, ABA has no effect on the germination of spores of *C herbarum* or the growth of their germ tubes (Plumbe and Willmer, 1985)

7.8 Other compounds

A variety of other naturally-occurring compounds of a diverse nature have been found to affect stomata Usually studies have not been detailed enough to determine the importance or significance of the compound in the control of stomatal behaviour Also, usually the effects on stomata have not been great and relatively high concentrations are needed to elicit a stomatal response

A number of phenolic compounds affect stomata. Ferulic acid, for example, was found to increase transpiration rates in barley leaves, but it had little effect on stomatal opening in *C. communis* epidermal strips (Plumbe and Willmer, 1986b). Michniewicz and Rozej (1974) also found that ferulic acid stimulated transpiration in cuttings of *P. vulgaris*. Sinapic acid and caffeic acid inhibit stomatal opening in epidermal strips of *C. communis* and reduce transpiration rates in excised leaves.

In general amino acids have little effect on stomatal responses (Plumbe and Willmer, 1985b, see Fig. 7.9) although proline has been reported to inhibit stomatal opening and cause closure of open stomata in epidermal strips of some species including *C. benghalensis* (Raghavendra and Reddy, 1987). Proline also accumulates in large amounts in leaves of some species exposed to water stress. This correlation between water stress, proline accumulation and inhibition of stomatal opening deserves further attention although it has been established that ABA accumulation by cells is not a prerequisite to proline accumulation in wilted leaves (Stewart and Voetberg, 1987).

Jasmonates occur naturally in plant tissues and have been considered as phytohormones. Raghavendra and Reddy (1987) found that methyl jasmonate was a potent inhibitor of stomatal opening and caused closure in epidermal strips of *C. benghalensis*. At the relatively high concentration of 1 mM, methyl jasmonate was observed to reduce transpiration rates in excised barley leaves though the effect was considered to be due to its toxicity (Horton, 1991).

Brassinosteroids, a relatively new group of biologically active natural products also affect stomata movements. Thus, 22*S*,23*S*-homobrassinolide and 24-epibrassinolide were relatively effective at inhibiting stomatal opening in epidermal strips of *C. communis* (Dahse *et al.*, 1991).

7.9 Artificial control of stomata

The artificial control of stomatal behaviour could be of use to man in a number of ways. Antitranspirants could be used over the long term or short term to close stomata, enabling a crop to withstand a period of drought. Such treatment would be particularly useful with perennial crops where survival may be more important than good yield for a particular season (since closing the stomata would prevent photosynthesis, and reduce yield as well as preventing water loss). Antitranspirants may also be sprayed on newly transplanted trees and other plants during dry periods to cut down water loss from the leaves and increase their chances of survival. It is also enticing to think that it may one day be possible to extend the opening period of stomata, possibly in late afternoon, under ideal photosynthesizing conditions when water was readily available. The daily period of photosynthesis would be extended and crop yields might, therefore, increase. For this sort of stomatal control protranspirants would be needed rather than antitranspirants and they would possibly have to overcome stomatal closing; in late afternoon due to endogenous rhythms. Fusicoccin, for example, may be used as a protranspirant and it has already been used at an experimental level to speed up the drying of newly cut grass to be used for hay. Compounds may also be applied to plants to close stomata and prevent entry of toxic pollutants which may damage tissues within the leaf.

Such compounds which will open or close stomata must be non-toxic to man and the plant, cheap to produce and must act specifically on guard cells without directly affecting photosynthesis or other plant processes in the underlying mesophyll. A danger of using antitranspirants may be one of overloading the cooling capacity of leaves since evaporation and escape of water vapour from the leaves will be reduced.

To date there is no substance (synthetic or natural) which has a higher permeability to CO_2 than to water vapour (Raschke, 1979). Also, because the water vapour gradient is steeper than the CO_2 gradient across the stomata more water molecules will be lost from the leaf than CO_2 will enter the leaf. Thus, no antitranspirant can reduce water loss without also reducing CO_2 uptake by the leaf. However, a reduction in stomatal conductance affects transpiration and assimilation to different degrees and the ratio between the rates of the two processes (transpiration/assimilation, the transpiration ratio) usually declines as stomata close.

There are basically two types of antitranspirants. One type may be classed as film antitranspirants. These are often silicone-based oils or emulsions of latex or waxy substances which are sprayed on crops to give a semi-waterproof coating. The other type of antitranspirants are sprays of either naturally occurring or synthetic compounds which directly affect stomata.

Many compounds have been considered as possible antitranspirants but most have deleterious side effects. Phenylmercuric acetate has been used in the past with some success, even in field trials, but, of course, it is a serious environmental pollutant. Some of the possible natural occurring antitranspirants unfortunately are likely to have side effects. ABA, however, is needed at very low concentrations to close stomata and, when sprayed on plants, appears to have minimal affects on other plant processes. The disadvantages of using ABA as an antitranspirant may be its short-lived effect and possibly its expense. Recently it has been found that acetylsalicylic acid (aspirin) prevents opening in epidermal strips. This observation may lend some support to the belief that the addition of aspirin to water bathing flowers preserves the flowers longer than in the absence of aspirin Although this needs further verification in whole plants, presumably the aspirin will be taken up in the transpiration stream and close the stomata in the foliage (and possibly petals and other floral parts), thereby preventing excessive water loss and wilting If this is the case, aspirin deserves further attention as a possible antitranspirant

Although there are a few impressive reports on the effects of film antitranspirants, results generally have not been successful Film antitranspirants are difficult to use with precision and may even cause leaf necrosis The search for ideal antitranspirants continues New, naturally occurring and synthetic compounds which affect stomata are still being discovered Some of these may prove to be useful for artificially controlling stomatal behaviour in the field

References

Allaway, WG and Mansfield, TA (1970) Experiments and observations on the after-effect of wilting on stomata of *Rumex sanguineus* *Can J Bot* , **48**, 513–521

Arntzen, CJ , Haugh, MF and Bobick, S (1973) Induction of stomatal closure by *Helminthosporium maydis* pathotoxin *Plant Physiol* , **51**, 569–574

Ayres, PG (1980) Stomatal behaviour in mildewed pea leaves solute potentials of the epidermis and effects of pisatin *Physiol Plant Pathol* , **17**, 157–165

Beardsell, MF and Cohen, D (1975) Relationship between leaf water status, abscisic acid levels, and stomatal resistance in maize and *Sorghum* *Plant Physiol* , **77**, 207–212

Behl, R and Hartung, W (1986) Movement and compartmentation of abscisic acid in guard cells of *Valerianella locusta*, effects of osmotic stress, external H concentration and fusicoccin *Planta*, **168**, 360–368

Bestman, HD (1993) The mode of action of the herbicide Topgun is linked to an effect on the integrity of plant cell membranes by the fatty acids nonanoic and decanoic acid *Plant Physiol Suppl* , **102**, 612

Biddington, NL and Thomas, TH (1978) Influence of different cytokinins on the transpiration and senescence of oat leaves *Physiol Plant* , **42**, 369–374

Blackman, PG and Davies, WJ (1983) The effects of cytokinins and ABA on stomatal behaviour of maize and *Commelina* *J Exp Bot* , **34**, 1619–1626

Blackman, PG and Davies, WJ (1984) Modification of the CO_2 responses of maize stomata by abscisic acid and by naturally-occurring and synthetic cytokinins *J Exp Bot*, **35**, 174–179

Bradford, KJ and Hsiao, TC (1982) Stomatal behaviour and water relations of waterlogged tomato plants *Plant Physiol*, **78**, 1508–1513

Brinckmann, E, Hartung, W and Wartinger, M (1990) Abscisic acid levels of individual leaf cells *Physiol Plant*, **80**, 51–54

Bunce, JA (1987) Species-specific responses to water stress of gas exchange parameters mimicked by applied abscisic acid *Can J Bot*, **6**, 103–106

Chang, T-Y, Senn, A and Pilet, P-E (1983) Effect of abscisic acid on maize root protoplasts *Z Pflanz*, **110**, 127–133

Cornic, G and Miginiac, E (1983) Nonstomatal inhibition of net CO_2 uptake by (+) abscisic acid in *Pharbitus nil* *Plant Physiol*, **73**, 529–533

Cornish, K and Zeevart, JAD (1986) Abscisic acid accumulation by in situ and isolated guard cells of *Pisum sativum* L and *Vicia faba* L in relation to water stress *Plant Physiol*, **81**, 1017–1021

Cowan, IR, Raven, JA, Hartung, W and Farquhar, GD (1982) A possible role for abscisic acid in coupling stomatal conductance and photosynthetic carbon metabolism in leaves *Aust J Plant Physiol*, **9**, 489–498

Cox, RC, Snaith, PJ and Mansfield, TA (1985) The significance of natural and synthetic auxins in the control of stomatal movements *Acta Hort*, **171**, 247–254

Creelman, RA, Gage, DA, Stults, JT and Zeevart, JAD (1987) Abscisic acid biosynthesis in leaves and roots of *Xanthium strumarium* *Plant Physiol*, **85**, 726–732

Dahse, I, Petzold, U, Willmer, CM and Grimm, E (1991) Brassinosteroid-induced changes of plasmalemma energization and transport and of assimilate uptake by plant tissues, in *Brassinosteroids Chemistry, Bioactivity, and Applications*, (eds HG Cutler, T Yokota and G Adam), American Chemical Society, Washington, DC, pp 176–188

Dale, J and Campbell, WH (1981) Response of tomato plants to stressful temperature increase in abscisic acid concentrations *Plant Physiol* **67**, 26–29

Das, VSR and Raghavendra, AS (1976) Reversal of abscisic acid induced stomatal closure by benzyladenine *New Phytol*, **76**, 449–452

Davies, FS and Lakso, AN (1978) Water relations in apple seedlings changes in water potential components, abscisic acid levels and stomatal conductance under irrigated and non-irrigated conditions *J Am Soc Hort Sci*, **103**, 310–313

Davies, WJ and Mansfield TA (1987) Auxins and stomata, in *Stomatal Function*, (eds E Zeiger, G D Farquhar and I R Cowan), Stanford University Press, Stanford, CA, pp 293–309

Davies, WJ and Zhang, J (1991) Root signals and the regulation of growth and development of plants in drying soil *Ann Rev Plant Physiol Plant Mol Biol*, **42**, 55–76

Davies, WJ and Jones, H G (1994) *Abscisic Acid Physiology and Biochemistry*, BIOS Scientific Publishers, Oxford

Dorffling, K, Streich, J, Kruse, W and Muxfeldt, B (1977) Abscisic acid and the after-effect of water stress on stomatal opening potential *Z Pflanz*, **81**, 43–56

Dorffling, K, Tietz, D, Streich, J and Ludewig, M (1980) Studies in the role of abscisic acid in stomatal movements, in *Plant Growth Substances 1979 (Proc 10th Int Conf Plant Growth Substances)*, (ed F Skoog), Springer, New York

Dorffling, K, Peterson, W, Sprecher, E *et al* (1984) Abscisic acid in phytopathogenic fungi of the genera *Botrytis*, *Ceratocystis*, *Fusarium*, and *Rhizoctonia* *Z Naturforsch*, **39**, 683–684

Downton, WJ S, Loveys, B R and Grant, WJ R (1988) Non-uniform stomatal closure induced by water stress causes putative non-stomatal inhibition of photosynthesis *New Phytol*, **110**, 503–509

Eamus, D (1986) Further evidence in support of an interactive model in stomata control *J Exp Bot*, **37**, 657–665

Eamus, D and Tomos, A D (1983) The influence of abscisic acid on the water relations of leaf epidermal cells of *Rhoeo discolor* *Plant Sci Lett*, **31**, 253–259

Ferri, M G and Rachid, M (1949) Further information on the stomatal behaviour as influenced by treatment with hormone-like substances *Anais Acad Brasil Cienc*, **21**, 155–166

Fitzsimons, PJ and Weyers, J D B (1987) Responses of *Commelina communis* guard cell protoplasts to abscisic acid *J Exp Bot*, **38**, 992–1001

Flores, A and Dorffling, K (1990) A comparative study of the effects of abscisic acid and new terpenoid abscisic acid analogues on plant physiological processes *J Plant Growth Reg*, **9**, 133–139

Grantz, D A, Ho, T-H, Uknes, S *et al* (1985) Metabolism of abscisic acid in guard cells of *Vicia faba* L and *Commelina communis* L *Plant Physiol*, **78**, 51–56

Harris, J M and Outlaw, WH, Jr (1991) Rapid adjustment of guard-cell abscisic acid levels to current leaf-water status *Plant Physiol*, **95**, 171–173

Harris M J, Outlaw, WH, Jr, Mertens R and Weiler, E W (1988) Water-stress-induced changes in the abscisic acid content of guard cells and other cells of *Vicia faba* L leaves as determined by enzyme-amplified immunoassay *Proc Natl Acad Sci USA*, **85**, 2584–2588

Hartung, W, Slovik, S and Baier, M (1990) pH change and redistribution of abscisic acid within the leaf under stress *Br Plant Growth Reg Group Monogr 21*

Hartung, W and Slovik, S (1991) Physicochemical properties of plant growth regulators and plant tissues determine their distribution and redistribution stomatal regulation by abscisic acid in leaves *New Phytol*, **119**, 361–382

Hartung, W, Gimmler, H, Heilmann, B and Kaiser, G (1980) The site of ABA metabolism in mesophyll cells of *Spinach oleracea Plant Sci Lett*, **18**, 359–364

Heilmann, B, Hartung, W and Gimmler, H (1980) The distribution of abscisic acid between chloroplasts and cytoplasm of leaf cells and the permeability of the chloroplast envelope for abscisic acid *Z Pflanz*, **97**, 67–68

Henson, I E (1981) Changes in abscisic acid content during stomatal closure in pearl millet (*Pennisetum americanum* (L) Leeke) *Plant Sci Lett*, **21**, 121–127

Henson, I E, Mahalakshmi, V, Alagarswamy, G and Bidinger, F R (1984) The effect of flowering on stomatal responses to water stress in pearl millet (*Pennisetum americanium* [L] Leeke) *J Exp Bot*, **35**, 219–226

Henson, I E and Mahalakshmi, V (1985) Evidence for panicle control of stomatal behaviour in water-stressed plants of pearl millet *Field Crops Res*, **11**, 281–290

Hite, D R C, Outlaw, W H, Jr and Seavy, M A (1994) Substitution of hydrazones for the 4′ carbonyl on abscisic acid rendered it ineffective in a rapid stomatal-opening-inhibition bioassay *Physiol Plant*, **92**, 79–96

Horton, R F (1971) Stomatal opening the role of abscisic acid *Can J Bot*, **49**, 583–585

Horton, R F (1991) Methyl jasmonate and transpiration in barley *Plant Physiol*, **96**, 1376–1378

Imber, D and Tal, M (1970) Phenotypic reversion of *flacca*, a wilty mutant of tomato, by abscisic acid *Science*, **169**, 592–593

Incoll, L O and Whitelam, G C (1977) The effect of kinetin on stomata of the grass *Anthephora pubescens* Nees *Planta*, **137**, 243–245

Incoll, L O and Jewer, P C (1987) Cytokinins and stomata, in *Stomatal Function*, (eds E Zeiger, G D Farquhar and I R Cowan), Stanford University Press, Stanford, CA, pp 281–292

Itai, C and Roth-Bejerano, N (1986) The effect of abscisic acid on epidermal cells protoplast swelling and ATPase activity *Physiol Plant*, **66**, 664–668

Iverson, T-H, Johnsson, A and Baggerud, C (1983) Effect of abscisic acid on leaf cell protoplasts *Z Pflanz* , **110**, 293–300

Jackson, M B and Hall, K C (1987) Early stomatal closure in water-logged pea plants is mediated by abscisic acid in the absence of foliar water deficits *Plant Cell Environ* , **10**, 121–130

Jewer, PC and Incoll, L D (1980) Promotion of stomatal opening in the grass *Anthephora pubescens* Nees by a range of natural and synthetic cytokinins *Planta*, **150**, 218–221

Jewer, PC and Incoll, L D (1981) Promotion of stomatal opening in detached epidermis of *Kalanchoe daigremontiana* Hamet et Perr *Planta*, **153**, 317–318

Jewer, PC , Incoll, L D and Shaw, J (1982) Stomatal responses of *Argenteum*–a mutant of *Pisum sativum* L with readily detachable epidermis *Planta*, **155**, 146–153

Jung, J and Grossmann, K (1985) Effectiveness of new terpenoid derivatives, abscisic acid and its methyl ester on transpiration and leaf senescence of barley *J Plant Physiol* , **121**, 361–367

Kays, S J and Pallas, J E (1980) Inhibition of photosynthesis by ethylene *Nature*, **285**, 51–52

Keck, R W and Boyer, J S (1974) Chloroplast response to low leaf water potentials III Differing inhibition of electron transport and phosphorylation *Plant Physiol* , **53**, 474–479

Kriedemann, PE , Loveys, B R and Downton, WJ S (1975) Internal control of stomatal physiology and photosynthesis II Photosynthetic response to phaseic acid *Aust J Plant Physiol* , **2**, 553–567

Krizek, D T and Milthorpe, F L (1966) Effect of photoperiodic induction on the transpiration rate and stomatal activity of debudded *Xanthium pennsylvanicum* *Plant Physiol Suppl* , **41**, xxviii

Lahr, W and Raschke, K (1988) Abscisic acid contents and concentrations in protoplasts from guard cells and mesophyll cells of *Vicia faba* L *Planta*, **173**, 528–531

Lancaster, J E Mann, J D and Porter, N G (1977) Ineffectiveness of abscisic acid in stomatal closure of yellow lupin, *Lupinus luteus* var Weiko III *J Exp Bot* , **28**, 184–191

Le Page-Degivry, M Th , Bidard, J M , Rouvier, E *et al* (1986) Presence of ABA, a phytohormone, in the mammalian brain *Proc Natl Acad Sci USA*, **83**, 1155–1158

Levitt, K L , Stein, D B and Rubinstein, B (1987) Promotion of stomatal opening by indoleacetic acid and ethrel in epidermal strips of *Vicia faba* L *Plant Physiol* , **85**, 318–321

Li, X Y, Wurtele, E S and LaMotte, C E (1994) Abscisic acid is present in liver worts *Phytochemistry*, **37**, 625–628

Little, C H A and Eidt, D C (1968) Effect of abscisic acid on bud break and transpiration in woody species *Nature*, **220**, 498–9

Livne, A and Vaadia, Y (1965) Stimulation of transpiration rate in barley

leaves by kinetin and gibberellic acid. *Physiol. Plant.*, **18**, 658–664.
Lohse, G. and Hedrich, R. (1992) Characterization of the plasma-membrane H^+-ATPase from *Vicia faba* guard cells. Modulation by extracellular factors and seasonal changes. *Planta*, **88**, 206–214.
Loveys, B.R. and Kriedemann, P.E. (1974) Internal control of stomatal physiology and photosynthesis. I. Stomatal regulation and associated changes of endogenous levels of abscisic and phaseic acid. *Aust. J. Plant Physiol.*, **1**, 407–415.
Loveys, B.R. and Robinson, S.P. (1987) Abscisic acid synthesis and metabolism in barley leaf protoplasts. *Plant Sci. Lett.*, **49**, 23–30.
Madhavan, S., Chrominski, A. and Smith, B.N. (1983) Effect of ethylene on stomatal opening in tomato and carnation leaves. *Plant Cell Physiol.*, **24**, 569–572.
Mansfield, T.A. (1967) Stomatal behaviour following treatment with auxin-like substances and phenylmercuric acetate. *New Phytol.*, **66**, 325–330.
Mansfield, T.A. (1976) Delay in the response of stomata to abscisic acid in CO_2-free air. *J. Exp. Bot.*, **27**, 559–64.
Marré, E. (1979) Fusicoccin: a tool in plant physiology. *Ann. Rev. Plant Physiol.*, **30**, 273–288.
Marten, I., Lohse, G. and Hedrich, R. (1991) Plant growth hormones control voltage-dependent activity of anion channels in plasma membrane of guard cells. *Nature*, **353**, 758–762.
Markhart, A.H. (1984) Amelioration of chilling-induced water stress by abscisic acid-induced changes in root hydraulic conductance. *Plant Physiol.*, **74**, 81–83.
Mawson, B.T., Colman, B. and Cummins, W.R. (1981) Abscisic acid and photosynthesis in isolated leaf mesophyll cells. *Plant Physiol.*, **67**, 233–236.
Meidner, H. (1967) The effect of kinetin on stomatal opening and the rate of intake of CO_2 in mature primary leaves of barley. *J. Exp. Bot.*, **18**, 556–561.
Meidner, H. (1970) Effects of photoperiod induction and debudding in *Xanthium pennsylvanicum* and of partial defoliation of *Phaseolus vulgaris* on rates of net photosynthesis and stomatal conductances. *J. Exp. Bot.*, **21**, 164–169.
Meidner, H. (1983) Our understanding of plant water relations. *J. Exp. Bot.*, **149**, 1606–1618.
Michniewicz, M. and Rosej, B. (1974) Stimulation of transpiration rate in bean plants by ferulic acid. *Naturwissenschaften*, **1**, 1–2.
Mittelheuser, C.J. and van Steveninck, R.F.M. (1969) Stomatal closure and inhibition of transpiration induced by (*RS*)-abscisic acid. *Nature*, **221**, 281–2.
Mott, K.A. and Takemoto, J.Y. (1989) Syringomycin, a bacterial phyto-

toxin, closes stomata *Plant Physiol*, **90**, 1435–1439

Munns, R and King, R W (1988) Abscisic acid is not the only stomatal inhibitor in the transpiration stream of wheat plants *Plant Physiol*, **88**, 703–708

Neill, S J and Horgan, R (1985) Abscisic acid production and water relations in wilty tomato mutants subjected to water deficiency *J Exp Bot*, **36**, 1222–1231

Neill, S J, McGaw, B A and Horgan, R (1986) Ethylene and 1-Aminocyclopropane-1-Carboxylic acid production in flacca, a wilty mutant of tomato, subjected to water deficiency and pre-treatment with abscisic acid *J Exp Bot*, **37**, 535–541

Ogunkanmi, A B, Tucker D J and Mansfield, T A (1973) An improved bioassay for abscisic acid and other antitranspirants *New Phytol*, **72**, 277–282

Ogunkanmi, A B, Wellburn, A R and Mansfield, T A (1974) Detection and preliminary identification of endogenous antitranspirants in water stressed *Sorghum* plants *Planta*, **117**, 293–302

Osborne, B A, O'Connell C, Campbell, G C and Weyers, J D B (1993) Stomata of *Gunnera tinctoria* do not respond to light, CO_2 or ABA *J Exp Bot Suppl*, **44**, 2

Pallaghy, C K and Raschke, K (1972) No stomatal response to ethylene *Plant Physiol*, **49**, 275

Pallas, J E and Box, I E (1970) Explanation for the stomatal response of excised leaves to kinetin *Nature*, **227**, 87–88

Pallas J E and Kays, J S (1982) Inhibition of photosynthesis by ethylene – a stomatal effect *Plant Physiol* **70**, 598–601

Paterson, N W, Weyers, J D B and A'Brook, R (1988) The effect of pH on stomatal sensitivity to abscisic acid *Plant Cell Environ*, **11**, 83–89

Pemadasa, M A (1982) Differential abaxial and adaxial stomatal responses to indole-3-acetic acid in *Commelina communis* L *New Phytol*, **90**, 209–219

Peng, Z -Y and Weyers, J D B (1994) Stomatal sensitivity to abscisic acid following water deficit stress *J Exp Bot*, **45**, 835–845

Pierce, M L and Raschke, K (1980) Correlation between loss of turgor and accumulation of abscisic acid in detached leaves *Planta*, **148**, 174–182

Plumbe, A M and Willmer, C M (1985) Phytoalexins, water-stress and stomata I Do phytoalexins accumulate in leaves under water-stress? *New Phytol*, **101**, 269–274

Plumbe, A M and Willmer, C M (1986a) Phytoalexins, water-stress and stomata II The effects of phytoalexins on stomatal responses in epidermal responses in epidermal strips and on guard cell protoplasts *New Phytol*, **102**, 375–384

Plumbe, A M and Willmer, C M (1986b) Phytoalexins, water-stress and stomata III The effects of some phenolics, fatty acids and some other compounds on stomatal responses *New Phytol*, **103**, 17–22

Raghavendra, A S and Reddy, K B (1987) Action of proline on stomata differs from that of abscisic acid, G-substances, or methyl jasmonate *Plant Physiol*, **83**, 732–734

Raschke, K (1975) Simultaneous requirement of carbon dioxide and abscisic acid for stomatal closing in *Xanthium strumarium* L *Planta*, **125**, 243–59

Raschke, K (1979) Movements of stomata, in *Encyclopedia of Plant Physiology, Vol 7, Physiology of Movements*, (eds W Hampt and M E Feinlieb), Springer, Berlin

Raschke, K (1982) Involvement of abscisic acid in the regulation of gas exchange evidence and inconsistencies, in *Plant Growth Substances 1982*, (ed PF Waring), Academic Press, London, pp 581–590

Raschke, K and Hedrich, R (1985) Simultaneous and independent effects of abscisic acid on stomata and the photosynthetic apparatus in whole leaves *Planta*, **163**, 105–118

Raschke, K, Firn, R D and Pierce, M (1975) Stomatal closure in response to xanthoxin and abscisic acid *Planta*, **125**, 149–60

Raschke, K, Pierce, M and Popiela, C C (1976) Abscisic acid content and stomatal sensitivity to CO_2 in leaves of *Xanthium strumarium* L after pretreatments in warm and cold growth chambers *Plant Physiol*, **57**, 115–121

Raskin, I and Ladyman, J A R (1988) Isolation and characterization of a barley mutant with abscisic-acid-insensitive stomata *Planta*, **173**, 73–78

Rubery, PH and Astle, M (1982) The mechanism of transmembrane abscisic acid transport and some of its implications, in *Plant Growth Substances 1982*, (ed PF Waring), Academic Press, London, pp 353–362

Schurr, U, Gollan, T and Schultze, E-D (1992) Stomatal response to drying soil in relation to changes in the xylem sap composition of *Helianthus annuus* II *Plant Cell Environ*, **15**, 561–567

Seemann, J R and Sharkey, T D (1987) The effect of abscisic acid and other inhibitors on photosynthetic capacity and the biochemistry of CO_2 assimilation *Plant Physiol*, **84**, 696–700

Sharkey, T D and Raschke, K (1980) Effects of phaseic acid and dihydrophaseic acid on stomata and the photosynthetic apparatus *Plant Physiol*, **65**, 291–297

Singh, B N, Galson, E, Dashek, W and Walton, D C (1979) Abscisic acid levels and metabolism in the leaf epidermal tissue of *Tulipa gesneriana* L and *Commelina communis* L *Planta*, **146**, 135–138

Slovik, S, Baier, M and Hartung, W (1992) Compartmentation and redistribution of abscisic acid in intact leaves I Mathematical formulation *Planta*, **187**, 14–25

Slovik, S and Hartung, W (1992) Compartmental distribution and redistribution of abscisic acid in intact leaves II Model analysis *Planta*, **187**, 26–36

Snaith, PJ and Mansfield, TA (1982) Control of the CO_2 responses of stomata by indol-3ylacetic acid and abscisic acid *J Exp Bot*, **33**, 360–365

Squire, G R, and Mansfield, TA (1972) Studies of the mechanism of action of fusicoccin, the fungal toxin that induces wilting, and its interaction with abscisic acid *Planta*, **105**, 71–78

Stålfelt, M G (1955) The stomata as a hydrophotic regulator of the water deficit of the plant *Physiol Plant*, **8**, 572–593

Stewart, C R and Voetberg, G (1987) Abscisic acid accumulation is not required for proline accumulation in wilted leaves *Plant Physiol*, **83**, 747–749

Stuart, K L and Coke, L B (1975) The effect of vomifoliol on stomatal aperture *Planta*, **122**, 307–310

Taiz, L and Zeiger, E (1991) *Plant Physiology*, Benjamin Cummings, San Francisco, CA

Tal, M and Imber, D (1970) Abnormal stomatal behaviour and hormonal imbalance in *flacca*, a wilty mutant of tomato II Auxin and abscisic acid-like activity *Plant Physiol*, **46**, 373–76

Tal, M, Imber, D, Erez, A and Epstein, E (1979) Abnormal stomatal behaviour and hormonal imbalance in flacca, a wilty mutant of tomato V Effect of abscisic acid on indoleacetic acid metabolism and ethylene evolution *Plant Physiol*, **63**, 1044–1048

Terashima, I, Wong, S-C, Osmond, C B and Farquhar, G D (1988) Characterisation of non-uniform photosynthesis induced by abscisic acid in leaves having different mesophyll anatomies *Plant Cell Physiol*, **29**, 385–394

Tietz, A and Kasprik, W (1986) Identification of abscisic acid in a green alga *Biochem Physiol Pflanz*, **68**, 115

Trewavas, A (1987) Sensitivity and sensory adaptation in growth substance responses, in *Hormone Action in Plant Development A Critical Appraisal*, (eds G V Hoad, J R Lenton, M B Jackson and R K Atkin), Butterworth, London, pp 19–38

Tucker, D J and Mansfield, TA (1971) A simple bioassay for detecting 'antitranspirant' activity of naturally occurring compounds such as abscisic acid *Planta*, **98**, 157–163

Ullrich, WR and Kuntz, G (1984) Effect of abscisic acid on nitrate uptake, respiration and photosynthesis in green algae *Plant Sci Lett*, **37**, 9–14

Vitagliano, C and Hoad, G V (1978) Leaf stomatal resistance, ethylene evolution and ABA levels as influenced by (2–chloroethyl) phosphonic acid. *Sci. Hortic.*, **8**, 101–106.

Walton, D.C. (1980) Biochemistry and physiology of abscisic acid. *Ann. Rev. Plant Physiol.*, **31**, 453–89.

Wardle, K. and Short, K.C. (1981) Responses of stomata in epidermal strips of *Vicia faba* to carbon dioxide and growth hormones when incubated on potassium chloride and potassium iminodiacetate. *J. Exp. Bot.*, **32**, 303–309.

Wellburn, A.R., Ogunkanmi, A.B., Fenton, R. and Mansfield, T.A. (1974) All-*trans*-farnesol: a naturally occurring antitranspirant? *Planta*, **120**, 255–263.

Weyers, J.D.B., Paterson, N.W. and A'Brook, R. (1987) Towards a quantitative definition of plant hormone sensitivity. *Plant Cell Environ.*, **10**, 1–10.

Willmer, C.M., Hamilton, K. and Pallas, J.E., Jr (1981) The use of an oxygen electrode for monitoring photosynthesis and respiration of leaves, in *Photosynthesis and Crop Productivity*, (ed. H. Metzner), Wissenschaftliche Verlagogesillschaft, Stuttgart, pp. 178–181.

Willmer, C.M., Don, R. and Parker, W. (1978) Levels of short-chain fatty acids and of abscisic acid in water-stressed and non-stressed leaves and their effects on stomata in epidermal strips and intact leaves. *Planta*, **139**, 281–287.

Wright, S.T.C. (1969) An increase in the 'inhibitor-B' content of detached wheat leaves following a period of wilting. *Planta*, **86**, 10–20.

Wright, S.T.C. (1977) The relationship between leaf water potential and the levels of abscisic acid and ethylene in excised wheat leaves. *Planta*, **134**, 183–189.

Wright, S.T.C. and Hiron, R.W.P. (1969) (+)-Abscisic acid, the growth inhibitor induced in wheat leaves by a period of wilting. *Nature*, **224**, 719–720.

Zeevart, J.A.D. and Creelman, R.A. (1988) Metabolism and physiology of abscisic acid. *Ann. Rev. Plant Physiol. Plant Mol. Biol.*, **39**, 439–473.

Zelitch, I. (1961) Biochemical control of stomatal opening in leaves. *Proc. Natl. Acad. Sci. USA*, **47**, 1423–1433.

Zhang, J., Schurr, U. and Davies, W.J. (1987) Control of stomatal behaviour by abscisic acid which apparently originates in the roots. *J. Exp. Bot.*, **38**, 1174–1181.

8 Ionic relations of stomatal movement and signal transduction in guard cells

8.1 Introduction

A variety of different hypotheses have been put forward over the last century and a half to explain how stomata open and close (see Chapter 1) All have proved inadequate to a greater or lesser degree Since the late 1960s, however, major advances in our understanding of how stomata function have been made with the application of diverse and new technologies, such as electrophysiology and ion imaging It is now clear that ion accumulation forms a major part of the osmotic increases observed during stomatal opening and recent experiments have started to link control of ion movement with transduction of physiological stimuli The conceptual framework for signal transduction relies heavily on models derived from animal cell systems, although sufficient data is beginning to accumulate to establish interactions in the signalling network that are unique to guard cells In this chapter the mechanisms leading to reversible ion accumulation and release by guard cells are considered, followed by a discussion of how such ion fluxes are regulated by internal and external signals

8.2 The involvement of ions in stomatal movements

The increases in guard cell osmotic pressure observed during opening (see Chapter 4) are a result of uptake of inorganic or organic solutes from the apoplast and/or synthesis of osmotically active molecules within the cell In this section the primary role of ion accumulation in stomatal movements and the contribution ions make to the overall osmotic changes are discussed

8.2.1 The participation of K^+ in stomatal opening and closing

In 1905, Macallum localized K^+ in guard cells of tulip using a histochemical test (Box 8 1), though a relationship between stomatal aperture and K^+ levels was not noted until nearly half a century later Imamura (1943) and Yamashita (1952) demonstrated K^+ accumulation in guard cells of a variety of species when stomata opened However it was not until Fujino (1967) published his work in English that scientists outside Japan became fully aware of the involvement of K^+ in stomatal function Fischer (1968a) and his colleagues independently confirmed the findings of the Japanese workers using epidermis of *Vicia faba* and concluded that the amount of K^+ and an accompanying anion were sufficient to account for the increased osmotic pressures of guard cells which occur when stomata open They also proposed that the uptake mechanism was specific for K^+ and that it was light activated As potassium is normally the most abundant inorganic ion in leaves, it is not surprising that K^+ is almost universally accumulated by guard cells regardless of the evolutionary level of the species or the location of the stomata on a plant (Willmer and Pallas, 1973, Dayanadan and Kaufman, 1975) For example, K^+ accumulates in guard cells during stomatal opening in leaves of *Equisetum* (horsetails), ferns, monocots and dicots (including CAM plants exhibiting night opening) and herbaceous and woody plants (including *Ginko biloba*) Additionally, when stomata open, K^+ also accumulates in guard cells located in awns, sepals and stems Contrary to the findings of Nelson and Mayo (1977) and Outlaw *et al* (1982), K^+ has been observed to increase in guard cells of *Paphiopedilum* (lady slipper orchid) species (Willmer *et al* , 1983) and a variegated cultivar of *Pelargonium zonale* (Jamieson and Willmer, 1984), both of which do not possess guard cell chloroplasts, but have functional stomata So far the only exception reported is the halophyte, *Cakile maritima*, which accumulates Na^+ in its guard cells (Eschel *et al* , 1974) It would be interesting, therefore, to determine whether other halophytes utilize Na^+ in a similar manner

Box 8.1. Histochemical test for potassium in leaf epidermis with stomata open or closed

Introduction

The location of K^+ in epidermis of leaves with open or closed stomata can be determined histochemically using an adaptation of a method developed by Macallum (1905) In the test sodium cobaltinitrite forms the triple salt, sodium potassium cobaltinitrite, in the presence of K^+ which crystallizes out at ice-cold' temperatures For this reason all staining procedures which require 'ice-cold' procedures should be carried out in a tray of crushed ice

$$Na_3Co(NO_2)_6 + 2K^+ \rightarrow NaK_2Co(NO_2)_6 + 2Na^+$$

sodium cobaltinitrite sodium potassium cobaltinitrite (yellow precipitate)

The yellow crystals which form are not easy to observe and so, after a suitable washing period, the triple salt is reacted with ammonium sulphide to form a black precipitate of cobalt sulphide

Procedure

1 To obtain closed stomata place plants in the dark for about 2 h, to open stomata very wide float leaves on water and illuminate for about 2 h Stomata from the leaf surface in contact with water will be wide open Epidermis with open or closed stomata is rinsed for 1–2 min in ice-cold distilled water to rid the tissue of K^+ from broken cells (either epidermal or adhering mesophyll cells) Calcium sulphate or calcium chloride (20 mM) has also been used as a washing medium with good results The Ca^{2+} exchanges for K^+ situated in the cell walls and cell debris The calcium salt is then rinsed out of the tissue by immersing the tissue in ice cold distilled water for 2 min The tissue should be agitated at all stages of the procedure so that air bubbles do not insulate the cells from contact with the incubation medium
2 The tissue is transferred to ice-cold , freshly-made sodium cobaltinitrite solution (the solution lasts for at least 2 weeks when kept in the refrigerator) Sodium cobaltinitrite solution made from commercially available solid does not always give good results and sodium cobaltinitrite solution prepared as follows is recommended add 75 ml of 13% acetic acid to 20 g cobalt nitrate and 35 g sodium nitrite Poisonous fumes of nitrogen dioxide are evolved and so the manipulation should be carried out in a fume cupboard, preferably with a stream of air bubbling through the solution to speed up evolution of the fumes For most types of epidermal tissue a 10 min exposure to the sodium cobaltinitrite is sufficient to give good results However, with some tissues particularly heavily cuticularized epidermis, longer periods of incubation up to 20 min may be necessary
3 The tissue is now washed in ice-cold distilled water until no more yellow stain flows from the tissue This usually entails washing the tissue for about 2 min in three changes of ice-cold distilled water
4 The tissue is next immersed in freshly-made 5% yellow ammonium sulphide solution at room temperature for about 2 min This step must be carried out in a fume cupboard
5 Finally, the tissue is briefly washed in water to get rid of surplus ammonium sulphide and mounted on a slide in a drop of water

Permanent mounts can be made but there does appear to be some slight loss of intensity of staining with time The procedures employed will close open stomata The pattern of localization obtained with the histochemical technique compares favourably with that obtained with the electron microprobe and it is considered that the histochemical test gives a reasonably accurate picture of the true K^+ location. Freeze-dried tissue does not give good results, unlike that apparently observed in animal tissue. It is also considered that the test is specific for K^+. Some results obtained using the procedures described above are shown in Fig. 8.B1 and are also presented in Willmer and Pallas (1973).

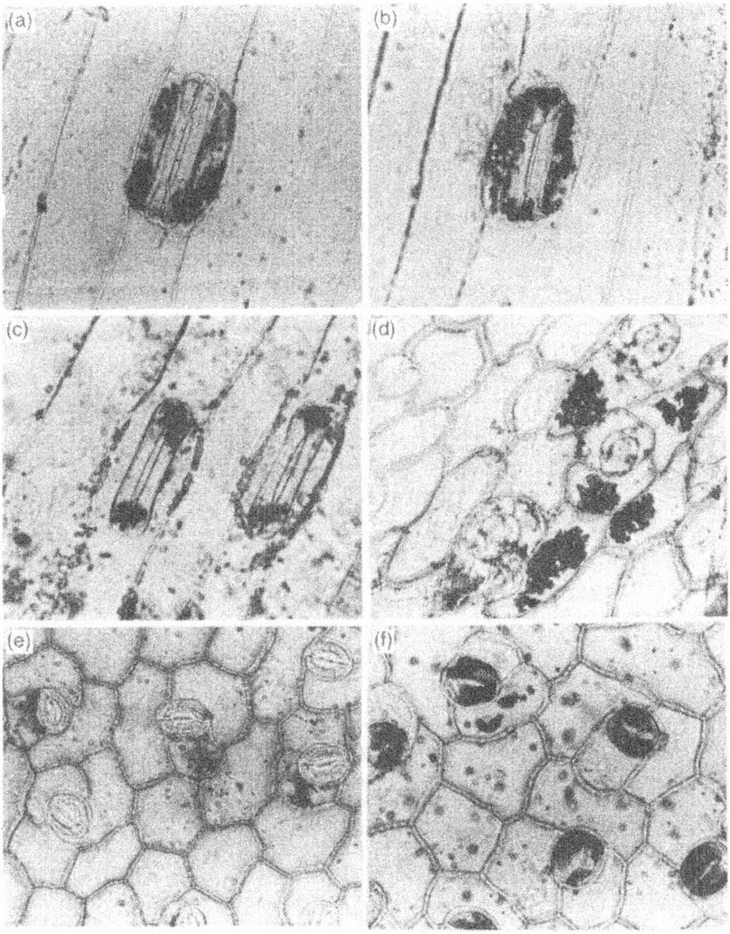

Figure 8.B1 Histochemical tests for K^+ in the epidermis of various species with open or closed stomata The test closes open stomata Blackened areas indicate detection of K^+ (a–c) Lower epidermis of *Avena sativa* (oat) showing the movement of K^+ from the subsidiary cells when stomata are closed to the guard cells when stomata are open, (d) K^+ storage in the epidermal cells of *Bryophyllum tubiflorum*, (e and f) epidermis from cotyledons of *Arachis hypogea* (peanut) showing K^+ in guard

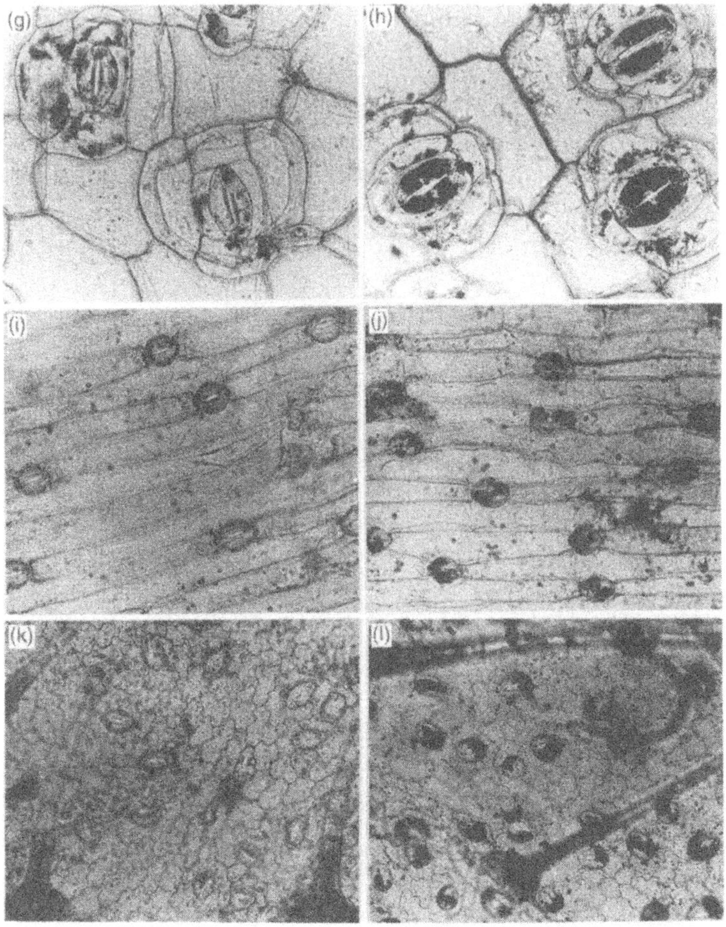

cells of open stomata (f) but not when they are closed (e), (g and h) lower epidermis of *Commelina communis* showing some K^+ in subsidiary cells when stomata are closed (g) and large accumulation of K^+ in guard cells of open stomata (h), (i and j) epidermis from onion showing accumulation of K^+ in guard cells of open stomata (j) but not in guard cells of closed stomata (i), (k) and (l) epidermis from the lower epidermis of *Pelargonium zonale* showing accumulation of K^+ in guard cells of open stomata (l) but not in guard cells of closed stomata (k) Also note that K^+ is present in the multicellular trichomes regardless of whether the stomata were closed (by a dark treatment) or opened (by a light treatment)

References

Macallum, A.B. (1905) On the distribution of potassium in animal and vegetable cells. *J. Physiol.*, **32**, 95–118.

Willmer, C.M. and Pallas, J.E. Jr (1973) A survey of stomatal movements and associated potassium fluxes in the plant kingdom. *Can. J. Bot.*, **51**, 37–42.

The potassium content of guard cells increases markedly with opening. Although there are substantial differences in the amount of K^+ accumulated between different species for the same aperture, values for an individual species are fairly consistent (see reviews by Raschke, 1975, 1979; Allaway and Milthorpe, 1976; Outlaw, 1983; Zeiger, 1983; MacRobbie, 1981, 1987, 1988). For example, typical levels of cell K^+ are 0.5 pmol per guard cell in closed stomata of *V. faba*, increasing to around 2–3 pmol per guard cell in wide open stomata, with average estimates of around 0.2 pmol μm^{-1} increase in aperture. These values translate into concentrations of about 200 and 500 mM K^+ for guard cells of closed and open stomata, respectively, although there is considerable variation in the guard cell volumes used for this calculation by the different groups of investigators (see Chapter 4). The presence of the neighbouring cells is also important as K^+ levels in guard cells, where all cells of the epidermal layer except the guard cells have been killed by a low pH pretreatment, are about 40% lower for an equivalent aperture range, than in guard cells with intact surrounding cells (MacRobbie, 1980). This observation was considered to be due to the guard cells not having to overcome the back-pressure of the neighbouring cells (see Fig. 8.1).

In Table 8.1 values for K^+ levels in guard cells are presented for a range of species and using a range of techniques for measurement of K^+. Concentrations have been calculated using recent estimates of cell volumes and taking into account the contribution of the non-osmotic volume (*NOV*, see Chapter 4) to facilitate comparisons between species.

The relationship between K^+ and aperture is linear for guard cells from a number of species including *Vicia* (e.g. Fischer, 1971), *Phaseolus* (e.g. Turner, 1973) and *Nicotiana* (e.g. Sawhney and Zelitch, 1969). However, in guard cells of *Commelina*, the relationship is non-linear and increases with increasing aperture (MacRobbie and Lettau, 1980a,b) (Fig. 8.1).

Zea mays is the only species with graminaceous-type guard cells where quantitative measurements of K^+ have been made (Raschke and Fellows, 1971; Lasceve *et al.*, 1987), and much lower levels of K^+ accumulate per micron aperture occur in this species (about 0.04 pmol μm^{-1} increase in aperture) compared to that in either *Vicia* or *Commelina* over the same aperture range. Schnable and Raschke (1980) also found that guard cells of wide open stomata of *Allium cepa* contained considerably lower amounts of K^+ than those of *Vicia* or *Commelina*, and only showed a change of 0.02–0.1 pmol μm^{-1} increase in aperture. In contrast, Losch (1985) found very high levels of both Na^+ and K^+ in guard cells of closed stomata of *Valerianella locusta*, but only K^+ increased during opening, at rates comparable to other species (0.23 pmol μm^{-1} increase in aperture).

Accumulation of high concentrations of salts in the cytoplasm are

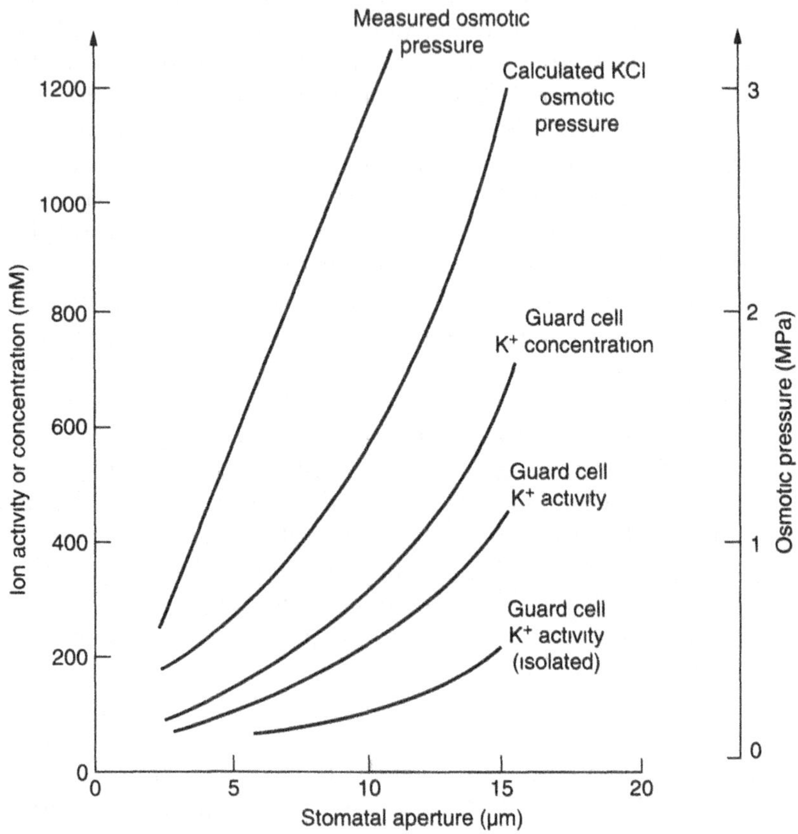

Figure 8.1 Changes in potassium levels and osmotic pressure with stomatal aperture in guard cells of *Commelina communis*. K^+ activities were measured using K^+-selective microelectrodes in guard cells with neighbouring cells alive (intact) or with only the guard cells alive (isolated). The electrode tip was probably located in the vacuole and measurements reflect vacuolar K^+ levels. Activities were converted to K^+ concentrations using a modified form of the Debye–Huckle equation (see Nobel, 1991). The osmotic contribution of these K^+ concentrations was estimated using the Van't Hoff relationship assuming all the K^+ was balanced by Cl^-. The measured osmotic pressure was determined using rapid plasmolysis techniques. Data redrawn from MacRobbie (1980) and MacRobbie and Lettau (1980a,b), with permission.

expected to be deleterious to cellular metabolism, suggesting that the majority of additional ions accumulated will be transferred to the vacuole and the cytoplasm will either shrink or accumulate a non-ionic osmoticum in parallel. There are no direct measurements of K^+ in the guard cell cytoplasm, but a value of 100–160 mM has been estimated from the potassium reversal potential in intact guard cells using electrophysiological techniques (e.g. Clint and Blatt, 1989). Unexpectedly, tracer flux measurements in *V. faba* gave estimated cytosolic K^+ concentrations that rose from about 50 to 350 mM during opening (Brindley, 1990a). Further studies are warranted in this area, given the importance of cytoplasmic K^+ in establishing the membrane potential and driving

Table 8.1 Levels of K^+ in guard cells of open and closed stomata from a variety of species

Species	Experimental material and measurement technique	Potassium salts added[a]	Amount of K^+ in guard cell[b] (pmol GC^{-1}) open	Amount of K^+ in guard cell[b] (pmol GC^{-1}) closed or relatively closed	Change in K^+ level per micron aperture change (pmol GC^{-1} μm^{-1})	K^+ concentration in guard cell[c] (mM) open	K^+ concentration in guard cell[c] (mM) closed or relatively closed	Change in K^+ concentration per micron aperture change (mM GC^{-1} μm^{-1})	Reference
Vicia faba (intact guard cells)									
	electron microprobe, intact leaf	–	2 12 (12)	0 10 (2)	0 202	454 (4 67)[d]	52 (1 92)[d]	40 2	Humble and Raschke, 1971
	flame photometry on 'rolled' epidermis from intact leaf	–	2 72[e] (16 5)	0 55[e] (6 5)	0 217	460 (5 91)[d]	174 (3 16)[d]	28 6	Allaway and Hsiao, 1973
	flame photometry on rolled epidermis from intact leaf	–	2 99[e] (15)	0 80[e] (7)	0 273	544 (5 49)[d]	243 (3 29)[d]	37 6	Allaway 1973
	micro-dissected guard cell pairs intact leaf	–	0 75 mol kg^{-1} 2 531[f] (4 12)[d] (10)	0 16 mol kg^{-1} 0 484[f] (1 37)[d] (0)	0 205	614	353	26 1	Outlaw and Lowry 1977
	electron microprobe, paradermal sections	–	1 68 (12)[g]	0 53 (2)[g]	0 115	360 (4 67)[d]	276 (1 92)[d]	8 4	Garrec *et al* , 1983
Vicia faba (isolated guard cells)									
	epidermal strip, $^{86}Rb^+$ uptake	RbCl	–	–	0 230[e]	–	–	–	Fischer, 1968
	epidermal strip, $^{86}Rb^+$ uptake	RbCl	2 00[e] (12)	0 714[e] (8)	0 321	428 (4 67)[d]	200 (3 57)[d]	57	Fischer, 1971
	^{42}K tracer accumulation	KCl	1 52[e] (10)	0 465[e] (6 7)	0 319	369 (4 12)[d]	145 (3 21)[d]	68	Pallaghy and Fischer, 1974
	electron microprobe, epidermal strips	KCl	0 94 (13 9)	0 235 (6 4)	0 094	181 (5 19)[d]	75 (3 12)[d]	14 1	Raschke and Schnabl, 1978
		KIDA	1 39 (11 7)		0 154	303 (4 59)[d]		30 4	

Table 8.1 *contd*

Species	Experimental material and measurement technique	Potassium salts added[a]	Amount of K⁺ in guard cell[b] (pmol GC 1)		Change in K⁺ level per micron aperture change (pmol GC 1 μm 1)	K⁺ concentration in guard cell[c] (mM)		Change in K⁺ concentration per micron aperture change (mM GC 1 μm 1)	Reference
			open	closed or relatively closed		open	closed or relatively closed		
	isotachophoretic analysis, epidermal strips	KCl	49 nmol mg 1 1 09[h] (9)	8 3 nmol mg 1 0 185[h] (1)	0 113	283 (3 845)[d]	112 (1 645)[d]	21 4	Shimada *et al* , 1979
	⁸⁶Rb tracer flux analysis	RbCl	0 896 (12 8)	0 218 (7 8)	0 136	183 (4 89)[d]	62 (3 515)[d]	24 2	Brindley, 1990
Commelina communis (intact guard cells)									
	ion-selective micro-electrodes, intact leaf	–	–	–	–	448	95	–	Penny and Bowling, 1974
	ion-selective micro-electrode[i], epidermal strip	KCl (two sets)	–	–	–	286–592 (11–14 μm)	85 (2 μm)	20 (2–10 μm) 102 (11–14 μm)	MacRobbie and Lettau, 1980b
						238–385 (11–14 μm)	75 (2 μm)	16 5 (2–10 μm) 49 (11–14 μm)	
	electron microprobe, paradermal section	–	7 5 (15)[g]	3 2 (7)[g]	0 537	1170 (6 41)[j]	665 (4 81)[j]	63 1	Garrec *et al* , 1983
Commelina communis (isolated guard cells)									
	ion-selective micro-electrode[i], epidermal strip	KCl	–	–	–	68 (10 μm) 157 (15 μm)	51 (5 μm)	3 4 (5–10 μm) 18 (10–15 μm)	MacRobbie and Lettau, 1980a
	⁸⁶Rb⁺ tracer flux analysis, epidermal strip	RbCl	1 195[k] (15)	0 192[k] (5)	0 1	186 (6 41)[j]	43 (4 41)[j]	13 2	MacRobbie and Lettau, 1980a

Species	Experimental material and measurement technique	Potassium salts added[a]	Amount of K^+ in guard cell[b] (pmol GC^{-1}) open	Amount of K^+ in guard cell[b] (pmol GC^{-1}) closed or relatively closed	Change in K^+ level per micron aperture change (pmol GC^{-1} μm^{-1})	K^+ concentration in guard cell[c] (mM) open	K^+ concentration in guard cell[c] (mM) closed or relatively closed	Change in K^+ concentration per micron aperture change (mM GC^{-1} μm^{-1})	Reference
	$^{86}Rb^+$ epidermal strip FC-induced opening	RbCl	0 869 (12)	0 337 (6 5)	0 097	98 (5 81)[j]	71 (4 71)[j]	4 9	Clint and MacRobbie, 1984
Zea mays (intact guard cells)									
	electron microprobe epidermis from intact leaf	–	0 4[l] (5)	0 15[l] (0)	0 04	359 (1 115)[m]	193 (0 777)[m]	33 2	Raschke and Fellows 1971
	electron microprobe intact leaf	–	–	–	–	317 (5 μm)	60 (0 μm)	51 4	Lasceve *et al* 1987
Tradescantia albiflora (intact guard cells)									
		–	–	–	–	633	152		Zlotnikova *et al* , 1977
Allium cepa (intact guard cells)									
	electron microprobe, epidermal strip	(–)	0 36 (5 5)	0 29 (2 6)	0 024	no volume data available	–	–	Schnabl and Raschke, 1980
		(KCl)	0 42 (8 7)	0 27 (3 6)	0 029				
		(K_2SO_4)	0 31 (8 8)	0 05 (2 3)	0 040				
Allium cepa (isolated guard cells)									
	electron microprobe, epidermal strip	KC1 (–)	0 43 (8 7)	0 22 (4 7)	0 052	–	–	–	Schnabl and Raschke, 1980
		(KCl)	0 46 (8 9)	0 34 (5 9)	0 040				
		(K_2SO_4)	0 23 (3 9)	0 12 (2 7)	0 092				

Table 8.1 *contd*

Species	Experimental material and measurement technique	Potassium salts added[a]	Amount of K^+ in guard cell[b] (pmol GC^{-1})		Change in K^+ level per micron aperture change (pmol GC^{-1} μm^{-1})	K^+ concentration in guard cell[c] (mM)		Change in K^+ concentration per micron aperture change (mM GC^{-1} μm^{-1})	Reference
			open	closed or relatively closed		open	closed or relatively closed		
Nicotiana tabacum (intact guard cells)									
	electron microprobe epidermal strip	–	–	–	–	500 (8)	210 (1)	41 4	Sawhney and Zelitch 1969
Valerianella locusta (intact guard cells)									
	physical separation of intact guard cells, atomic absorption spectroscopy	–	3 5 (6)	2 1 (0)	0 233	875 (4)[n]	525 (4)[n]	58	Losch, 1985

a Salts in parentheses indicate the nutrient supplement of the medium the plants were grown in
b Values in brackets are stomatal apertures in µm
c Values in brackets are osmotic volumes of guard cells in pl
d Volume for guard cells of *Vicia faba* calculated using an average of values determined by Raschke (1979), Allaway and Milthorpe (1974) and Brindley (1990) with subtraction of the non-osmotic volume determined by Weyers and Fitzsimons (1985) V=2 42+0 275A–1 05 (*NOV*)
e Stomatal frequency of individual guard cells in *Vicia faba* taken as 112 mm^{-2} (Fischer, 1971)
f Calculated using 3 375 ng GC^{-1} for light treated and 3 025 ng GC^{-1} for dark treated (Outlaw and Lowry, 1977)
g Authors original volume estimates used
h Authors give 4 cm^2 epidermis as equivalent to 1 mg dry weight Stomatal guard cell density 112 mm^{-2} from Fischer (1971)
i Potassium activity measurements have been converted to concentrations using an approximate form of the Debye–Huckel equation (Nobel, 1991)
j Volume for guard cells of *Commelina communis* taken from MacRobbie and Lettau (1980b) with subtraction of the non-osmotic volume determined by Weyers and Fitzsimons (1985) 4+0 2A–0 59 (*NOV*) Stomatal frequency of individual guard cells taken as 120 mm^{-2} (MacRobbie and Lettau, 1980b)
k Represents average of two regressions quoted as both similar
l Taken from fig 7 of Raschke and Fellows (1971) assuming opening to about 5 µm and expressed per individual guard cell
m Volume estimates taken from Raschke and Dickerson published in Raschke (1979) expressed per guard cell with subtraction of 0 2 pl as the nominal *NOV*
n Insufficient data to calculate volume changes – average volume given by author used

forces for K^+ uptake and transfer to the vacuole as well as its effect on the ionic strength of the cytoplasm.

Changes in vacuolar K^+ activities (the free ionized K^+ as opposed to the total K^+ concentration) have been measured using ion-selective electrodes in the guard cells and surrounding cells in *Commelina communis* (Penny and Bowling, 1974; MacRobbie and Lettau, 1980b) and *Tradescantia albiflora* (Zlotnikova *et al.*, 1977). In general, the subsidiary cells and epidermal cells adjacent to the stomatal complex showed an inverse relationship in vacuolar K^+ accumulation compared to the guard cells, but to varying degrees (Fig. 8.2). In *Tradescantia*, the lateral subsidiary cells showed the greatest change, with little difference in either the terminal cells or epidermal cells further away (Zlotnikova *et al.*, 1977). In *Commelina*, Penny and Bowling (1974) and Willmer and Pallas (1984) found changes throughout the complex, while in the study by MacRobbie and Lettau (1980b) changes occurred mainly in the terminal subsidiary cell and the epidermal cells.

Interestingly, the levels of K^+ in the guard cell, subsidiary cell and epidermal cell vacuoles were not enough to account for the total amount of K^+ in the epidermis (MacRobbie and Lettau, 1980b; Penny and Bowling, 1974). In addition, although the changes in K^+ concentration in the epidermal cells were less than that in the guard cells, the total volume of the epidermal cells was at least 10-fold greater. Thus, the total change in the amount of K^+ in cells other than the guard cells was considerably more than the amount of K^+ that ever appeared or disappeared from the guard cells, suggesting a significant proportion of the K^+ was located in the apoplast (MacRobbie and Lettau, 1980b).

A proportion of the K^+ in the apoplast will be bound to negative charges on the wall polymers to an extent dependent on the apoplastic pH and the presence of other cations, such as Ca^{2+}. The guard cell wall is characterized by a high pectin content (see Chapter 3) and a high cation binding capacity (0.3–0.5 M; Saftner and Raschke, 1979). Although the ratio of wall volume to protoplast volume is much larger in guard cells than other cell types, the total apoplast volume as a repository for ions is relatively small, similar to that of the guard cell lumen, at about 5 pl in *V. faba* (Raschke, 1979). The amount of unbound K^+ surrounding guard cells during stomatal movements remains uncertain. According to Blatt (1985), the apoplastic levels were submillimolar throughout the epidermis of *Commelina*, *Vicia* and *Pisum*. However, Bowling (1987) found that the K^+ and Cl^- activities in guard cell walls of *C. communis* varied according to whether stomata were open or closed. Thus at 12 μm aperture, the K^+ and Cl^- activities were 3 and 5 mM, respectively, and when stomata closed apoplastic levels rose to 100 and 33 mM, respectively (Fig. 8.2). In a subsequent study following changes in apoplastic K^+ with time during opening, an increase in apoplastic K^+

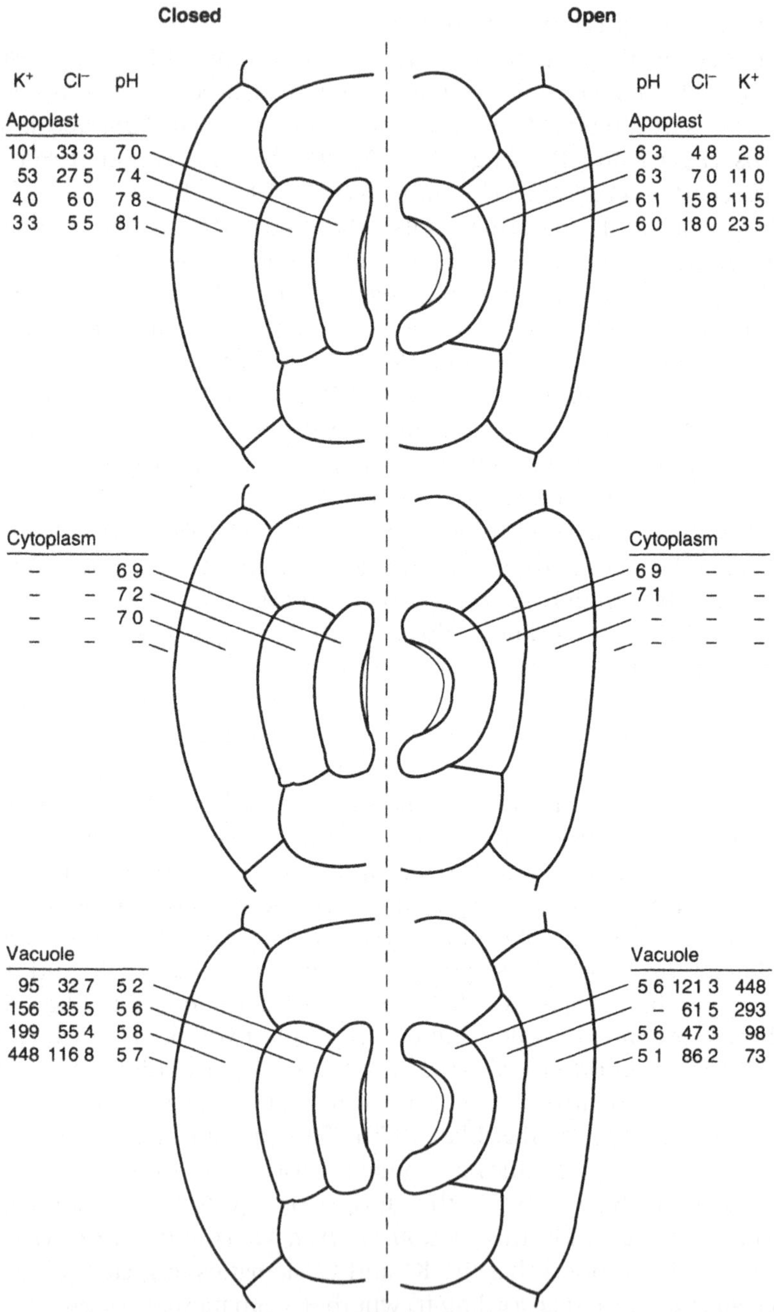

Figure 8.2 Ion concentrations in the apoplast, cytoplasm and vacuole of cells in the stomatal complex of *Commelina communis* measured using ion-selective electrodes Data taken from Penny and Bowling (1974, 1975), Penny *et al* (1976), Bowling (1987) and Edwards and Bowling (1986) with permission

was observed to follow the initial rapid decline, which appeared to spread as a wave from the walls of the epidermal cells to the guard cells (Bowling and Smith, 1990) This may have originated from exchange of H^+ for K^+ from fixed binding sites in the wall during activation of the guard cell plasma membrane H pump (see Section 8 5 3) and/or pH-stimulated release of K^+ by the epidermal cells (Bowling and Smith, 1990) From Bowling s results guard cell walls appear to behave like cell walls of pulvinar motor cells which also act as reservoirs of K for reversible cell volume changes during rhythmic leaf movements (Freudling *et al* , 1988)

The external K activity is also important in defining the electrochemical gradient for K movement and influences both the membrane potential and the gating characteristics of the plasma membrane ion channels (Section 8 4 1) The amount of ions in the apoplast may also have a significant effect on the osmotic pressure in the wall, against which the guard cell protoplast (GCP) competes for water This may also account in part for the difference in measured guard cell turgor pressure and guard cell osmotic pressure recorded by Meidner and Edwards (1975, 1979) (see Chapter 4)

In grass stomata, K fluxes of the guard cells appear to be closely coupled to K^+ fluxes of the subsidiary cells Microprobe and histochemical evidence indicates that virtually all the K^+ involved in stomatal movements shuttles between these two cell types (Raschke and Fellows, 1971), although this may depend on the precise experimental conditions (Lasceve *et al* , 1987) This may represent the most advanced ion uptake system in guard cells, as the inverse correlation between guard cell and subsidiary cell turgor might facilitate the more rapid stomatal movements generally observed in this grass species compared to other species The morphology of the dumb-bell-shaped guard cell also enables wider apertures for a lower K^+ accumulation in comparison to elliptical guard cells

The level of apoplastic K^+ is also important in epidermal strip experiments where the apoplastic K concentration equilibrates with the bulk medium K concentration (see Weyers and Meidner, 1990) Marked differences are observed between species in the amount of K^+ required to support stomatal opening in these systems Thus, in *Vicia*, near maximal opening can occur in 1 mM KCl, while in *Commelina*, higher concentrations, from 30 to 50 mM, are required for substantial opening (e g Travis and Mansfield, 1979a) Increasing concentrations of KCl or other cations, up to 100–150 mM, in the medium increase the stomatal aperture, but decrease the ability of guard cells of *Commelina* (Travis and Mansfield, 1979a) and *Vicia* (e g Fischer and Hsaio, 1968) to respond to closing stimuli Under certain conditions, such as treatment with fusicoccin, stomata from *Commelina* can open in low or zero external K^+

concentrations (e g Squire and Mansfield, 1974, Reddy *et al* , 1984), and stomata from *Z mays* can also open in the absence of exogenous K^+ (Pallaghy, 1971) It is possible that local release of K^+ from the subsidiary cells and preferential uptake by the guard cells is more rapid than diffusion from the apoplast to the medium in these conditions

8.2.2 Effects of other cations on stomatal movements

Other monovalent cations will support opening in epidermal strip experiments to varying degrees, even though they are not normally accumulated by guard cells Thus Imamura (1943) found that cations stimulated opening at low concentrations in the following order of activity K > Na > Li while at higher concentrations the order was reversed Iljin (1957) observed that monovalent cations stimulated opening in the following order of activity Li^+ > Na > Cs^+ > K > Rb^+ Willmer and Mansfield (1969) found NaCl could open stomata of *C communis*, but such cells were less sensitive to closing stimuli (e g Jarvis and Mansfield, 1980) More recently, Thiel and Blatt (1991) have shown that Na , when present at the cytoplasmic side, blocks potassium channels that would allow cation efflux and stomatal closure Thus Na^+ effectively locks the stomata open once it has been accumulated

Divalent cations do not support stomatal opening, but may affect the ability and selectivity of the monovalent cations accumulated by the guard cells At low concentrations, calcium increased the selectivity of the guard cells for K over other cations (e g Humble and Hsiao, 1969), but at higher concentrations Ca prevented opening and stimulated closure in a number of species The Ca + concentration that caused closure was also markedly different between species, thus 50% closure in *Commelina* occurred at 0 1 mM (De Silva *et al* , 1985a), while 10- to 100 fold higher concentrations were needed to effect closure in *V faba* The effectiveness of Ca however, depends on the K^+ concentration of the incubation medium increasing K^+ decreasing the effects of Ca^{2+} The level of Ca in the apoplast *in vivo* has not been measured directly, but endogenous differences in apoplastic Ca may be physiologically important Thus De Silva *et al* (1986) found that as chelation of the apoplastic Ca occurred the difference in apertures between adaxial and abaxial stomata of *C communis* disappeared (De Silva, 1986), the bulk Ca levels in both sets of epidermis, however, were similar (Atkinson, 1991) Atkinson *et al* (1992) showed that the amount of calcium entering the xylem was controlled in a range of species and the level of Ca was further tightly regulated *en route* to the guard cells in *Commelina* (Atkinson *et al* , 1989, Ruiz *et al* , 1993) This buffering effectively isolated the guard cells from slow changes in rhizosphere Ca levels, although experimentally, rapid increases in Ca^{2+} caused partial stomatal closure (Atkinson *et al* , 1990)

8.2.3 The involvement of anions in stomatal movements

Virtually every potassium ion accumulated in the vacuole or cytoplasm of a guard cell requires an accompanying negative charge to maintain charge balance In principle, the negative charge could be derived from simultaneous uptake of an inorganic ion, such as Cl , across the plasma membrane or synthesis of an organic anion, such as malate , from an uncharged precursor, such as starch, with concurrent expulsion of the additional protons generated, or a combination of both systems

In *C communis*, MacRobbie (1980) found that the level of Cl in guard cells increased during opening using tracer flux analysis with $^{82}Br^-$ as a Cl analogue, but the overall level of Cl was not sufficient to balance the K accumulated The predicted subcellular distribution of Cl determined from the efflux curves indicated that the cytoplasmic levels of K and Cl were similar, with a much larger difference in the relative vacuolar levels (MacRobbie, 1981) Using ion-selective electrodes Penny and Bowling (1974) and Penny *et al* (1976) also found that the ratios of vacuolar Cl to K^+ levels in guard cells of open and closed stomata of *C communis* decreased from 0 34 (closed stomata) to 0 27 (open stomata)

The early X-ray microprobe studies also highlighted the discrepancy between the changing levels of K accumulated and the relatively constant levels of inorganic anions such as SO_4^{2-} and PO_4^{3-} or minor changes of Cl during stomatal opening in *V faba* (Humble and Raschke, 1971) For example, the extent that Cl balances K in *Vicia* varies from 6 to 45%, depending on the level of Cl supplied in the growth medium or the incubation buffer (Humble and Raschke, 1971, Pallaghy and Fischer, 1974, Raschke and Schnabl, 1978, Van Kirk and Raschke, 1978a) Allaway (1973) showed that part of the difference could be accounted for by increases in organic anions, principally malate , and this observation has been made in several other species (see Chapter 9) The balance between malate and Cl may also depend on the pH of the apoplast solution Thus Dittrich *et al* (1979) found that Cl uptake was reduced at high external pH values as would be predicted from operation of a Cl /nH^+ symport mechanism

In contrast to *Vicia*, Schnabl and Raschke (1980) found that the ratio of Cl to K^+ was near unity in *Allium cepa* across a broad range of treatments, with the notable exception of incubation in K_2SO_4, when little change in Cl occurred Guard cells of *Allium* lack starch and may be deficient in malate synthesis, unless mobilization of low molecular weight fructans occurs (see Chapter 9)

Parallel uptake of K and Cl movement was observed by Raschke and Fellows, (1971) for *Z mays*, where Cl consistently accounted for 40% of the K^+ and occasionally more Lasceve *et al* (1987), however, found that the Cl /K^+ ratio shifted from about 0 3 for closed stomata of *Z mays* to about 1 in fully open guard cells

8.2.4 The osmotic contribution of potassium salts during stomatal movements

There is considerable controversy in the literature on whether the osmotic contribution from the measured or predicted levels of potassium salts are sufficient to account for the increases in osmotic pressure measured during stomatal opening (see Chapter 4) Approximately 40% more K^+ as K_2malate has to be accumulated to give the same change in osmotic potential as KCl, on account of both the lower absolute number of osmotically active anions present and the lower dissociation of K_2malate Most early reports concluded that the levels of KCl matched the osmotic pressure for wide stomatal opening in *V faba* (Fischer, 1968a,b, 1971, Fischer and Hsaio, 1968, Humble and Raschke, 1971, Allaway and Hsaio, 1973, Pallaghy and Fischer, 1974) and *Nicotiana tabacum* (Sawhney and Zelitch, 1969) However, guard cell osmotic pressures were determined plasmolytically in these studies and would underestimate the solute potential due to salt leakage for the guard cells during the measurement (see Chapter 4) More recent values using a modified plasmolysis technique (Willmer and Beattie, 1978, Raschke, 1979) or cryoscopic techniques (Bearce and Kohl, 1970) have indicated much higher osmotic pressures are associated with stomatal opening

Several investigators consider that, while a significant proportion of the osmotic pressure is attributable to potassium salts, additional osmotica are required, particularly at low apertures or the early stages of opening (MacRobbie and Lettau, 1980a,b, Brindley, 1990a, Laffray *et al* , 1984) (Fig 8 1) The most likely candidate is currently considered to be sucrose (see Chapter 9) However, it is also worth noting that Fitzsimons and Weyers (1986a,b) found that volume increases in GCPs can be accounted for entirely by the uptake of K^+ plus an accompanying anion (having a mean elementary charge of 1 63)

Thus K uptake and to a lesser extent Cl uptake and synthesis of malate form the basis of the osmotic changes occurring during stomatal movements Additional solutes may either be required under some conditions, however, as the levels of ions accumulated in some species do not precisely match the measured osmotic pressures or the time course over which uptake occurs may lag behind the opening response

8.3 Ion fluxes and the driving forces for ion movement

Ion accumulation during opening is an active process requiring an input of energy To generate favourable thermodynamic gradients for ion accumulation, the plasma membrane and tonoplast are 'energized' by proton-pumping ATPases (and possibly redox systems), that establish a proton motive force (p.m.f.) comprising both a proton concentration gradient and an electrical gradient:

$$\text{p.m.f.} = 0.203RT\log_{10}\frac{[\mathrm{H}^+]_{\mathrm{out}}}{[\mathrm{H}^+]_{\mathrm{in}}} - z_{\mathrm{H}}FV_{\mathrm{m}} \tag{8.1}$$

Where R is the universal gas constant (J mol^{-1} K^{-1}), T is the absolute temperature (K), z_H is the electronic charge on a proton (+1), F is a Faraday (96 500 coulomb mol^{-1}) and V_m is the membrane potential (V).

Ions may move passively through channels in response to the electrical component of the p.m.f. or be driven by the pH gradient via a proton-coupled co-transport system. During closure, the direction of the ion fluxes are reversed, but a different set of transport proteins appear to be involved. The following sections discuss the ion transport events associated, first, with stomatal opening and, second, with stomatal closure. The sections on stomatal opening consider the energization of the plasma membrane and tonoplast, followed by discussion of K^+ and anion uptake across these membranes. The sections on stomatal closure consider the release of K^+ and then anions across both the tonoplast and plasma membrane.

8.3.1 Characteristics and activity of the plasma membrane proton pumping system(s)

The electrogenic activity of the plasma membrane proton pump has been investigated using microelectrode measurements in intact guard cells and using patch-clamp techniques with GCPs. The first microelectrode measurements by Pallaghy (1968) in guard cells of *Nicotiana* indicated a negative membrane potential. Treatments that open stomata would be expected to activate the pump and hyperpolarize the membrane potential. However, Pallaghy (1968) also found that the membrane potential was insensitive to light (an opening stimulus) or HCO_3^- (a closing stimulus). Likewise, Saftner and Raschke (1981) found little evidence for an electrogenic component to membrane potentials in guard cells from *Commelina*, *Allium*, *Vicia* and *Zea*. Similar results were reported by Edwards and Bowling (1984) for *Tradescantia*, although small (10–40 mV) light-induced hyperpolarizations had been previously reported for guard cells of this genus by Gunar *et al.* (1974). Light induced hyperpolarizations were also reported for guard cells from *Allium* (Zeiger *et al.*, 1977; Moody and Zeiger, 1978) and *Vicia* (Ishikawa *et al.*, 1983). Blatt (1987a) has suggested that salt-leakage from the measuring electrode containing 3 M KCl may have shifted the membrane potential towards the equilibrium potential for Cl^- in some of these experiments and therefore obscured changes in membrane

Table 8.2 Properties of the K^+ inward rectifier at the guard cell plasma membrane

	Channel characteristics	Reference
Conductance	5–10 pS	j, k
Activation voltage	activates on hyperpolarization more negative than −100 to −120 mV	a, b, c, d, e, g, f, h, i, j, k, l
Modulation by ions and second messengers		
$[K^+]_o$	(i) independent of $[K^+]_o$ above 1 mM, inactivated below	a, h
	(ii) $[K^+]_o$ dependent decrease in activation potential	l
$[Ca^{2+}]_i$	increasing $[Ca^{2+}]_i$ shifts activation voltage to more negative values and decreases current amplitude	c, d, e, g, f, j
$[Ca^{2+}]_o$	(i) increasing $[Ca^{2+}]_o$ has no direct effect	a
	(ii) increasing $[Ca^{2+}]_o$ decreases current amplitude	d, e, f
pH_i	internal alkalinization inactivates	b, f, l
pH_o	external acidification shifts activation voltage to more positive values and increases current amplitude	a
Time dependence	(i) *Vicia faba* – exponential activation τ=100–400 ms, exponential deactivation τ=10–120 ms (voltage dependent)	a, e, g, f, h, i, j, k
	(ii) *Zea mays* – exponential activation τ=41 ms, exponential deactivation τ=10–50 ms (voltage dependent)	d
	(iii) no time-dependent inactivation	a, h
Selectivity	selectivity sequence $K^+>Rb^+>Na^+>Li^+>>Cs^+$	h
	$K^+>Rb^+>Cs^+>>Na^+$	a
	Ca^{2+} may permeate the open K^+ channel	d, e
Inhibitors	TEA (I_{50} 9 mM)	a
	Ba^{2+} (I_{50} 0 8 mM)	g, f, k
	Al^{3+} (I_{50} 15 μm)	h
	charybdotoxin (I_{50} << 50 nM)	a
	α-dendrotoxin (I_{50} 0 2 nM)	a

[a] Blatt, 1992, [b] Blatt and Armstrong, 1993, [c] Fairley-Grenot and Assmann, 1991, [d] Fairley-Grenot and Assmann, 1992a, [e] Fairley-Grenot and Assmann, 1992b, [f] Lemtiri-Chlieh and MacRobbie, 1994, [g] Luan *et al.*, 1993, [h] Schroeder, 1988, [i] Schroeder and Fang, 1991, [j] Schroeder and Hagiwara, 1989, [k] Schroeder *et al.*, 1987, [l] Thiel *et al.*, 1993

potential due to pump activity. Thus, more recent investigations have concentrated on measurement of the current passing through the pump using voltage clamp techniques in intact cells with multi-barrelled electrodes or by whole-cell patch-clamping in GCPs (see Box 8.2).

Blatt (1987b) has estimated that one proton is pumped per ATP hydrolysed at pH 7.4. The free energy available from hydrolysis of ATP depends on the energy charge in the cytoplasm (see Chapter 9), but is about 42 kJ mol^{-1}. Thus, assuming the largest pH gradient experienced by the plasma membrane of the guard cell *in vivo* is about 2.5 units (pH

Box 8.2. Patch-clamp techniques

Introduction

The patch-clamp technique enables recording of net ion flow through ion transporters in membranes. The measurements can be made from entire cells or from small excised membrane patches. The orientation of the membrane patch can be controlled to some extent (Fig. 8.B2a). This is advantageous during perfusion experiments as potentially either side of the membrane can be exposed to changing conditions

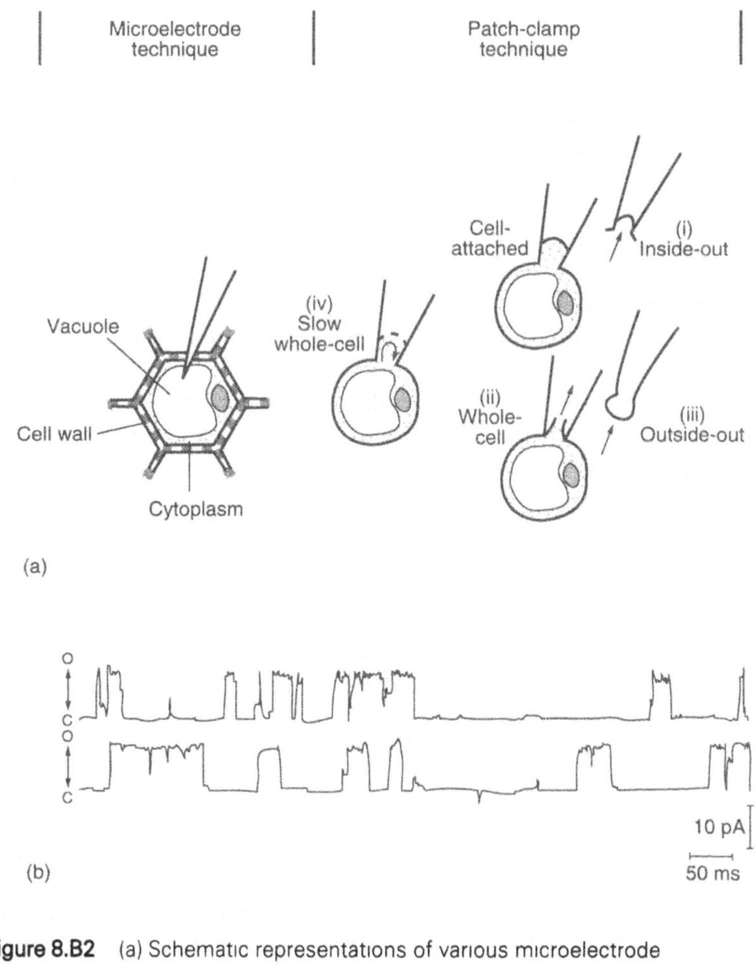

Figure 8.B2 (a) Schematic representations of various microelectrode configurations (Left) Classical impalement technique applied to a plant cell (Right) Patch-clamp technique applied to a plant protoplast (b) Single-channel currents recorded with an isolated membrane patch Upward deflections indicate channel opening and downward deflections channel closure Redrawn from Hedrich and Schroeder (1989), with permission

in the bathing medium, whilst the composition of the pipette solution remains effectively constant Recent developments also allow micro-perfusion of the pipette solution itself Description of a variety of electrophysiological techniques can be found in Standen *et al* (1987) and results of patch-clamp experiments specifically in plants are reviewed in Hedrich and Schroeder (1989) and Tester (1990)

Procedure

1 A heat polished micropipette is pressed against the plasma membrane or vacuolar membrane and spontaneously forms a high resistance seal (10^9–$10^{12}\Omega$ seal) if gentle suction is applied
2 The mechanical stability of this arrangement allows
   i Withdrawal of the pipette encircling the patch (termed an inside-out patch)
   ii Rupture of the patch by suction with the cell attached (termed whole cell mode)
   iii Rupture of the patch and withdrawal of the electrode to allow the peripheral membrane to reseal (termed outside-out patch)
   iv Slow whole cell measurements with the membrane patch permeabilized to allow electrical access without loss of large diffusible molecules
3 The membrane forms part of an electrical circuit between an electrode in the patch-pipette and an electrode in the bathing medium The current flowing through individual channels is typically in the pA range (about 10^7 ions s^{-1}) and can be measured by sensitive amplifiers The flow of ions through pumps is much lower (10^3–10^4 ions s^{-1}) and can only be measured in the whole cell configuration as the sum of many pumps acting together
4 Single channel openings are recognized as step increases in the current flowing (Fig 8 B2b)
5 Several characteristics of single channels are normally measured
   (i) The conductance under defined ionic conditions Typically these values are expressed for symmetric or asymmetric ion concentrations on either side of the membrane
   (ii) The open time or closed time of the channel
   (iii) The frequency of channel opening
   (iv) The ion selectivity
   (v) Modification of the above in response to pharmacological agents
6 Traces from multiple channels or pumps, usually in the whole cell mode, yield other parameters, in particular the time dependence of activation and inactivation kinetics of the channels or pumps
7 There are a number of limitations to the patch-clamp technique in plants, in particular the technique requires access to the plasma membrane, which normally involves preparation of protoplasts,

although laser microsurgery techniques can cut holes in walls of intact cells In addition, the volume of the lumen of the patch electrode is massive compared to the cell volume, except in the slow whole cell mode The composition of the cytoplasm is effectively diluted by the pipette solution and may lead to loss of diffusable activators/inhibitors and perturbation of metabolism

References

Hedrich, R and Schroeder, J I (1989) The physiology of ion channels and electrogenic pumps in higher plants *Ann Rev Plant Physiol Plant Mol Biol*, **40**, 539–569

Standen, N B, Gray, P T A and Whitaker, M J (1987) *Microelectrode techniques The Plymouth Workshop Handbook,* Company of Biologists Ltd, Cambridge

Tester, M (1990) Plant ion channels whole-cell and single-channel studies *New Phytol*, **114**, 305–340

5 0 in the apoplast and a cytoplasmic pH of 7 5), the maximum membrane potential that could be expected *in vivo* is about –290 mV (calculated from equation 8 1) This potential is termed the reversal potential of the pump when there is no net H^+ flux through the pump At voltages more positive than the nominal reversal potential the current through the pump increased and then saturated at around –150 mV (e g Blatt, 1987b, Becker *et al*, 1993) However, the maximum pump currents recorded were highly variable depending on the experimental system used and the time of year, usually with a minimum in the winter months (Thiel *et al*, 1992, Becker *et al*, 1993) In intact cells Blatt (1987b) recorded currents of around 320 pA per guard cell, whilst early patch-clamp studies only recorded currents in the region of 1 6–5 5 pA (Assmann *et al*, 1985, Serrano *et al*, 1988) These low currents may have arisen from dilution of essential cytoplasmic factors by the large volume of solution in the patch-electrode More recently, currents averaging around 18–25 pA per guard cell were found under conditions that minimized washout of cytoplasmic components (Schroeder, 1988, Lohse and Hedrich, 1992) For comparison, currents accountable to K^+ fluxes during opening are in the region of 20–24 pA per guard cell averaged over a 3 h period (Outlaw, 1983, Raschke *et al*, 1988)

8.3.2 Biochemical properties of the plasma membrane H^+-ATPase

Biochemical characterization of membrane transporters in guard cells has been difficult as only a small amount of guard cell material can be isolated free from contamination from other cell types in comparison with other plant cells or tissues. As discussed in detail in Chapter 9, early work using intact epidermal strips is unlikely to provide data specific to guard cells which comprise only a small percentage of the epidermis, thus more recent studies have used GCPs or epidermal tissues with guard cells as the only intact cells, followed by varying degrees of membrane isolation, ranging from crude homogenates to purified plasma membranes. There is a broad concensus on the properties of the plasma membrane H^+-ATPase between reports for both *Vicia* and *Commelina*, the only species so far studied (Table 8.3). In all reports there was an ATPase activity that was dependent on Mg^{2+}, with a pH optimum between pH 6.4 and 6.8; activity was sensitive to vanadate and stimulated by K^+ by about 30–40%. These features and the general kinetics (V_{max} and $K_{m,ATP}$), are characteristic of plasma membrane P-type H^+-ATPases in other species (Serrano, 1989). The maximum ATPase activity was significantly higher in guard cells compared to other cells on protein, chlorophyll or plasma membrane surface area bases, although there was also a marked seasonal variation, similar to that observed in electrophysiological studies. More recently Kinoshita *et al.* (1995) have shown that the ATPase is reversibly inhibited by Ca^{2+} at physiological concentrations (I_{50} = 0.3 μM). A 100 kDa molecular weight band has also been detected, using antibodies raised against the H^+-ATPase from *Arabidopsis*, in extracts of guard cells of *Z. mays* (Villalba *et al.*, 1991) and *C. communis* (Fricker and Askerlund, unpublished), but the same antibody gave variable staining of guard cells in sections from leaves of *Avena sativa* and *Pisum sativum* (Parets-Soler *et al.*, 1990).

8.3.3 H^+ fluxes in epidermal strips and GCPs

The rate of proton efflux from the guard cell must be sufficient to account for the observed rates and duration of K^+ and Cl^- fluxes. Continuous measurements of external pH changes in the bathing medium around epidermal strips or GCPs provide an indication of the net proton efflux, minus the contribution of secondary proton coupled transport events, such as Cl^- uptake via a Cl^-/nH^+ symporter. Light stimulates proton efflux from guard cells into the bathing medium from epidermal strips of both *Vicia* (Raschke and Humble, 1973; Gepstein *et al.*, 1982/83) and *Commelina* (Inoue and Katoh, 1987; Fricker and Willmer, 1990a). The measured rates and the total amount of H^+ efflux were similar to the rate and levels of K^+ accumulated (e.g. Fricker and Willmer, 1990a).

Although the H^+ efflux is most probably due to the activity of the H^+-ATPase, there are conflicting results regarding the effectiveness of the plasma-membrane H^+-ATPase inhibitor, vanadate, at preventing both H^+

Table 8.3 Biochemical properties of the guard cell plasma membrane H^+-ATPase

Species	Tissue preparation	pH optimum	V_{max}	V_{max} (pmol GC^{-1} h^{-1})	$K_{m\ ATP}$ (mM)	K^+ stim (%)	Inhibitors I_{50}, µm	Activators	Reference
Vicia faba	GCP membrane pellet	6 8	800 (µmol mg Chl^{-1} h^{-1})	1 6[a]	0 5	none	Ca^{2+}(2000)		Shimazaki and Kondo, 1987
Vicia faba	Acid-washed epidermis, two-phase polymer purification	6 4	510 (nmol mg $protein^{-1}$ min^{-1})	5 1[b]	0 67	–	VO_4^{3-}(15) DCCD (20) Ca^{2+} (800)	FC slight stimulation	Blum *et al* , 1988
Vicia faba	Acid-washed epidermis two-phase polymer purification	6 4 (ATPase) 6 8 (H^+pump)	17 (nkat mg $protein^{-1}$)	10 2[b]	~1	–			Becker *et al* , 1993
Commelina communis	GCP membrane pellet	6 5	5 7 (µmol mg $protein^{-1}$ h^{-1})	0 9[c]	ND	30–40	VO_4^{3-}(10–100) DCCD (5) Ca^{2+}(~150) ABA (13)[d]	Red light pre-illumination ~50%[d]	Nejidat *et al* , 1986, Nejidat *et al* , 1989
Commelina communis	GCP crude homogenate	6 6	16 4 (µmol mg $protein^{-1}$ h^{-1})[c]	2 5	0 33–1 12	25			Fricker and Willmer, 1987, Fricker and Willmer, 1990a

[a] Assuming a chlorophyll content of 2 pg $protoplast^{-1}$ for *Vicia faba* (Shimazaki *et al* , 1983)
[b] Assuming a protein content of 166 pg $protoplast^{-1}$ for *Vicia faba* (Outlaw *et al* , 1981)
[c] Assuming a protein content of 152 pg $protoplast^{-1}$ for *Commelina communis* (Fitzsimons and Weyers, 1983)
[d] Only the K^+-stimulated ATPase activity was affected

extrusion from guard cells and stomatal opening. The different results are probably related to the level of vanadate uptake by guard cells, which is reduced in the presence of high level of permeant anions such as Cl^- (Schwartz *et al.*, 1991). Thus, vanadate, at concentrations up to 2 mM, did not inhibit stomatal opening in *C. communis* in the presence of 50 mM KCl (Karlsson and Schwartz, 1988), but it was effective when applied as part of pre-treatment in the absence of Cl^- (Fricker and Willmer, 1990a). Likewise, when vanadate was supplied with low chloride concentrations, it inhibited light-stimulated stomatal opening in epidermal peels of *C. communis* L (Schwartz *et al.*, 1991), H^+ extrusion from epidermal strips of *V. faba* (Gepstein *et al.*, 1982–83) and blue light or fusicoccin induced swelling of GCPs from *Vicia* (Amodeo *et al.*, 1992).

Activation of the proton pump at the plasma membrane of guard cells would be expected to result in proton efflux and acidification of the apoplast *in vivo*. The extent of the acidification will depend on a range of factors including the rate of H^+ efflux, the number of protons returned to the cytoplasm by co-transport systems (e.g. for Cl^-), the buffering capacity of the cell wall polymers, the rate of diffusion of protons away from the guard cells and any related H^+ transport activities of the neighbouring cells. In an intact leaf, where the proton efflux would occur into the relatively small volume of the guard cell apoplast, these fluxes would be expected to generate very large, localized pH changes. However, Bowling and co-workers found that the pH only fell from about pH 7 around closed stoma to pH 6.3 after opening in both *Tradescantia virginiana* and *C. communis* (Bowling and Edwards, 1984; Edwards and Bowling, 1986; Edwards *et al.*, 1988) (Fig. 8.2). Transitory changes during opening responses, however, were much greater, being up to two pH units over 10 min in *C. communis* (Edwards *et al.*, 1988). These changes are still considerably less than expected from the level of H^+ efflux into the bathing medium from epidermal strips, given an estimated guard cell apoplast volume of 5 pl (Raschke, 1979). Even with a total ion-buffering capacity of 0.3–0.5 M, pH changes in excess of three pH units might be anticipated from a proton efflux of 2 pmol per guard cell or more. However, the fluorescent pH indicator, primulin, which binds to the cell wall, showed that the H^+ efflux spreads as a wave away from the guard cells through the apoplast of the neighbouring cells as stomata open (Edwards *et al.*, 1988). This increases the effective buffering volume in the apoplast. According to Bowling and Edwards (1984) cytoplasmic pH values across the stomatal complex (guard, subsidiary) and adjacent epidermal cells of *Tradescantia* remain between pH 6.8 and 7.0, whilst the epidermal cell vacuoles become more acidic as the guard cell vacuole alkalinizes during opening (see Fig. 8.2). Thus apoplast pH values may be influenced by a number of cells in the stomatal complex in the intact leaf Changes in apoplastic pH have several potential effects on guard cells including alterations in activity of the plasma membrane ion channels (see Section 8 3 5) and of the plasma membrane H -ATPase The decrease in external pH during opening reduces the activity of the H^+-ATPase, but activates K^+ channels allowing K^+ influx (see Section 8 3 5) and thus maintains a high capacity for K^+ uptake (Blatt, 1992) Decreases in apoplastic pH will have other effects, e g low pH will neutralize the negative binding sites of the wall polymers and increase the level of protonation of weak acids such as abscisic acid (ABA) and indole-3-acetic acid (IAA) (see Chapter 7)

8 3 4 Plasma membrane redox systems

Redox systems in the plasma membrane capable of pumping protons have been demonstrated for some types of plant cells (e g Crane *et al* , 1985) and may provide an alternative or extra means to the H^+-ATPase to generate a proton motive force The presence of such a system in guard cell plasma membranes has been suggested to account for blue light stimulated proton efflux (see Raghavendra, 1990), but the available evidence is controversial Basically, three features have been used as evidence for a redox H^+-pump, i e the reduction of exogenous ferricyanide by intact cells (ferricyanide does not readily cross the plasma membrane), the stimulation of the reduction of ferricyanide by NAD(P)H (which also does not cross membranes readily) and an increased oxygen consumption by cells in the presence of NAD(P)H Furthermore, ferricyanide inhibits stomatal opening in intact epidermal strips and inhibits swelling of GCPs of *Commelina* (Vavasseur *et al* , 1994), although Roth-Bejerano *et al* (1988) found the opposite effect with GCPs Blue-light stimulated proton extrusion from GCPs was also stimulated by exogenous NADH and inhibited by ferricyanide (Gautier *et al* , 1992) Moreover, the rate of blue light stimulated H^+ efflux was greater in CO_2-free conditions than in the presence of CO_2 (Lasceve *et al* , 1993) and it was suggested that there was competition for reducing power between CO_2 fixation and the redox system Further evidence for the participation of a redox system in blue light stimulated opening was found by Vani and Raghavendra (1989) They observed that the action spectrum for blue light stimulated opening was similar to the absorption spectrum of flavins and redox activity was inhibited by flavin-quenching reagents

However, Pantoja and Willmer (1988, 1991) showed that GCPs of *C communis* secreted peroxidase which was also capable of reducing ferricyanide and caused increased rates of NAD(P)H-stimulated oxygen consumption This indicates that a substantial proportion of ferricyanide reduction and O_2 consumption measured experimentally is not related to a redox chain, but equally does not exclude the presence of such a system, particularly at pH values below pH 6.0 when peroxidase activity is low (Pantoja and Willmer, 1991). Furthermore, the initial report by Shimazaki *et al.* (1986) indicating blue light stimulated H^+ efflux was insensitive to vanadate is now thought to reflect a lack of vanadate uptake as found by Amodeo *et al.* (1992), rather than to provide evidence for an alternative redox pump. Pantoja (in Fricker and Willmer, 1990a) found that vanadate partially inhibited the blue light stimulated acidification of media by GCPs of *C. communis*, directly demonstrating that at least part of the blue light stimulated H^+ pumping was due to P-type ATPase activity. In addition, Assmann *et al.* (1985) reported that blue light did not increase pump currents if ATP was omitted from the patch electrode in the whole cell mode.

Nevertheless, it is possible that redox events also affect the H^+-ATPase directly or indirectly via alteration in the level of H^+ as its substrate. Thus, the effects of artificial electron acceptors, such as ferricyanide or neutral red, may be complex, causing membrane depolarization via e^--transport, charge compensating K^+ efflux and intracellular acidification (Ullrich *et al.*, 1990). Cousson *et al.* (1994) also believe that the putative redox system and the H^+-ATPase system are intimately linked, a view partly inferred from their finding that fusicoccin overpowers the inhibitory effects of ferricyanide on stomatal opening and swelling of GCPs.

8.3.5 Transport of K^+ across the plasma membrane during stomatal opening

Unaided diffusion of charged ions through biological membranes is very slow and could not account for the large and rapid ion fluxes observed during stomatal movements. Thus specific transport systems must exist in both the plasma membrane and tonoplast that provide a route for ion movement coupled to a favourable thermodynamic gradient to drive accumulation of K^+ and its accompanying anion.

Uptake of K^+ across the plasma membrane is thought to occur passively through a channel(s) in response to the electrical gradient established by the H^+-ATPase that extrudes protons across the plasma membrane and establishes an inside negative membrane potential (Zeiger *et al.*, 1978). Membrane energization may also occur through the redox based system described previously (see Section 8.3.4). It is possible that guard cells may also possess a secondary coupled, active K^+-uptake system at the plasma membrane, perhaps involving a co-transport system similar to that identified in other cells types by Schachtman and Schroeder (1994). This system would be required for K^+ uptake if apoplastic K^+ concentrations were submillimolar (Blatt, 1987a) or when the Nernst K^+ diffusion potential is positive of the membrane potential, which can occur experimentally during treatment with fusicoccin, for example (Clint and Blatt, 1989)

The major ion channel allowing K^+ uptake in guard cells is called the K^+ inward rectifier, according to the predicted direction of K^+ flux under physiological conditions The K^+ inward rectifier is only activated at membrane potentials more negative than –100 mV (Schroeder *et al* , 1987), when the net driving force for K^+ would typically be into the cell The properties of the K inward rectifier from *V faba* have been measured by a number of groups using electrophysiological techniques (see Box 8 2) In contrast, only one group has reported measurements on K channel activity in *Z mays* (Fairley-Grenot and Assmann, 1992a), and channels with comparable properties may exist in guard cells of *Arabidopsis* and *Nicotiana* (Armstrong and Blatt, personal communication) Detailed reviews describing channels in guard cells and other cell types can be found in Hedrich and Schroeder (1989), Tester (1990), Blatt (1991), Assmann (1993) and Blatt and Thiel (1993) A summary of channel properties in guard cells are given in Table 8 4 for the K^+ inward rectifier and its gating characteristics presented schematically in Fig 8 3

The K^+ inward rectifier has a conductance of 5–10 pS (Siemens is the unit of conductance and is equivalent to the reciprocal of the resistance in Ohms), opens at voltages more negative than –100 mV and allows K influx if the apoplastic K concentration is greater than 1 mM At apoplastic K levels lower than 1 mM, the thermodynamic driving force for K^+ is likely be out of the cell The channel appears to inactivate at these low apoplastic concentrations and would thus not participate in K^+ efflux (Schroeder, 1988, Blatt 1992, see also Schroeder and Fang, 1991) The gating characteristics of the inward rectifier are essentially insensitive to external potassium levels higher than a few mM (Schroeder, 1988, Blatt 1991, 1992)

In addition to regulation by voltage, the K^+ inward rectifier is remarkably sensitive to changes in potential signalling intermediates such as cytosolic calcium, cytosolic pH and external pH (Fig 8 3) Physiological increases in cytosolic calcium ranging from 100 nM to 1 5 μM inactivated the channel, with a shift in the gating potential to more negative values and a reduction in the current amplitude (Schroeder and Hagiwara, 1989, Fairley-Grenot and Assmann, 1992) Decreasing external pH from pH 8 1 to 5 5 had the reverse effect, shifting the activation potential to more positive voltages and increasing the conductance to a maximum at pH 5 5 (Blatt, 1992) This pH range spans the changes in apoplastic pH that occur on opening (see Section 8 3 3) Increases in cytoplasmic pH can also inactivate the K inward rectifier (Blatt, 1992, Blatt and Armstrong, 1993, Lemtiri-Chlieh and MacRobbie, 1994, Thiel *et al* , 1993)

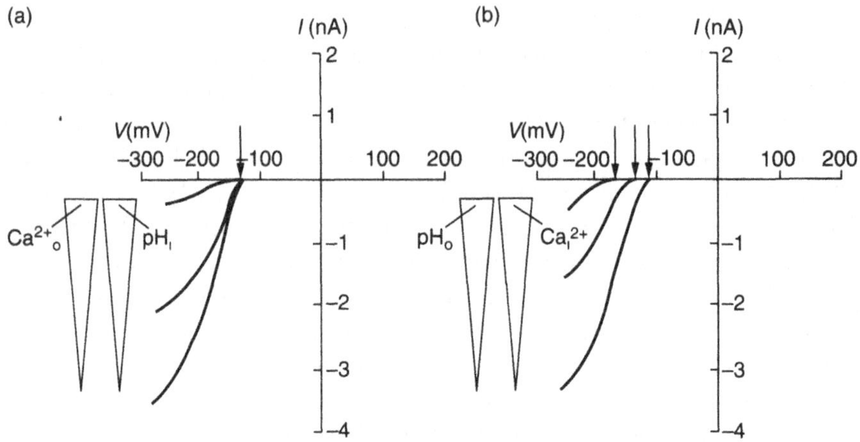

Figure 8.3 Current–voltage (*I*–*V*) characteristics of the K^+ inward rectifier at the plasma membrane of guard cells from *Vicia faba* The channel opens at around –100 mV and the current flowing through the channel increases with hyperpolarization, i e as the potential becomes more negative The current flowing through the channel decreases as either the external Ca^{2+} or internal pH are increased as indicated by the triangles (a) Increasing internal Ca^{2+} or external pH also reduces the current and, in addition, shifts the gating potential to more negative values indicated by arrows (b) Note that the currents measured in guard cell protoplasts using patch-clamp techniques are about 10-fold lower than those shown here for intact guard cells, but the majority of the voltage-gating characteristics are similar Data taken from reports in Table 8 3

8.3.6 Transport of Cl^- across the plasma membrane during stomatal opening

The apoplastic Cl concentration around guard cells of closed stomata is about 33 mM (Bowling, 1987) while the cytoplasmic concentration is estimated to be between 50 to 120 mM (e.g. MacRobbie, 1987), thus uptake of Cl is opposed by both the Cl concentration gradient and the prevailing plasma membrane potential and accumulation of the Cl^- requires coupling to a second system to achieve a net negative ΔG. Dittrich *et al* (1977) were the first to suggest that a nH$^+$/Cl symport system operated in guard cells since stomatal opening was enhanced as the external pH surrounding epidermal strips was decreased. This system remains the most probable candidate. Further details of Cl transport in plant cells are discussed by Sanders (1984).

8.3.7 Properties and activity of the tonoplast H^+ pumps

During opening, it is anticipated that the majority of ions accumulated will be transferred to the vacuole, with generation of a compatible solute in the cytoplasm, such as sucrose (see Chapter 9), to increase the osmotic pressure in the cytoplasm However, there has been remarkably little work on the pumps that might energize the tonoplast in guard cells Fricker and Willmer (1990b) demonstrated proton pumping activity in crude homogenates of GCPs of *C communis* that was dependent on Mg-ATP and had a pH optimum at pH 8 0 Pumping was inhibited by nitrate and *N*, *N'*-dicyclohexyl carbodiimide, but not vanadate These features are characteristic of a vacuolar (V-type) proton pump (Sze, 1985) Mg-ATP also stimulated an inwardly directed current in isolated guard cell vacuoles (Hedrich *et al* , 1988, Pantoja *et al* , unpublished) The V-type H^+-ATPase from GCP of *Commelina* has been further characterized by Willmer *et al* (1995) ATPase activity was higher in GCP than mesophyll cell protoplasts on protein, chlorophyll or membrane area bases The enzyme was stimulated by Cl and inhibited by high concentrations of Ca (I_{50} = 1 8 mM), whilst K^+ had no effect Recent studies have indicated the presence of several isoforms of V-type ATPase in plants and that multiple genes encode a major subunit of the enzyme (see Sze *et al* , 1992) It is intriguing to speculate that the high V-type ATPase activities in guard cell are related to expression of a specific isoform However, Willmer *et al* (1995) could not find evidence for specific control of the V-type ATPase in guard cells, as incubation of GCP in light, fusicoccin or ABA had no effect on the subsequent ATPase activity measured

The activity of the tonoplast H^+-ATPase would be expected to acidify the vacuole lumen It has long been known that large pH changes occur in different cell types of the epidermal layer during stomatal movements (e g Scarth, 1929, 1932, Pekarek, 1936, Small and Maxwell, 1939, Pallas, 1966, Dayadan and Kaufman, 1975) Although in early work, the cellular compartment being measured was uncertain, it was probably the vacuole as it is the largest compartment within the cells In general, values of pH 4 0–5 5 were reported for guard cells of closed stomata rising to between pH 5 6 to 6 5 after opening in a variety of species Equally large pH changes moving in the opposite direction have been observed in epidermal and subsidiary cells as stomata open and close More recently, Penny and Bowling (1975) and Bowling and Edwards (1984) used pH-sensitive microelectrodes to measure vacuolar pH values in *C communis* and *Tradescantia virginiana* They found the guard cell vacuolar pH was between pH 5 2–5 3 and increased by 0 4–0 7 pH units upon opening The pH of the subsidiary cell vacuole followed in the same direction, but to a lesser extent, whilst the pH in the vacuole of the epidermal cell acidified during opening by 0 4–0 6 pH units The cytoplasmic pH remained around pH 7 0 in all three cell types during opening, at least in *Tradescantia* (Bowling and Edwards, 1984) (Fig 8 2) The vacuolar alkalinization observed during opening in all these data apparently contradicts the expected acidification from activation of an inwardly directed H pump as the primary event energizing the tonoplast One possibility is that accumulation of malate rather than Cl may increase the buffering capacity of the vacuole, combined with recycling of the protons via the *n*H /K antiport to give net alkalinization

8.3.8 Accumulation of K^+ by the vacuole

Potassium accumulation into the vacuole cannot occur passively against an estimated tonoplast membrane potential of +20 to +40 mV, with the vacuole more positive than the cytoplasm (We have not adopted the recent change in convention that advocates that the tonoplast membrane potential should be referenced to the extra-cytosolic medium, i e the vacuole lumen, by analogy to the plasma membrane potential, such that $V_m = V_{cytosol} - V_{vacuole}$, Bertl *et al* , 1992 Using this convention, the cytoplasm would be -20 to -40 mV more negative than the vacuole Unfortunately this convention introduces confusion with the earlier literature and is particularly inappropriate for discussion of ion accumulation by vacuoles)

K^+ uptake may be mediated by a K^+/nH^+ antiport system or by a tonoplast pyrophosphatase which is thought to pump H^+ and K simultaneously into the vacuole There is currently no experimental support for the tonoplast cation exchanger in experiments designed to assay for its presence (Fricker and Willmer, 1990b) and there is only limited evidence for the tonoplast pyrophosphatase from pyrophosphate induced proton pumping in vesicles (Fricker, unpublishe

d) In addition, operation of the H^+/K *symport* driven by the tonoplast-pyrophosphatase is more difficult to reconcile with the vacuolar alkalinization observed, without an additional mechanism to recycle the stoichiometric increase of protons in the vacuole

Despite the relatively small tonoplast pH gradient and membrane potential, the thermodynamic driving forces would be sufficient to accumulate at least a tenfold excess of K^+ if coupled to H^+ exchange Ward and Schroeder (1994) have suggested that continued H pumping may also provide the driving force for rapid K^+ *efflux* during closure by maintaining the membrane potential positive inside the vacuole and this might explain the observed vacuolar acidification on closure

8.3.9 Accumulation of anions by the vacuole

The predicted tonoplast membrane potential (+20 to +40 mV) would be sufficient to drive a limited (5-fold) accumulation of anions passively via channels A category of channels that activate rapidly (and hence termed Fast-activating Vacuolar (FV)-channels) have been characterized in the tonoplast of some cells (see Hedrich and Schroeder, 1989) In principle, FV-channels could account for anion influx at positive tonoplast membrane potentials, but there have been no detailed characterization of FV-channels in guard cell vacuoles Guard cell vacuoles can accumulate significant concentrations of malate , and Pantoja *et al* (unpublished) found evidence for a 40 pS, voltage-gated malate channel in guard cell vacuoles from *C communis*, although this channel awaits detailed characterization

8.3.10 Efflux of K^+ from the vacuole during stomatal closure

K^+ efflux from the vacuole is favoured by the prevailing membrane potential and assumed vacuole to cytoplasm concentration gradient of K^+, and could thus occur passively through activation of an appropriate ion channel There have been relatively few studies and no concensus at present as to the identity of these channels Hedrich *et al* (1988) reported a high-conductance channel at the tonoplast from a range of cell types, that showed little discrimination between cations and anions This channel showed slow activation kinetics and was therefore termed a Slow-Vacuolar (SV)-type channel The SV-channel in GCPs from *Allium cepa* has recently been characterized in more detail by Amodeo *et al* (1994) They found the half time to activate the channel was in excess of 1 s and the channel was strongly selective for cations over anions, but showed little discrimination between the monovalent cations ($Na^+ > K^+ > Rb^+ > Cs^+$) Ward and Schroeder (1994) also found SV-channels in GCPs from *V faba* that were activated at high (10–100 μM) cytoplasmic Ca^{2+} levels and negative tonoplast potentials, similar to the SV-channel reported by Amodeo *et al* (1994) and SV-channels in vacuoles from other cell types (e g Hedrich and Neher, 1987) They demonstrated that these channels were also permeable to Ca^{+}, as well as to K^+, but not significantly to Cl It is not clear, however, whether the negative tonoplast potentials required to activate SV-channels occur *in vivo*, thus the significance of these channels in stomatal function is not fully understood (although see below)

In addition to the SV-channel in the tonoplast of GCPs from *V faba*, Ward and Schroeder (1994) also found a second, voltage-independent, channel with a high conductance (70 pS) and high selectivity for K^+ ($K^+ > Rb^+ > NH_4 >> Cs^+ = Na^+ = Li^+$), which they termed a Vacuolar K^+ (VK)-channel The VK-channel was also strongly activated by cytoplasmic Ca^{2+} over a physiological concentration range (Ward and Schroeder, 1994) This channel could thus provide a route for K^+ efflux at the predicted vacuole-positive potentials *in vivo*, in response to an elevation of cytoplasmic Ca^{+} Intriguingly, VK-channels also showed a strong dependence on cytoplasmic pH, activating with decreasing pH This is exactly the opposite response to that observed for cytoplasmic pH activation of the plasma membrane K outward channel (see Section 8 3 12)

One way of integrating the mode of action of the SV- and VK-channels was suggested by Ward and Schroeder (1994) as follows elevated cytoplasmic Ca could activate VK-channels allowing K^+ efflux and depolarizing the tonoplast membrane potential sufficiently to activate SV-channels The resulting efflux of Ca from the vacuole through the SV-channel would further depolarize the potential and maintain the Ca^{2+} activation of both VK-channels and SV-channels To maintain the driving force for continued K^+ efflux would require a mechanism to repolarize the membrane, possibly through parallel anion efflux, through an as yet uncharacterized pathway, or continued activity of the tonoplast H^+ pumps

8.3.11 Efflux of anions from the vacuole during stomatal closure

Anion efflux from the vacuole could occur via channels in conjunction with depolarization of the tonoplast potential during stomatal closure However, the appropriate channels have yet to be identified in guard cells

8.3.12 Efflux of K$^+$ across the plasma membrane during stomatal closure

Release of K^+ during closure requires depolarization of the plasma membrane potential to remove the driving force for K^+ uptake and allow K^+ efflux At membrane potentials more positive than –100 mV the K^+ inward rectifier closes and will not, therefore allow any K efflux However, a second K channel, termed the K^+ outward rectifier, opens as the membrane potential falls more positive than the K^+ equilibrium potential (e g Blatt, 1988, 1991) Under these conditions the driving force for K is out of the cell The K^+ outward rectifier has a slightly higher conductance (10–25 pS) than the K^+ inward rectifier There are, again, some differences between results for guard cells in epidermal strips and patch-clamped GCPs, notably a faster rate of current activation as the voltage is stepped to a range where the channel opens, and a stronger voltage dependence with the use of epidermal strips (reviewed by Blatt, 1991, Blatt and Thiel, 1993) It is possible that the properties of the K^+ outward rectifier are altered during protoplast isolation or a number of different outward rectifying K^+-channels exist The properties of the K outward rectifier are summarized in Table 8 4 and its voltage-gating properties depicted in Fig 8 4

In contrast to the K inward rectifier, the activation voltage of the K^+ outward rectifier alters markedly with external K^+, shifting to more positive values with increasing K concentrations This effectively tracks the K^+ equilibrium potential, and means the channel will only be open under conditions that favour K^+ efflux, even in the face of increasing

Table 8.4 Properties of the K^+ outward rectifier at the guard cell plasma membrane

	Channel characteristics	Reference
Conductance	10–65 pS	f, m, i
Activation voltage	activates at potentials more positive than −120 to −80 mV in intact cells	a–c, n
	activates at potentials more positive than −60 to −40 mV in patch-clamp experiments	d, f–m
	region of negative slope conductance at potential more positive than +50 to +80 mV	a, n
Modulation by ions and second messengers		
$[K^+]_o$	increasing $[K^+]_o$ shifts the activation voltage to more positive values	a, b, j
$[Ca^{2+}]_i$	(i) insensitive to $[Ca^{2+}]_i$	f, h, l
	(ii) increasing $[Ca^{2+}]_i$ reduces current amplitude	d
$[Ca^{2+}]_o$	increasing $[Ca^{2+}]_o$ reduces current amplitude	d, h
pH_i	current amplitude increased by increasing pH_i with no shift in activation voltage	c, h
pH_o	current amplitude increased by increasing pH_o and the activation voltage shifts to more positive values	g
Time dependence	(i) intact cells of *Vicia faba* – activation $t_{½}$ ~ 100 ms, deactivation $t_{½}$ ~ 11 ms	a, b
	(ii) patch-clamp of GCPs from *Vicia faba* – sigmoid or double exponential activation τ=100–600 ms, exponential deactivation τ=20–200 ms	e, g, i–k, m
	(iii) patch-clamp of GCPs from *Zea mays* – sigmoidal activation τ=41 ms, exponential deactivation mean τ=852 ms	d, e
	(iv) no time or voltage inactivation	j, k
Selectivity	$K^+ > Rb^+ > Na^+ > Li^+ >> Cs^+$	j
Inhibitors	TEA^+ (I_{50} 1 mM)	a, i, n
	Ba^{2+} (I_{50} 0 1 mM)	h, m
	Na^+ blocks when applied to the cytoplasmic side (I_{50} 80 mM)	n
	insensitive to 4-aminopyridine	i
	NH_4^+, benzylamine, trimethylamine and procaine (I_{50} 3–10 mM)	c
	insensitive to Al^{3+}	i

[a] Blatt, 1988b, [b] Blatt, 1990, [c] Blatt and Armstrong, 1993, [d] Fairley-Grenot and Assmann, 1992, [e] Fairley-Grenot and Assmann, 1993, [f] Hosoi, Iino and Shimazaki, 1988, [g] Ilan *et al*, 1994, [h] Lemtiri-Chlieh and MacRobbie, 1994, [i] Schauf and Wilson, 1987, [j] Schroeder, 1988, [k] Schroeder, 1989, [l] Schroeder and Hagiwara, 1989, [m] Schroeder *et al*, 1987, [n] Thiel and Blatt, 1991

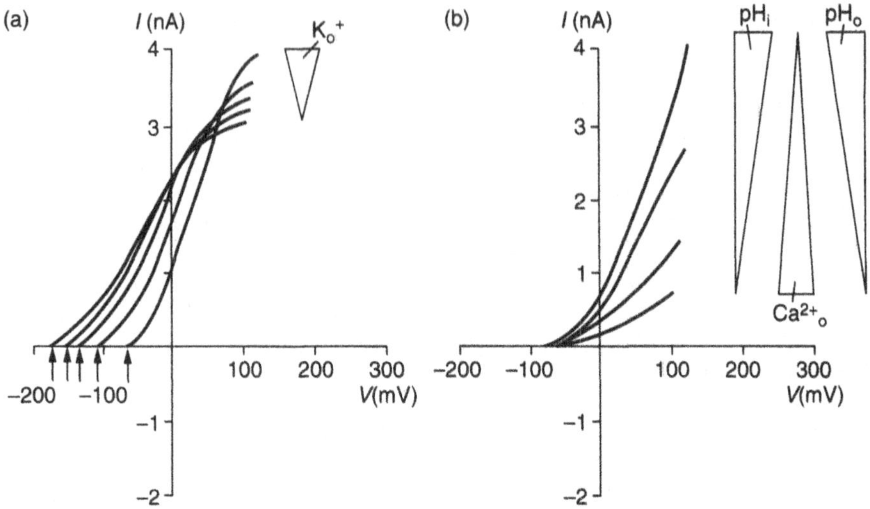

Figure 8.4 Current–voltage *(I–V)* characteristics of the K^+ outward rectifier at the plasma membrane of guard cells from *Vicia faba*. The channel always opens at voltages more positive than the K^+ equilibrium potential (indicated by arrows) as external K^+ concentrations ($[K^+]_o$) are increased (a). The current flowing through the channel also decreases as external Ca^{2+} ($[Ca^{2+}]_o$) is increased or as either internal pH (pH_i) or external pH (pH_o) is decreased (b). Note that the currents measured in guard cell protoplasts using patch-clamping techniques were about 5- to 10-fold lower than those shown here for intact guard cells. Data taken from reports in Table 8.4.

external potassium levels. This may be particularly pertinent to guard cells, when increases in apoplastic K^+ occur during closure (Bowling, 1986; Bowling and Smith, 1990).

Most significantly, in studies so far the K^+ outward rectifier was not affected by increases in cytosolic Ca^{2+} from physiological resting levels between 100 and 200 nM, although an increase in cytoplasmic Ca^{2+} from 2 to 200 nM caused a decrease in current amplitude in GCPs from *Zea* (Fairley-Grenot and Assmann, 1992b). In addition, there are some effects of external Ca^{2+} at high (10 mM) concentrations which also cause a reduction in current amplitude. However, the K^+ outward rectifier is dramatically activated by both cytoplasmic alkalinization (Blatt, 1992; Blatt and Armstrong, 1993; Lemtiri-Chlieh and MacRobbie, 1994) and apoplastic alkalinization (Ilan *et al.*, 1994). For example, Blatt and Armstrong (1993) found a change in pH of only 0.16 units from 'resting' pH around pH 7.8 was sufficient to increase K^+ currents by 250%.

Additional K^+ conductances have been described at the plasma membrane of guard cells of unknown function (e.g. Thiel *et al.*, 1992). Hosoi *et al.* (1988) also report a rare 4–5 pS channel that was voltage insensitive. These conductances are part of an ensemble waiting detailed char-

Table 8.5 Properties of the R-type anion channel at the guard cell plasma membrane

	Channel characteristics	Reference
Conductance	25–40 pS	b, c, f, g
Activation voltage	activates at voltages more positive than −80 mV	a–g
Modulation by ions and second messengers		
$[Cl^-]_o$	increasing $[Cl^-]_o$ shifts activation voltage to slightly more negative potentials and increases single channel conductance	b
$[Ca^{2+}]_i$	increased $[Ca^{2+}]_i$ increases current amplitude with no change in activation voltage	a, c, g
$[Ca^{2+}]_o$	increasing $[Ca^{2+}]_o$ increases current amplitude via increasing $[Ca^{2+}]_i$	a
Nucleotides	ATP, ATPγS, GTPγS all stimulate current synergistically with Ca^{2+}	a
Time dependence	activation $t_{½}$<10–>30 ms (voltage dependent)	a, f
	deactivation $t_{½}$=20–100 ms	a, g
	time-dependent inactivation $t_{½}$=10–20 s, slow recovery from inactivation	a, g
Selectivity	NO_3^- ≥ I^- > Br^- > Cl^- >> malate^{2-} impermeable to gluconate and glutamate	a–c
Inhibitors and activators	(i) Zn^{2+} blocks channel	c
	(ii) auxin, malate and a number of externally-applied channel blockers cause a shift in the activation voltage to more negative potentials, reduce the current at positive potentials (anion influx) and have time- and concentration-dependent effects on peak current amplitude, see below	
	Auxins 1-NAA>2,4-D>IAA	a
	Channel blockers NPPB>IAA-94>niflumic acid>ethacrynic acid>A-9-C> probenecid	d
	Stilbenes DIDS (I_{50}=0.2 μm)>DNDS (I_{50}=0.5 μm)>SITS (I_{50}=4 μm)	e
	malate (I_{50}=0.4 mM)	b

[a] Hedrich *et al.*, 1990, [b] Hedrich and Marten, 1993, [c] Keller *et al.* 1989, [d] Marten *et al.*, 1992, [e] Marten *et al.*, 1993, [f] Marten *et al.*, 1991, [g] Schroeder and Keller, 1992

acterization that are often grouped under the heading of a 'leak' conductance. A separate category of channels have been characterized by Cosgrove and Hedrich (1991) that are activated by suction applied to the patch-electrode and are thus called stretch-activated channels. The major stretch-activated K^+-channel in guard cells showed little voltage dependence or selectivity between monovalent cations and had a conductance of 25 pS for inward currents and 45–50 pS for outward currents. The channel did not immediately inactivate after release of suction and could remain active for up to 5 min. A second stretch-activated K^+-channel with lower conductance was also observed occasionally. At the moment it is not clear how to integrate these conductances in models of stomatal behaviour based on the non stretch-activated channels already characterized.

8.3.13 Efflux of anions across the plasma membrane during stomatal closure

The route for anion efflux at the plasma membrane has been characterized in some detail. Two main anion channels have been found, that differ primarily in their activation kinetics. One of these channels opens rapidly as the voltage is stepped to an activating range ($t_{1/2}$ 10–30 ms) and also spontaneously inactivates after a few seconds ($t_{1/2}$ 10–12 s) (Hedrich *et al.*, 1990) (see Table 8.5 and Fig. 8.5a and b). This channel has been named according to a variety of acronyms including GCAC1, standing for the guard cell anion channel (Hedrich *et al.*, 1990), R-type for rapid activating (Schroeder and Keller, 1992) and QUAC for quick activating anion channel (Linder and Raschke, 1992). We have adopted the R-type nomenclature as it parallels other existing nomenclature for ion channels (see Hedrich and Schroeder, 1989). R-type channels show a pronounced peak current amplitude between –10 to –80 mV, but are completely closed at more negative potentials (Fig. 8.5d and e).

The other major type of anion efflux channel required several tens of seconds to reach full open probability and has therefore been called S-type (Schroeder and Keller, 1992) or SLAC for slow activating anion channels (Linder and Raschke, 1992) and it exhibits little time-dependent inactivation over tens of seconds (see Table 8 6 and Fig. 8 5a). The S-type channels showed a current maximum at –20 to –50 mV (Schroeder and Hagiwara, 1989; Linder and Raschke, 1992; Schroeder and Keller, 1992; Schroeder *et al.*, 1993), but remained open at more negative membrane potentials, including the normal range of the resting membrane potential in guard cells (Linder and Raschke, 1992; Schroeder *et al.*, 1993) (Fig. 8.5c).

Both R- and S-type channels were activated by increases in cytoplasmic Ca^{2+} (Schroeder and Hagiwara, 1989; Keller *et al.*, 1989) (Fig. 8.5d) and the R-type channel required binding of a nucleotide for full activity (Hedrich *et al*, 1990). The two channels have been distinguished further on the basis

Figure 8.5 Time dependence and current–voltage (*I*–*V*) relations of the anion channels at the plasma membrane of guard cells from *Vicia faba* measured using patch-clamp techniques The S-type anion channel activates slowly over tens of seconds during continuous stimulation (a) whilst the R-type anion channel activates rapidly, but subsequently inactivates over a few seconds as stimulation is continued (b) Both channels showed a maximum conductance at around –50 mV (c–e), although only the S-type channel typically remains open at conductances more negative than –100 mV (c)

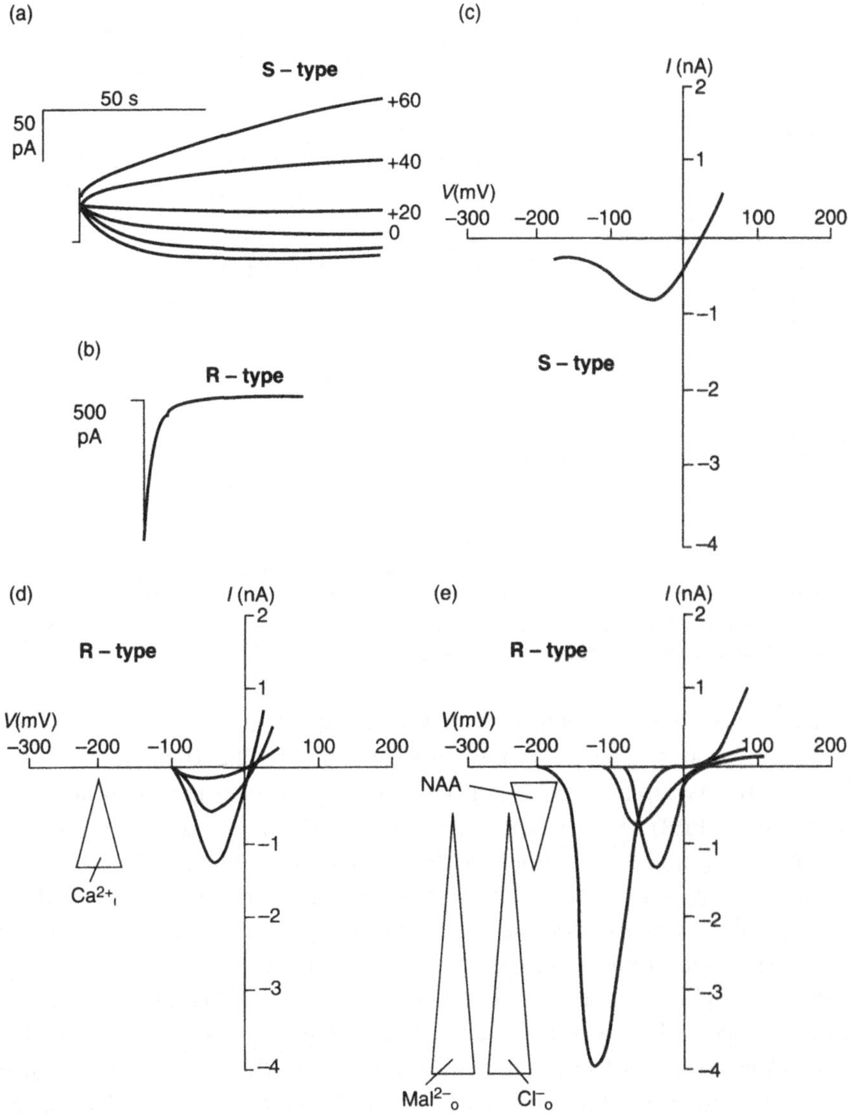

The voltage-gating characteristics of the R-type channel are affected by a variety of ligands Increasing cytoplasmic Ca^{2+} ($[Ca^{2+}]_i$) stimulates current flow through the channel without altering its voltage dependence (d) A range of channel blockers, including malate^{2-} and auxin, gave a complex voltage-dependent block of the channel Channel conductance was reduced at potentials positive of the Cl^- equilibrium potential in all cases, however, malate^{2-} caused an additional increase in conductance and shifted the gating-voltage to more negative values NAA also shifted the gating voltage to more negative values, but caused a decrease in conductance that developed over time (e) *Note that anion currents are more difficult to characterize in intact guard cells and tend to* be much lower in magnitude than those shown here, measured in guard cell protoplasts under optimized conditions Data taken from reports in Tables 8 5 and 8 6

Table 8.6 Properties of the guard cell plasma membrane S-type anion channel

	Channel characteristics	Reference
Conductance	1–36 pS	a, c, d, e
Activation voltage	(i) little voltage dependence	b
	(ii) peak current amplitude at –10 to –70 mV	a, c, d, e
Modulation by ions and second messengers		
$[Ca^{2+}]_i$	increased $[Ca^{2+}]_i$ increases current amplitude	b
$[Ca^{2+}]_i$	no effect	e
pH_o	no effect	e
Time dependence	(i) slow activation and deactivation ($t_{½}$>5–10 s)	a, c, d, e
	(ii) little inactivation	
Selectivity	NO_3 >Br >F ~Cl^-~I > $malate^{2-}$	e
Inhibitors and activators	Voltage-independent block by external antagonists NPPB (I_{50} 7 μm)>IAA-94 (I_{50} 10 μm)>>DIDS	d

[a] Linder and Raschke, 1992, [b] Schroeder and Hagiwara, 1989, [c] Schroeder and Keller, 1992, [d] Schroeder *et al*, 1993, [e] Schmidt and Schroeder, 1994

of their inhibition by pharmacological agents: the S-type channels was fairly insensitive to 4,4′-diisothiocyanostilbene-2,2′-disulphonic acid (DIDS), but completely inhibited by 5-nitro-2-(3-phenylpropylamino)benzoic acid (NPPB) (Schroeder *et al.*, 1993), whilst R-type channels were affected by a range of channel blockers, including DIDS at low concentrations (Marten *et al.*, 1992, 1993). The extent of the inhibition for both channel types was also dependent on the voltage, with complete inhibition at positive membrane potentials and variable effects at negative membrane potentials. There appeared to be two potential binding sites for the inhibitor accessible from the external face of the R-type channel. Binding of channel blockers to one site caused a shift in the activation voltage to more negative potentials and binding to the second site caused a voltage-dependent alteration in the current amplitude. In the latter case, all the ligands inhibited ion flux at potentials more positive than the equilibrium potential for Cl , but had concentration-dependent and time-dependent effects on the peak current in the range –10 to –80 mV (Marten *et al*, 1992, 1993). A similar voltage-dependent block of the R-type channel occurred in response to two physiological important substances, i.e. IAA (Marten *et al*, 1991) and $malate^2$ (Hedrich and Marten, 1993) (Fig. 8.5e). The possible consequences of $malate^2$ and IAA action on stomatal responses are considered further in Sections 8.6.2 and 8.6.5, respectively.

Specific binding of IAA-23, an auxin analogue, allowed Marten *et al* (1992) to purify a 60 kDa protein by affinity chromatography, which cross-reacted with antibodies to anion channels from kidney membranes This polypeptide is thus a possible candidate for at least one subunit of the R-type anion channel

8.4 Summary of ion movement events during stomatal movements

The putative transporters involved in ion fluxes during stomatal movements are depicted in Fig 8 6 Stomatal opening requires activation of the plasma membrane H^+-ATPase to hyperpolarize the membrane potential negative of the K^+ equilibrium potential Activation of a K^+ inward rectifying channel or a voltage-insensitive K^+-channel would allow K^+ uptake, whilst balancing anions could be accumulated through a putative $2H^+/Cl^-$ symporter or stimulation of malate^{2-} synthesis Equally important will be energization of the tonoplast by the tonoplast H^+-ATPase and activation of a putative tonoplast nH^+/K^+ antiporter or tonoplast H^+/K^+-pyrophosphatase and activation of an anion uptake channel This pattern of ion accumulation is not unique to stomatal opening, but appears to follow general models for salt accumulation in most plant tissues and has been termed the chemiosmotic hypothesis by Zeiger *et al* (1978), by analogy with the model first put forward by Peter Mitchell for transport coupling across mitochondrial membranes via the proton motive force (Mitchell, 1966) The more specialized function of guard cells is manifest in the variety of stimuli that promote stomatal opening and presumably the complexity of the signal transduction systems that must converge at some point to ensure integration of all these disparate events

Stomatal closure requires depolarization of both the plasma membrane and tonoplast, which may or may not involve inhibition of the primary pumps, and activation of appropriate ion efflux pathways These include activation of VK-channels, SV-channels and an anion efflux system at the tonoplast, and the K^+ outward rectifier, S-type and/or R-type anion channels at the plasma membrane

8.5 Signal perception and transduction by guard cells

Our understanding of the molecular events that initiate and co-ordinate the ion transport processes are based on models of signal perception and transduction developed extensively for animal systems It is likely that there are a number of the basic principles and components in com-

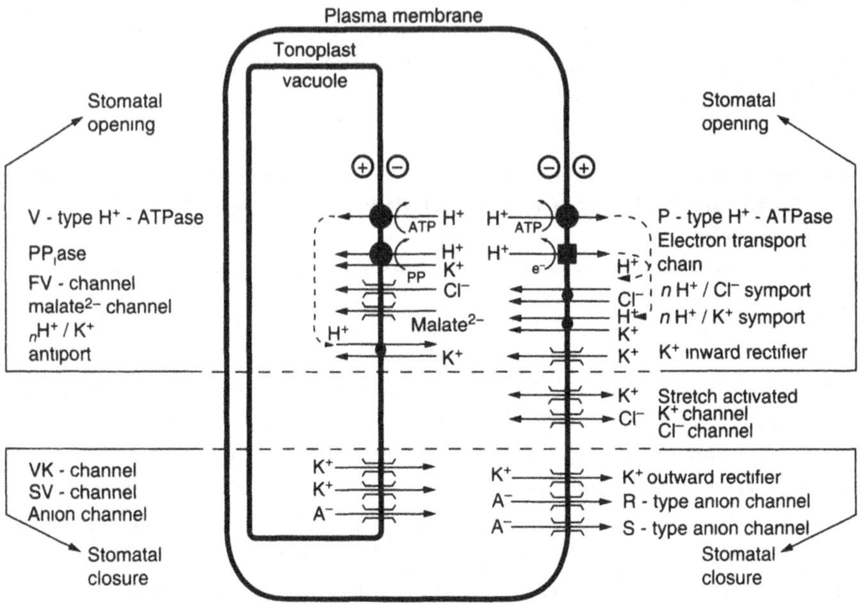

Figure 8.6 Summary of the ion transport systems associated with stomatal movements During stomatal opening K^+ movement at the plasma membrane occurs in response to the membrane potential established by either a proton pumping ATPase or a redox chain Cl uptake is achieved via a Cl /nH^+ symport system An additional K^+/nH^+ symport system may operate at low levels of external K^+ in the apoplast At the tonoplast, anion accumulation occurs in response to the membrane potential established by either a proton pumping ATPase or a pyrophosphatase K^+ movement occurs in response to the pH gradient via a K^+/nH^+ antiporter or concurrently with H^+ as part of the symport activity of the pyrophosphatase During stomatal closure, efflux of both cations and anions (A) at the tonoplast and plasma membrane is thought to occur via a variety of channels

mon and this has provided a useful stimulus to investigations in plants. Equally, however, it is recognized that plants have developed solutions to a rather different set of problems and may have developed a unique array of signalling pathways or utilize similar motifs in alternative configurations (Assmann, 1993). Much of the current evidence has focused attention on the role of cytoplasmic free Ca^{2+} as a signalling intermediate and how changes in cytoplasmic Ca^{2+} may be brought about in response to specific stimuli (e.g. Gilroy *et al.*, 1993). A brief description of the general model for Ca^{2+}-based signal transduction in animals is given below and in Fig. 8.7 (see Alberts *et al.*, 1994) and is reviewed in detail by Berridge (1993).

In animal cells, external signals interact with surface receptors to initi-

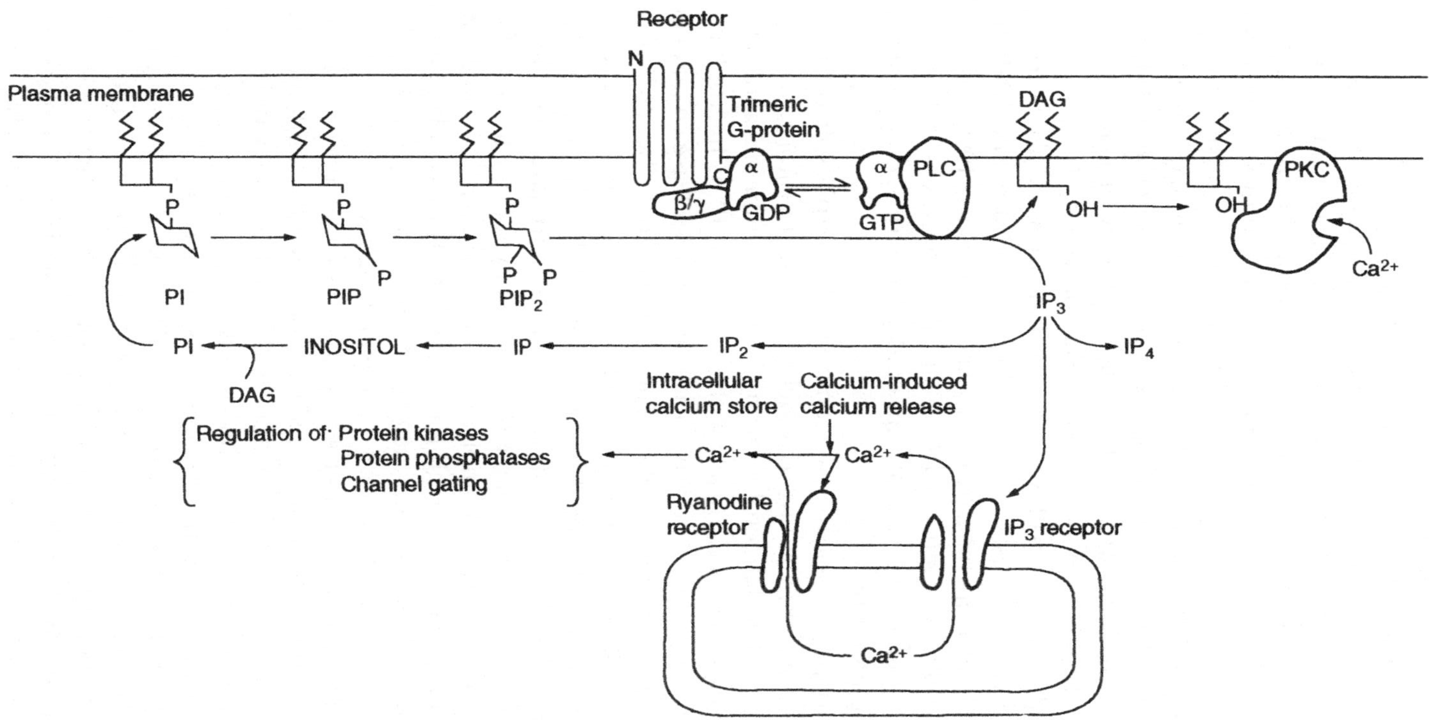

Figure 8.7 A schematic representation of the typical phosphatidyl inositol signalling pathway, thought to operate in animal cells, that is used as a model for signalling in plant cells Key GDP, guanosine diphosphate, GTP = guanosine triphosphate, PLC = phospholipase C, DAG = diacylglycerol, PKC = protein kinase C, IP_4 = inositol-1 3,4,5-tetraphosphate, IP_3 = inositol-1,4,5-triphosphate, IP_2 = inositol-1,4-biphosphate, IP = inositol-4-monophosphate, PI = phosphatidylinositol, PIP = phosphatidylinositol-4-monophosphate, PIP_2 = phosphatidylinositol-4,5-biphosphate

ate signal cascades, often operating through reversible protein phosphorylation. The receptors may be enzymatically active, such as tyrosine kinases, or act via association with an intermediate trimeric G-protein complex. On binding to the activated receptor, the α-subunit (Gα) of the G-protein exchanges bound GDP for GTP and dissociates from the βγ subunits. Gα remains active for about 1 s before hydrolysing the bound GTP, and diffuses in the membrane, where it may stimulate phospholipase C or adenylate cyclase. Adenlate cyclase cleaves ATP to give cyclic AMP, which in turn activates protein kinase A. Protein kinase A has not been identified in plants and it was previously thought that this rapid cAMP signalling pathway was not functional in plants. However, recent evidence suggests a role for adenylate cyclase and cAMP in promoting stomatal opening (Curvetto and Delmastro, 1990, Morsucci *et al.*, 1991, 1992, Curvetto *et al.*, 1994) although cyclic AMP does not stimulate opening of stomata in epidermal strips of *C. communis* (Willmer, unpublished).

Phospholipase C cleaves a membrane lipid (phosphatidyl-4,5-bisphosphate or PIP_2) to give diacylglycerol and inositol-1,4,5-trisphosphate (IP_3). Diacylglycerol remains in the membrane and activates protein kinase C synergistically with Ca^{2+}, whilst IP_3 is hydrophilic and diffuses through the cytoplasm to activate a Ca^{2+}-release channel in the endoplasmic reticulum termed the IP_3-receptor. The phosphatidyl inositol signalling system is considerably more complex however, as additional inositol metabolites, such as inositol-1,3,4,5-tetraphosphate and inositol-1,3,5-triphosphate are also known to have signalling functions. Elevations in cytoplasmic Ca^{2+} caused by release from intracellular stores may also show complex dynamics, ranging from single peaks to waves propagating through the cytoplasm. This complex behaviour is thought to allow a greater degree of control and spatial localization of the Ca^{2+} signal. A second endomembrane Ca^{2+}-release channel, termed the ryanodine receptor, can also be triggered by the original increase in cytoplasmic Ca^{2+} as part of a feedforward mechanism known as calcium-induced calcium release. Increased cytoplasmic Ca^{2+} may directly modulate other ion channels or protein kinase C activity, or act via interaction with the ubiquitous Ca^{2+} binding protein, calmodulin (CaM), and stimulation of Ca^{2+}-CaM dependent protein kinases. Propagation of the original signal will thus depend on the range of kinases present, their substrate specificity and the activity of protein phosphatases that reverse the kinase-mediated protein phosphorylation. Some classes of protein phosphatases are themselves activated by Ca^{2+}-CaM, making interpretation of some experiments, such as inhibitor studies, rather difficult.

8.5.1 Perception and transduction of ABA signals

There is much information on the action of ABA in guard cells where it is known to inhibit stomatal opening and promote stomatal closure (see Chapter 7) It has also been known for many years that exogenous Ca^{+} can inhibit opening and bring about closure in the absence of ABA (e g Schwartz, 1985, Schwartz *et al* , 1988, see Section 8 2 2) However, a link between ABA and Ca remained undetected until De Silva *et al* (1985a) observed that there appeared to be a synergistic effect of ABA and exogenous Ca on the inhibition of stomatal opening in epidermal strips of *C communis* (Fig 8 8) Curvetto and Delmastro (1990) also found that ABA or Ca prevented stomatal opening in *V faba*, but no evidence for a synergistic interaction was found Further reports indicated that exogenously applied ABA or Ca^{+} could contract swollen GCPs and inhibit swelling of GCPs independently (Smith and Willmer, 1987) In principle, external Ca could act as a signal influencing stomatal movements, however, there is evidence in *Commelina* that apoplastic Ca^{+} levels around guard cells are well buffered from changes in Ca^{+} levels in the rhizosphere or xylem (e g Ruiz *et al* , 1993, Ruiz and Mansfield, 1994, also see Section 8 2 2) Apoplastic Ca^{+} levels may affect the wall elasticity (see Bittisnich *et al* , 1987), but most effects of changing exogenous Ca on stomatal movements are thought to occur through Ca influx across the plasma membrane and modulation of cytoplasmic free calcium levels The electrochemical gradient driving Ca^{+} into the cytoplasm is very large, with both an estimated 10 000-fold concentration gradient and an inside negative membrane potential Rapid calcium influx could therefore occur passively on opening of a Ca^{2+}-channel at the plasma membrane So far four possible channels have been identified that could conduct Ca one is a non-selective plasma membrane cation channel that closes on depolarization of the plasma membrane (Schroeder and Hagiwara, 1990), according to Fairley-Grenot and Assmann (1992b) the K^{+} inward rectifier would also allow some influx of Ca a third possibility is a stretch-activated Ca^{+} channel identified by Cosgrove and Hedrich (1991), and there is evidence for another Ca channel that opens on depolarization of the plasma membrane (MacRobbie, 1989)

Although ABA can effect stomatal movements in the absence of external Ca^{+}, there is good evidence indicating cytosolic Ca^{+} levels change in response to ABA under some conditions Thus, McAinsh *et al* (1990) found that ABA induced a slow increase in cytoplasmic free Ca^{+} in guard cells of *C communis* from resting levels around 70–250 nM to over 1 μM and the stomata subsequently closed (Fig 8 9) (see Box 8 3) Increases in cytoplasmic Ca could modulate a number of ion transporters directly and constitute an important signal in guard cells, how

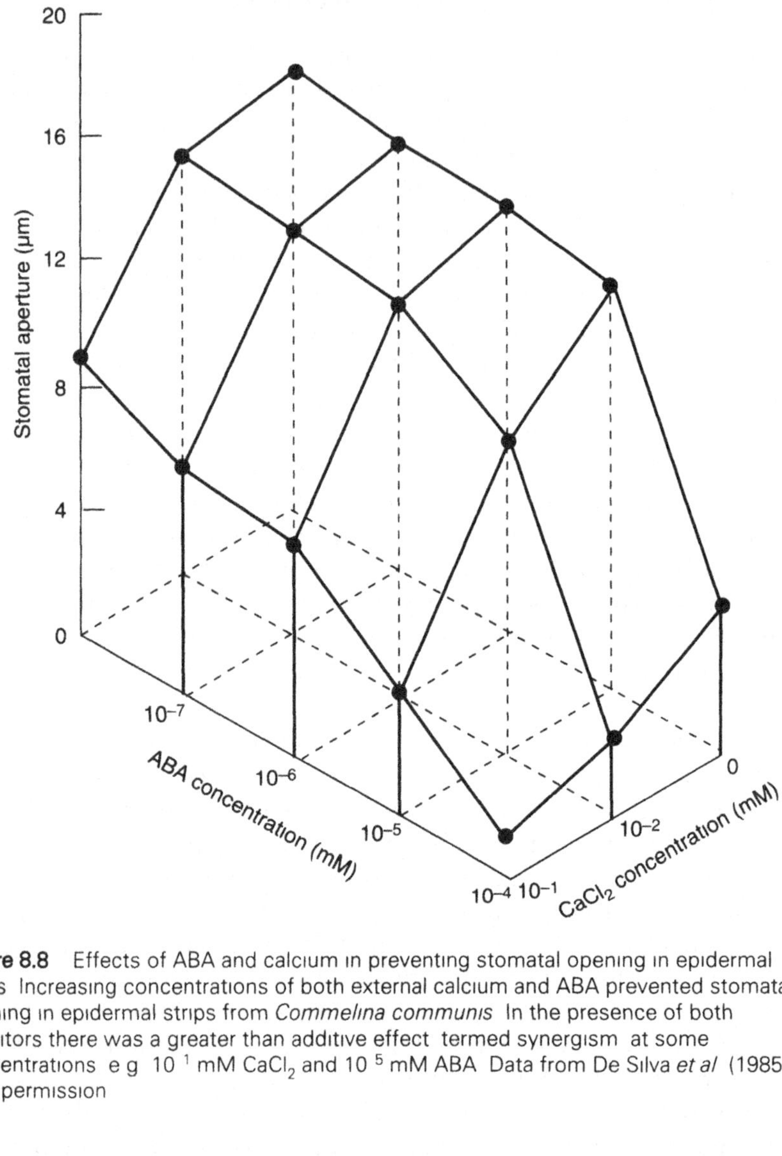

Figure 8.8 Effects of ABA and calcium in preventing stomatal opening in epidermal strips Increasing concentrations of both external calcium and ABA prevented stomatal opening in epidermal strips from *Commelina communis* In the presence of both inhibitors there was a greater than additive effect termed synergism at some concentrations e g 10^{-1} mM $CaCl_2$ and 10^{-5} mM ABA Data from De Silva *et al* (1985a) with permission

ever, the extent to which cytoplasmic Ca^{2+} levels alter in response to ABA is observed to be highly variable, even though stomata close in all cases Thus a variety of responses in terms of cytoplasmic calcium levels have been reported, including no detectable change of guard cell Ca^{2+} (Gilroy *et al* , 1991, Irving *et al* , 1992), transient spikes (Schroeder and Hagiwara, 1990, Gilroy *et al* , 1991, McAinsh *et al* , 1992), sustained elevations varying from 100 to 1000 nM above resting levels (McAinsh *et al* , 1990, 1992, Gilroy *et al* , 1991, Irving *et al* , 1992) or high amplitude oscillations (Gilroy *et al* 1991, McAinsh *et al* , 1992) The work of Allen *et al.* (1994) indicates that part of this variation may be due to the temperature the plants were grown at. Plants pre-conditioned at high temperatures consistently showed sustained high increases or oscillations of cytoplasmic Ca^{2+} in guard cells, whilst plants grown at relatively low temperatures showed no changes in cytoplasmic Ca^{2+} (Fig. 8.10). In all cases the stomata closed. Thus the extent to which the Ca^{2+}-signalling pathway operates depends on the previous environmental conditions experienced by the plants amongst other factors. As a result of such studies correlating ABA treatments with changing cytoplasmic Ca^{2+} levels and of the studies by MacRobbie (1990) examining the effects of Ca^{2+} on ion fluxes in guard cells induced by ABA, a view has developed that there are both Ca^{2+}-dependent and Ca^{2+}-independent components in the signal transduction pathway leading to closure (see reviews by MacRobbie, 1992; Schroeder, 1992; Assmann, 1993).

A second controversial area is whether ABA is sensed at the outer surface of the plasma membrane or inside the guard cell. There is evidence to support both contentions. Support of the view that the ABA receptor is located in the plasma membrane with an external binding site is as follows. If ABA is applied to stomata at pH 8.0 they close. Since, at this pH, ABA is fully ionized and in such a form it may not be able to diffuse across membranes, it was suggested that its effect was via an externally

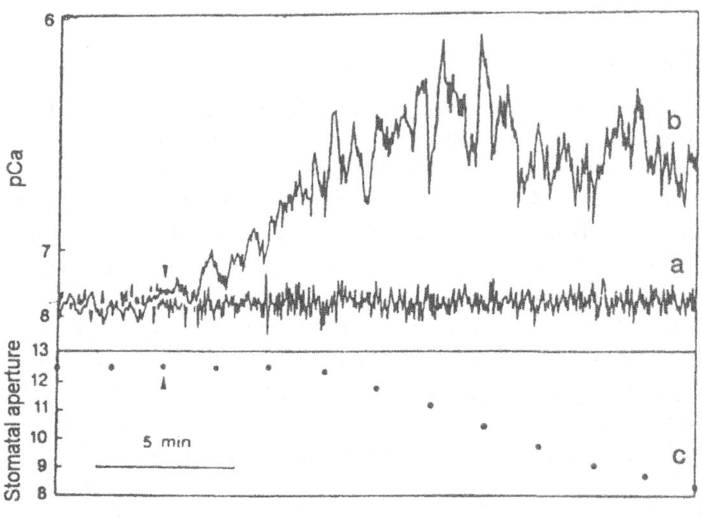

Figure 8.9 Changes in the free calcium level in the cytoplasm of guard cells from *Commelina communis* with time, measured using ratio photometry of the fluorescent calcium indicator dye, Fura-2. The calcium level in resting cells was about 25 nM (a), but increased over a 10 min period to about 600 nM with addition of ABA at the time point indicated by arrows (b). Stomatal closure ensued after a delay of about 5 min (c). Data from McAinsh *et al.* (1990)

Box 8.3 Measurement of ion activities using fluorescence techniques

Introduction

Dynamic changes of specific ions in the cytoplasm of living cells can be directly quantified and mapped using a combination of specific fluorescent probes and sensitive detectors such as photomultipliers or cameras The basic principle involves introduction of a fluorescent dye into the cytoplasm which binds to the ion of interest A variety of dyes have been synthesized that respond to Ca^{2+}, H^+, K^+, Na^+ and Mg^{2+} Although measurements can be made with dyes that exhibit a simple change in intensity on binding to the ion of interest, the most useful dyes have a shift in either the excitation or the emission spectrum A spectrum for a typical emission ratio dye for calcium is shown in Fig 8 B3a The extent of the spectral shift is related to the concentration of the ion, but independent of the concentration of the dye (Figs 8 B3a and b) An estimate of the shift in spectrum is obtained from the ratio of the intensity at two wavelengths, normally the peak for the bound form divided by the peak for the unbound form (Fig 8B3b) This can be calibrated against ratio values from known ion concentrations As the ratio is independent of dye concentration, the values are not affected by leakage of dye, bleaching of dye or changes in cell shape

Procedure

1 The dyes are highly charged in order to interact with the ion of interest and are membrane impermeant as a consequence Thus a range of specialized techniques have been developed to introduce the dyes into cells Ester derivatives of the dyes are uncharged and hence membrane permeant In the cytoplasm the ester groups are enzymatically cleaved off to release the free, active dye This approach potentially loads a large population of cells, but does not work well in most plant tissues as the esters appear to be hydrolysed externally in the wall or not at all The dyes are also uncharged at low pH (about pH 4 5) due to protonation of the carboxyl residues Cells can therefore be directly loaded with the dye at these pH values At pH 7 in the cytoplasm the protons dissociate to

Figure 8 B3 (a) Emission spectrum for the Ca^{2+} sensitive dye Indo 1 showing the shift in spectrum upon binding Ca^{2+} (b) Ratio measurements at the peaks for the Ca^{2+} free and Ca^{2+} bound form of the dye are insensitive to the amount of dye present and therefore accurately report the level of Ca^{2+} even if the dye concentration changes during the experiment (c) Schematic diagram of the typical system used to make fluorescence measurements from guard cells using imaging techniques or photometry

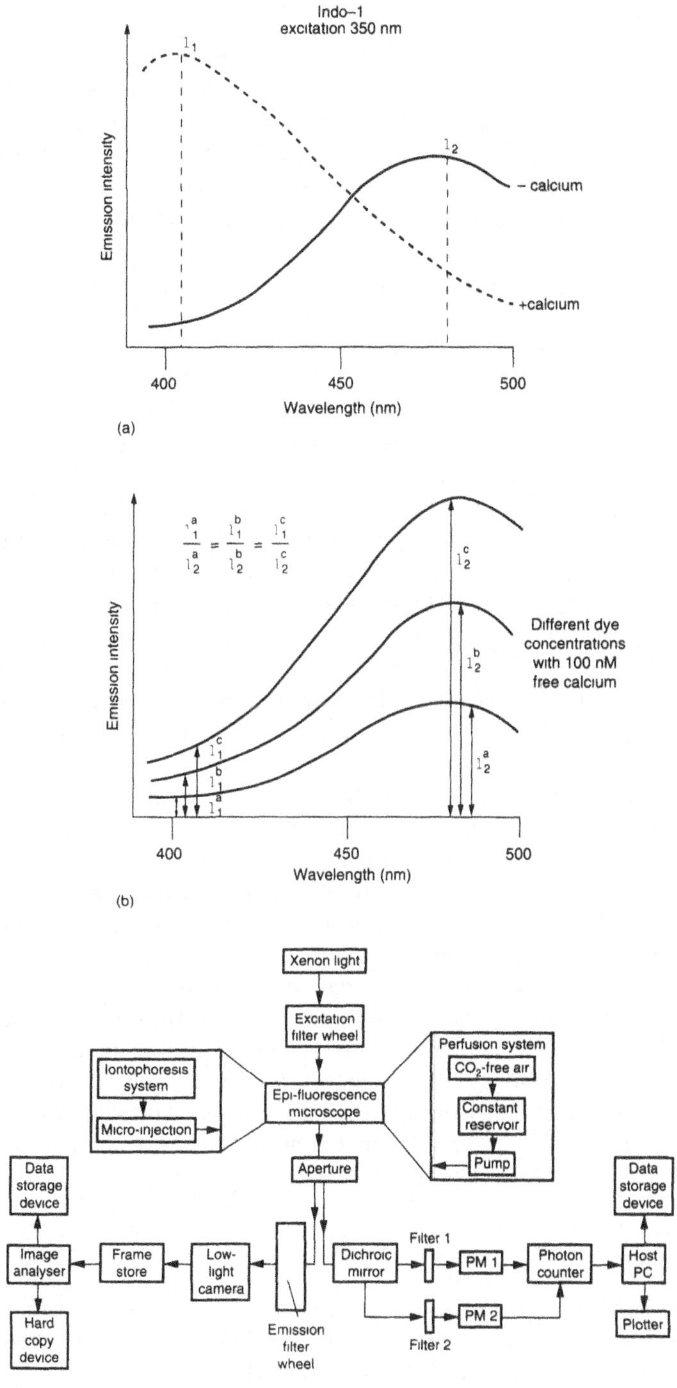

Indo–1
excitation 350 nm
Emission intensity
l1
l2
– calcium
+calcium
400
450
500
Wavelength (nm)
(a)
Emission intensity
Different dye concentrations with 100 nM free calcium
400
450
500
Wavelength (nm)
(b)
Xenon light
Excitation filter wheel
Iontophoresis system
Micro-injection
Epi-fluorescence microscope
Perfusion system
CO2-free air
Constant reservoir
Pump
Aperture
Data storage device
Image analyser
Hard copy device
Frame store
Low-light camera
Emission filter wheel
Dichroic mirror
Filter 1
Filter 2
PM 1
PM 2
Photon counter
Host PC
Data storage device
Plotter
(c)

give the free, active dye Some plant cells have been loaded by this technique, but in many, the uncharged dye binds very strongly to the cell wall Dyes can be loaded by microinjection using pressure or a small electric current directly into the cytoplasm from the barrel of a microelectrode used to penetrate the cell wall This technique is extensively used in the plant world, but suffers the drawback that only a single cell can be studied at any given instant

2 The fluorescence signal from the dye within a single cell is very weak and requires sophisticated detection apparatus Two basic approaches exist at present, depending on whether an average is made over a region of the sample or an actual image is taken Both operate on the same principle (Fig 8 B3c) Some ion indicators can also be imaged using confocal techniques described in Box 3 1
   i The dye is excited at the appropriate wavelength Dual excitation dyes require alternation of the excitation wavelength using spinning filter wheels or chopping between two light sources
   ii The epi-fluorescent signal from the dye is transmitted to the detector, being either a photomultiplier tube or an ultra-low light level camera Dual emission dyes require alternating wavelength selection or double measurement systems at this stage
   iii The ratio of the intensities at the two wavelengths is presented as a trace (photomultipliers) or an image Typically the image is colour coded to aid interpretation with high concentrations red and low concentrations blue
   iv Photomultiplier systems are potentially more sensitive and faster, but have poorer spatial resolution than camera systems

3 To quantify the response, the ratio values obtained are compared with ratios measured from known concentrations of the ion concerned The simplest calibration is performed in media designed to resemble the cytoplasm, with respect to ionic strength, pH, etc (external calibration) An alternative calibration can be performed *in situ* by allowing rapid equilibration of the internal concentration of the ion with a defined external concentration using a specific ion carrier (ionophore) Both methods have their problems and absolute values must be treated with caution An extensive description of the dyes available, instrumentation and calibration techniques can be found in Mason (1993)

References

Mason, W T (1993) *Fluorescent and Luminescent Probes for Biological Activity A Practical Guide to Technology for Quantitative Real-Time Analysis*, Academic Press, New York

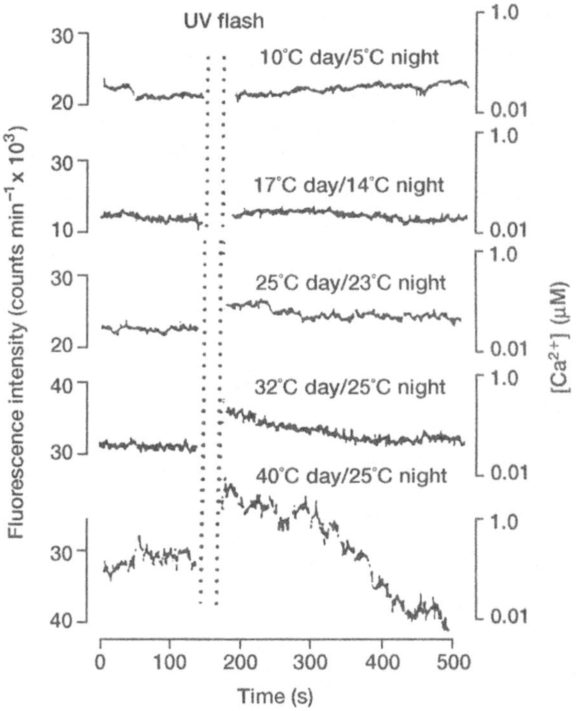

Figure 8.10 Calcium transients induced by photolysis of caged-ABA in the cytoplasm of guard cells of *Commelina communis* grown at varying temperatures Fluorescence of the calcium indicator, Calcium Green, was measured photometrically from individual guard cells simultaneously loaded with caged-ABA by microinjection Upon UV photolysis of the caged-ABA by a 30 s UV pulse calcium was observed to increase reproducibly in plants previously exposed to day temperatures of 25°C or higher No such changes were observed in plants grown at lower day temperatures, although a loss of turgor equivalent to stomatal closure always occurred in that guard cell Data from Allan *et al* (1994), with permission

facing receptor (e.g. Hartung, 1983). Another finding in support of an externally facing ABA receptor is that ABA binding proteins have been located that are accessible from the outside of the plasma membrane in *V. faba* (Hornberg and Weiler, 1984). A third piece of evidence is that when ABA is injected into guard cells of closed stomata, they still open in response to light (Anderson *et al.*, 1994). The data supporting the view that ABA receptor(s) are located inside the guard cell are equally numerous. Thus stomata respond to much lower concentrations of ABA at acid pH values than alkali ones, when ABA will be in the protonated form and readily enters guard cells. Furthermore, application of ABA directly to the cytoplasm by microinjection (Schwartz *et al*, 1994) or photo-release of ABA from an inactive 'caged' precursor (Allan *et al*,

1994) (see Box 8.4), causes closure of open stomata. Further evidence for internally located ABA receptors comes from the finding that when ABA is applied to the cytoplasmic face of an isolated patch of plasma membrane, outward rectifying K^+ channels are activated (Schauf and Wilson, 1987). Thus there is no clear concensus on this aspect at present. Indeed, there may be a number of different receptors for ABA some of which are located within the cell and some facing the apoplast (see Fig. 8.11).

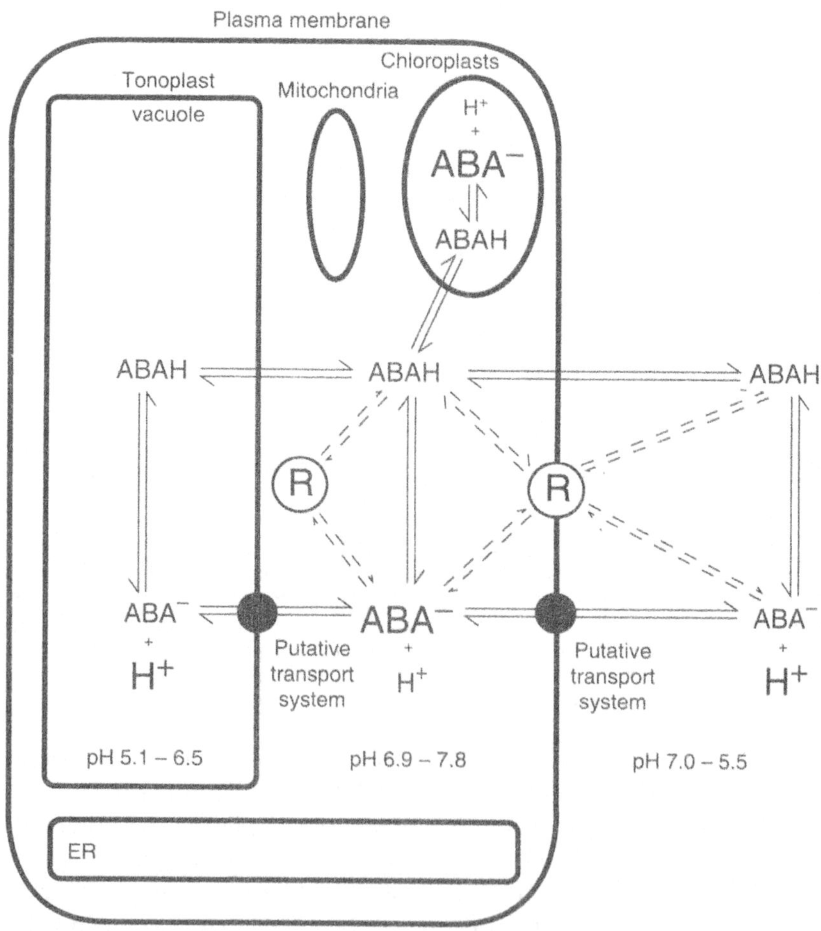

Figure 8.11 Predicted subcellular distribution of ABA based on the prevailing pH gradients in different compartments of a guard cell. The pH values are taken from a number of published sources and indicate the range of pHs that may occur during stomatal movements The protonated form of ABA is thought to be membrane permeable, although additional transporters may exist at the plasma membrane and/or tonoplast in *Commelina communis* The potential receptor sites (R) for ABA are also indicated.

Box 8.4 Photoreactive caged compounds

Introduction

Compounds or ions are complexed to 'cage' molecules that render them inert, before introduction into the cell, usually by microinjection. A short flash of high intensity UV light causes the cage to dissociate and release the compound in an active form. By controlling the intensity and duration of the illumination the amount of compound released can be finely regulated. This approach allows subtle and rapid perturbation of specific target systems within the cytoplasm. Most attention has focused on introduction of signalling molecules such as caged-Ca^{2+} and caged-IP_3. The technique can also be combined with electrophysiological measurements.

Procedure

The compound forming the cage can be derived from a variety of organic compounds by a series of simple reactions. The major problem is separation of the derivatized form of the compound from the uncaged and therefore active compound to prevent introduction of the active molecule. The coupling molecule for IP_3 is shown below:

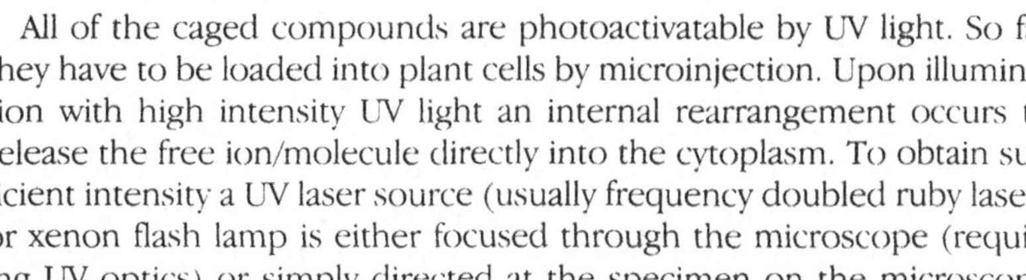

All of the caged compounds are photoactivatable by UV light. So far they have to be loaded into plant cells by microinjection. Upon illumination with high intensity UV light an internal rearrangement occurs to release the free ion/molecule directly into the cytoplasm. To obtain sufficient intensity a UV laser source (usually frequency doubled ruby laser) or xenon flash lamp is either focused through the microscope (requiring UV optics) or simply directed at the specimen on the microscope stage. A pulse of illumination lasting from milliseconds to seconds is sufficient to photolyse a defined amount of the cage and release known amounts of compound, although quantitation of the amount of compound released is not always straightforward within biological tissues, when the local intensity of illumination is variable and the volume of the compartment poorly defined. A number of controls are required to show that the potential for UV damage to the cell is minimized by the short duration of illumination and that the other products of photolysis, such as H^+ or acetate ions, do not initiate responses.

References

McCray, J A and Trentham, D R (1989) Properties and uses of photoreactive caged compounds *Ann Rev Biophys Biochem* , **18**, 239–270

While the actual receptors for ABA remain to be characterized, there is growing evidence that the link between the receptor and the Ca^{2+}-dependent part of the signalling pathway is mediated by the classical signal transduction pathway found in animal cells, involving the phosphatidyl inositol pathway (see Fig 8 7) Parmar and Brearley (1993) characterized turnover of inositol phospholipids in guard cells and detected a wide range of different phosphoinositides, including 3-phosphorylated phosphoinositides, but the intracellular Ca^{2+} channel activator, IP_3, was not unequivocally identified Parmar (discussed in MacRobbie, 1992) also showed that the turnover of inositol phospholipids was stimulated in response to ABA Moreover, Gilroy *et al* (1990) found evidence that release of Ca^{2+} from an internal Ca^{2+} pool in guard cells could be triggered by photorelease of IP_3 from an inactive 'caged' form which subsequently caused stomatal closure (Fig 8 12) (see Box 8 4) Blatt *et al* (1990) also showed that photorelease of IP_3 was associated with inactivation of the K^+ inward rectifier and activation of a putative anion efflux channel However, at the moment the link between results showing that activation of IP_3-gated Ca^{2+}-release can trigger stomatal closure and the natural elevation of guard cell IP_3 by a physiological signal remains somewhat circumstantial

The ability of IP_3 to stimulate release of Ca^{2+} from an internal store in guard cells is also the focus of another debate In other plant cells, Ca^{2+}-release channels that are activated by IP_3 are located on the tonoplast, with the vacuole as the major intracellular Ca^{2+} pool (Johannes *et al* , 1991) This is in contrast to animal cells, where the endoplasmic reticulum is the primary intracellular Ca^{2+} pool that is mobilized during signal transduction (Berridge, 1993) In addition, the IP_3-gated Ca^{2+} channel may not be the only intracellular Ca^{2+} release pathway in guard cells as Allen and Sanders (1994) have identified two additional vacuolar calcium release channels in GCPs from *V faba* Furthermore, Ward and Schroeder (1994) suggest that Ca^{2+} efflux from the vacuole can also occur through the non-selective SV-type channels on depolarization of the membrane to positive values The physiological significance of these various channels remains to be determined, but the range of putative Ca^{2+}-mobilizing channels at the plasma membrane and tonoplast are depicted in Fig 8 13

One consequence of an increase in cytoplasmic Ca^{2+}, arising by whatever means, would be modulation of the ion transport systems at the

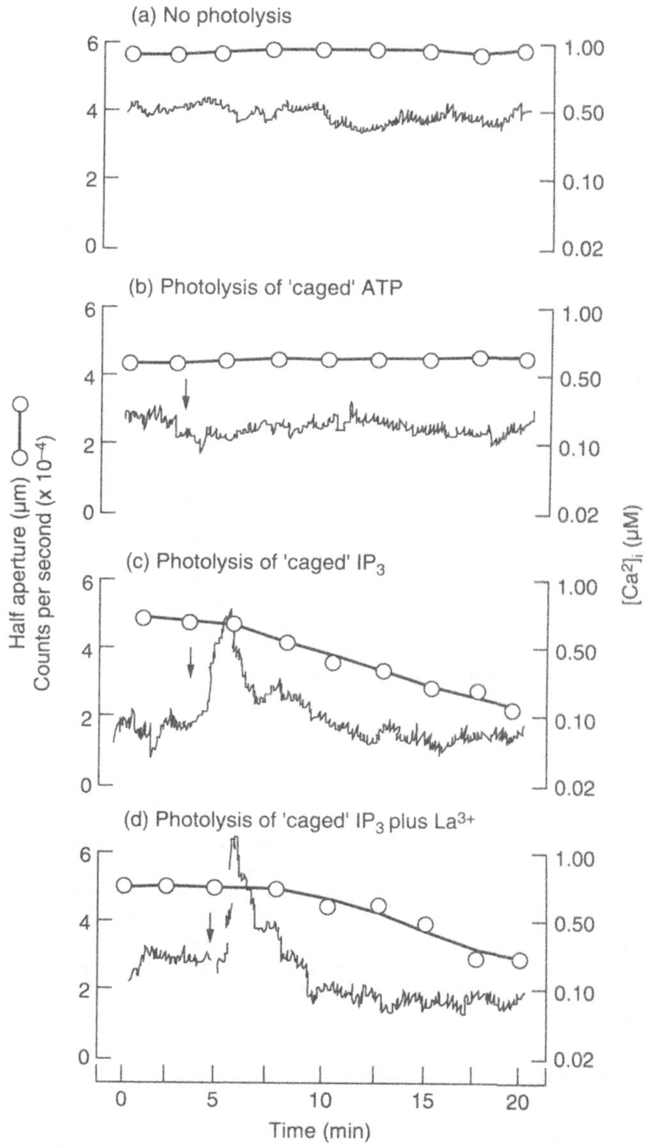

Figure 8.12 Photometric traces showing the effects of caged IP_3 on cytoplasmic calcium levels and stomatal apertures (○) from guard cells of *Commelina communis* previously microinjected with the calcium indicator, Fluo-3 Cytosolic calcium levels and stomatal apertures do not alter in cells where the caged-IP_3 is not photolysed (a) or where caged-ATP is photolysed as a control (b) UV photolysis of caged-IP_3 caused a transient increase in cytosolic calcium to about 1 μM and triggered stomatal closure (c) A similar response was observed in the presence of the impermeant calcium-channel blocker, lanthanum, indicating the source of the calcium was from internal stores (d) Arrows indicate where the cells were exposed to a 30 s UV flash Half aperture values are presented as only one guard cell of the pair was microinjected with caged compounds Data from Gilroy *et al* (1990), with permission

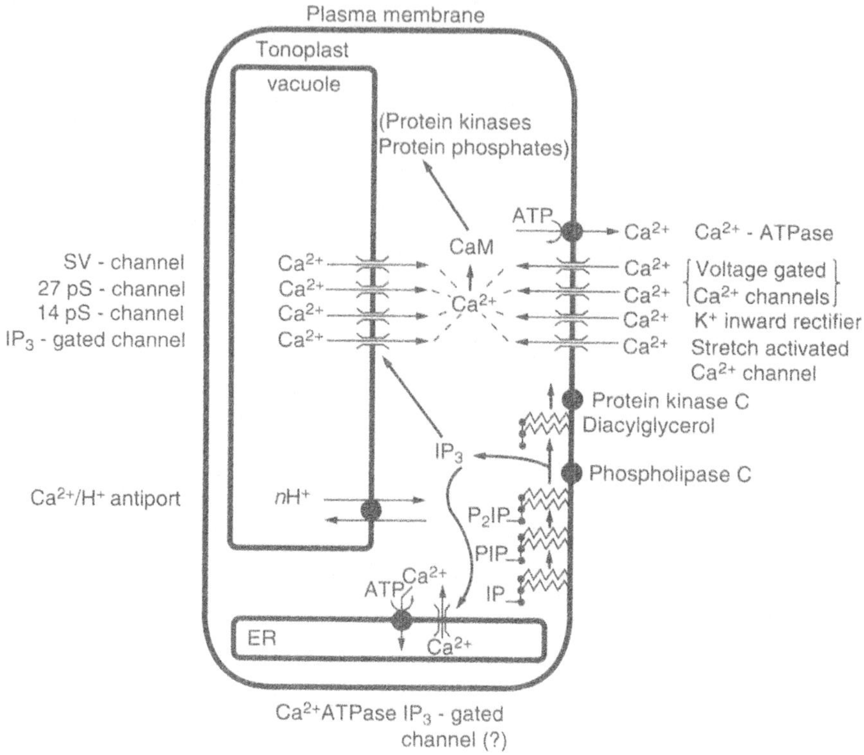

Figure 8.13 Summary of the potential ion-transport and signalling systems associated with cytosolic calcium in stomatal guard cells. Increases in cytosolic calcium may arise from influx of calcium through channels located at the plasma membrane, endoplasmic reticulum or tonoplast. Calcium is subsequently removed from the cytoplasm by active transport systems at one or more of these three membranes. Some of the effects of elevated calcium are thought to be mediated by reversible protein phosphorylation involving a number of calcium or Ca^{2+}-CaM-dependent protein kinases and protein phosphatases

plasma membrane and tonoplast. Thus elevated Ca^{2+} is known to inactivate the K^+ inward rectifier at the plasma membrane (see Section 8.3.5) which would prevent K^+ uptake necessary for stomatal opening. Elevated cytoplasmic Ca^{2+} is also known to activate both VK- and SV-type channels at the tonoplast (see Section 8.3.10) and S- and R-type anion channels at the plasma membrane (see Section 8.3.13). It would be expected that such channel activity would increase K^+ and anion efflux from the vacuole and increase anion efflux at the plasma membrane with concomitant stomatal closure. Notably, however, the K^+ outward rectifier, which is thought to be the major route for K^+ efflux at the plasma membrane, is not activated by increased Ca^{2+} above resting levels (see Section 8.3.12).

The increased cytoplasmic Ca^{2+} could also stimulate stomatal closure or prevent opening in other ways. For example, CaM is small protein which binds calcium and then activates a variety of different enzymes. CaM is present at high levels in guard cells and a number of CaM-binding proteins are also detectable in guard cells, at least two of which are enriched in these cells (Ling and Assmann, 1992). Furthermore application of CaM inhibitors prevented stomatal closing in response to ABA, CO_2 and elevated Ca^{2+} (De Silva *et al.* 1985b, Donovan *et al.*, 1985; Nejidat, 1987). At least some of the proteins activated by Ca^{2+}-CaM are likely to be protein kinases and protein phosphatases which modulate the activity of other enzymes by covalent phosphorylation. A specific link has been suggested between one such Ca^{2+}-CaM dependent protein phosphatase, called calcineurin, and the calcium-sensitivity of the K^{+} inward rectifier. Thus, Luan *et al.* (1993) showed that specific inhibitors of calcineurin could block the Ca^{2+}-dependent inactivation of the K^{+} inward rectifier. This suggests that elevated Ca^{2+} normally interacts with CaM, which then activates calcineurin causing dephosphorylation of the channel protein itself or an associated regulatory protein, and subsequent channel inactivation. The inhibitors used by Luan *et al.* (1993) are formed from complexes between the protein cyclosporin and a cyclosporin binding protein (called cyclophilin). Luan *et al.* (1993) also demonstrated that guard cells contained endogenous cyclophilin, providing further evidence for participation of this pathway in guard cells.

There is also evidence that other types of protein phosphatases are important in guard cell responses. Thus, Thiel and Blatt (1994) found that application of another phosphatase inhibitor called okadaic acid, which affects protein phosphatases type 1 and 2A but not calcineurin, also reduced currents through both the K^{+} inward rectifier and the K^{+} outward rectifier in guard cell of *V. faba*. Furthermore, in the *abi1* mutant of *Arabidopsis thaliana*, which has a wilty phenotype and stomata insensitive to ABA, the wild type gene involved normally encodes a protein serine threonine phosphatase type 2C (Meyer *et al.*, 1994; Leung *et al.*, 1994).

To make matters more complex, increases in cytoplasmic Ca^{2+} in guard cells have also been reported for stomatal *opening* in response to IAA, cytokinins and fusicoccin (Irving *et al.*, 1992). In addition, Ca^{2+}-CaM has been shown to activate some enzymes that result in stomatal opening. Thus, Shimazaki *et al.* (1992, 1993) found blue light dependent H^{+} pumping and stomatal opening were blocked by inhibitors of myosin light chain kinase which is activated by Ca^{2+}-CaM. Furthermore, a range of CaM inhibitors reduced blue light stimulated H^{+} pumping and stomatal opening (Shimazaki *et al.*, 1992, 1993) suggesting Ca^{2+}-CaM complexes are important in regulation of opening as well as ABA-induced closure. This level of ambiguity in results may reflect a genuine complexity in the signal transduction pathway, as the Ca^{2+}-CaM complex activates both protein kinases and protein phosphatases The net balance of these activities and hence phosphorylation state for any potential target is thus hard to predict

Following the scheme outlined in Fig 8 7, the link between the putative receptor for ABA at the plasma membrane and the phosphatidyl inositol pathway leading to elevation of Ca^{2+} from intracellular stores may be mediated by trimeric G-proteins Thus, Fairley-Grenot and Assmann (1991) found that continued activation of G-proteins using, for example non-hydrolysable analogues of GTP such as GTPγS, inhibited the K^+ inward rectifier As this effect was prevented by buffering the cytoplasmic calcium level, it was suggested an increase in cytoplasmic Ca^{2+}, possibly via the phosphatidyl inositol pathway, was involved In a subsequent paper, Wu and Assmann (1994) showed that changes in K^+ channel activity still occurred upon activation of G-proteins in membrane patches, suggesting that G-proteins can activate K^+ channels by a membrane delimited pathway However, G-protein activators can also stimulate stomatal opening (Curvetto and Delmastro, 1990, Lee *et al* , 1993), indicating the role of G-protein coupling in guard cells is complex and not yet fully understood

As indicated earlier, ABA is not always seen to elevate cytoplasmic Ca^{2+} levels, suggesting that part of the mode of action of ABA is independent of increases in Ca^{2+}, although a basal level of cytoplasmic Ca^{2+} is probably required (e g Lemtiri-Chlieh and MacRobbie, 1994) One possible way that this may happen is through changes in cytosolic pH Evidence in support of this view comes from the work of Irving *et al* (1992) who found ABA treatment of guard cells caused alkalinization of the cytoplasm by 0 04–0 3 pH units with a time lag of about 2 min Activation of the K^+ outward rectifier during exposure to ABA also occurred in response to alkalinization with a similar time lag (Blatt, 1990, Thiel *et al* , 1992, Blatt and Armstrong, 1993) Furthermore, activation of the K^+ outward rectifier in response to ABA did not occur if cytoplasmic pH was experimentally maintained constant (Blatt and Armstrong, 1993, Lemtiri-Chlieh and MacRobbie, 1994) Equally, a role for decreased cytoplasmic pH in opening responses was also suggested in the work of Irving *et al* (1992), who found that the cytoplasm acidified by 0 27 pH units in guard cells exposed to fusicoccin This change in pH may be responsible for inactivation of the K^+ outward rectifier during fusicoccin treatments reported by Blatt and Clint (1989)

Electrophysiological measurements of ion channel activities provide a good indication of the routes that ions may move during stomatal movements but do not give a clear picture of the magnitude of the ion fluxes that occur *in vivo* Tracer flux experiments allow measurement of unidirectional ion fluxes and provide complementary data to the electrophysiological studies Thus MacRobbie (1981, 1983), using acid treated epidermis to isolate guard cells, found that the rate constants for efflux of both K^+ and Cl were stimulated by ABA, reaching a peak rapidly before dropping to initial values after about 60 min (Fig 8 14a) MacRobbie (1983) and Brindley (1990b) also observed similar flux transients on transfer of epidermis from light to dark (Fig 8 14b) and such transients are probably a characteristic feature of stomatal closing responses Further analysis of the ABA-stimulated $^{86}Rb^+$ efflux showed that short pulses (less than 2 min) of ABA were sufficient to trigger a full efflux transient In addition, MacRobbie (1990) found that the putative receptors for ABA were rapidly de-sensitized and unable to respond to subsequent applications of ABA The effect of ABA on anion fluxes was similar to that on the cation fluxes (MacRobbie, 1984) there was a transient stimulation of Br (a Cl analogue) efflux with little or no reduction in influx The majority of this current is probably carried by S-type anion channels as R-type channels inactivate within tens of seconds and would not remain open over the 60 min closing period In support of a major role for S-type channels, Schroeder *et al* (1993) found that DIDS, a potent inhibitor of R type anion channels, did not aleviate ABA and malate induced stomatal closure In contrast, NPPB, which inhibits both S- and R-type anion channels, completely blocked the closing response

The responses to ABA described above relate to rapid alterations in guard cells ion content and the potential signal transduction systems that may be involved There is yet another possible way in which ABA may bring about its effects on stomata, particularly longer-term effects such as those seen in the after-effect of wilting (see Section 7 2 6) Preliminary experiments indicate that when GCPs are exposed to ABA the patterns of expression of certain proteins are changed Thus there is an increased synthesis or reduced degradation of certain proteins and some entirely new ones may be expressed Some of these proteins may be similar to stress proteins induced in other tissues However, none of these proteins have been specifically identified in guard cells and how they may regulate stomatal movements currently remains a mystery Taylor *et al* (1995) have also shown that reporter genes fused with ABA-regulated promotors were activated in guard cells of *Arabidopsis thaliana* and *Nicotiana tabacum* which indicated that guard cells are competent to relay an ABA signal to the nucleus (see Chapter 10)

8.5.2 Perception and transduction of CO_2-closing responses

Despite the central role of CO_2 in influencing stomatal behaviour, there are remarkably few papers dealing with the site of perception and

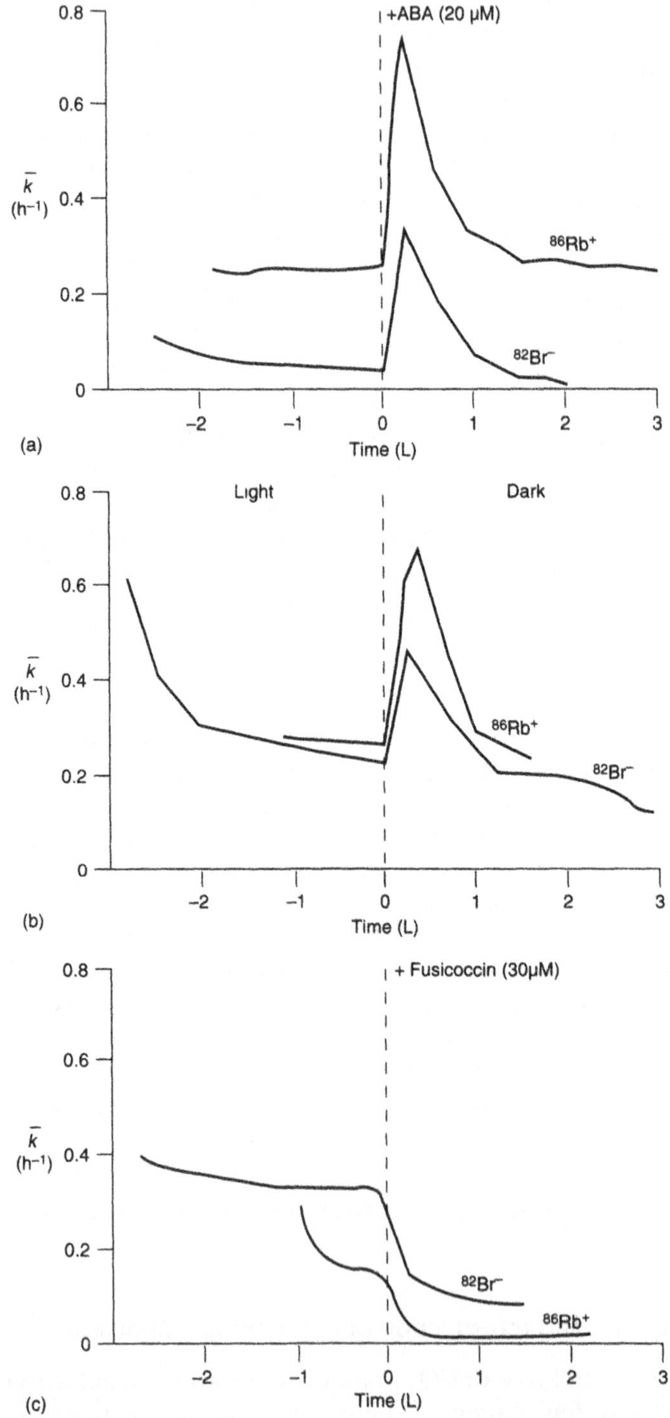

0.8
0.6
0.4
0.2
0
k̄
(h⁻¹)
+ABA (20 µM)
86Rb+
82Br−
−2
−1
0
1
2
3
(a)
Time (L)
Light
Dark
86Rb+
82Br−
(b)
Time (L)
+ Fusicoccin (30µM)
82Br−
86Rb+
(c)
Time (L)

mode of action of CO_2 (reviewed in Mansfield *et al.*, 1990). The physicochemical properties of CO_2 further complicate any simple model for its action (Fig. 8.15). It dissolves in water and is membrane permeable, although the measured plasma membrane conductance is lower than theoretically expected (Gimmler *et al.*, 1990), and CO_2 also dissociates to HCO_3^- to an extent depending on pH. The rate of equilibration, particularly of the reverse reaction, may be catalysed by carbonic anhydrase which is present in guard cells (Willmer and Birkenhead, unpublished). Guard cells appear to be able to discriminate small changes in external CO_2 in the range 0.01–0.036%, despite the changing pH regimes in the apoplast and cytoplasm, and varying levels of CO_2 production in guard cells by respiration and depletion by CO_2 fixation. According to Raschke (1977) full stomatal opening in the light requires a minimum level of CO_2 (but see Travis and Mansfield, 1977, and Chapter 6) and stomatal opening driven by fusicoccin increases with high levels of CO_2, in parallel with increases in epidermal malate levels (Travis and Mansfield, 1979b). These data would suggest that CO_2 has a dual action in guard cells, i.e. a stimulatory effect on opening through promotion of malate synthesis, which is normally masked by a second as yet uncharacterized closing signal (Mansfield *et al*, 1990). Recently, Hedrich and Marten (1993) have linked the two systems and put forward a hypothesis that CO_2 levels are sensed via increased synthesis of malate^{2-} in the guard cells or other leaf tissues under high CO_2 conditions. Excess malate^{2-} would be lost from the guard cell (Van Kirk and Raschke, 1978b) or may diffuse through the apoplast from other tissue to the guard cell apoplast. As external malate^{2-} effectively opens the R-type anion channel by shifting the activation voltage to more negative values and increasing the peak current amplitude (Fig. 8.5), there would be further release of anions A proportion of the anions released would be malate^{2-} which in turn, would generate a positive feedback signal to further activate the channel. Release of Cl^- also affects the channel behaviour, increasing the current conductance and maintaining the capacity for anion efflux as the anion concentration gradient declines. Such channel behaviour would result in stomatal closure.

Figure 8.14 The effects of ABA (a), light to dark transitions (b) and fusicoccin (c) on the efflux of $^{86}Rb^+$ and $^{82}Br^-$ from isolated guard cells of *Commelina communis* after labelling to steady state *k* is the apparent rate constant for exchange and is equivalent to the rate of tracer lost per time period divided by the average tracer content of the tissue in that period Both ABA (a) and light to dark transitions (b) stimulated a transient increase in the apparent influx rate constant for both $^{86}Rb^+$ and $^{82}Br^-$ Under these conditions there was little change in the apparent efflux rate constants In contrast, fusicoccin caused a marked and irreversible decrease in both efflux rate constants (c) Data redrawn from MacRobbie (1981 1983, 1984) Clint (1987) and Clint and MacRobbie (1984), with permission

8.5.3 Perception and transduction of light signals in stomatal opening

The main requirement to stimulate ion accumulation is hyperpolarization of the plasma membrane and tonoplast through energization of the primary proton pumps. Provided the plasma membrane potential is driven negative of both the potassium equilibrium potential and the activation voltage of K^+ inward rectifier, K^+ accumulation can occur. However, in many epidermal strip experiments using external K^+ levels in the range 30–75 mM, and possibly also *in vivo* where K^+ levels up to 100 mM have been reported around closed stomata (Bowling, 1987), the membrane potential is likely to be positive of the activation voltage for the K^+ inward rectifier. Under these conditions, K^+ uptake may be mediated by voltage-insensitive K^+ channels (e.g. Hosoi *et al.*, 1988). Increases in plasma membrane H^+-ATPase pump currents have been observed for a variety of light opening stimuli, but with markedly different kinetics (Serrano and Zeiger, 1988; Assmann, 1993; Kearns and Assmann, 1993). For example, perception of low intensity blue light by an unknown receptor activates the H^+-ATPase with a characteristic delay of about 30 s (Assmann *et al.*, 1985) and triggers H^+ efflux in GCPs (Shimazaki *et al.*, 1986; Willmer and Pantoja, 1992). H^+ extrusion is maintained for several minutes after the end of pulse of blue light, indicating that continuous input of blue light energy is not required to maintain the response. GCPs swell by 27% after similar blue light pulses (Amodeo *et al.*, 1992). In contrast, the red light effect is probably related primarily to energy supply from the chloroplasts in the form of ATP, which also stimulates proton pumping, but without a time delay. The current is not sustained after the light is removed. However, additional photosynthetic products other than ATP may also be involved in transmission of the light response, as the current is stimulated further by P_i in the presence of ATP (Serrano *et al.*, 1988).

The transduction pathway between opening stimuli, such as blue light, and activation of the pump has not been determined in detail. External application of synthetic diacylglycerols that activate protein kinase C in animals, induced H^+ pumping in GCPs and stomatal opening in epidermal strips (Lee and Assmann, 1991). Stimulation by the synthetic diacylglycerols was not additive to that of light, suggesting both processes acted through a common intermediary at some stage. Evidence for the presence of protein kinase C is lacking in plants, however, H-7, a relatively specific inhibitor of protein kinase C, prevented stomatal opening and stimulated stomatal closure (Lee and Assmann, 1991). Diacylglycerol may be produced by hydrolysis of a number of different membrane lipids including PIP_2 which also yields IP_3 but it is worth noting that if endogenous diacylglycerol were derived from hydrolysis of PIP_2 as part of the inositol pathway, the effect of diacylglycerol in stimulating opening is

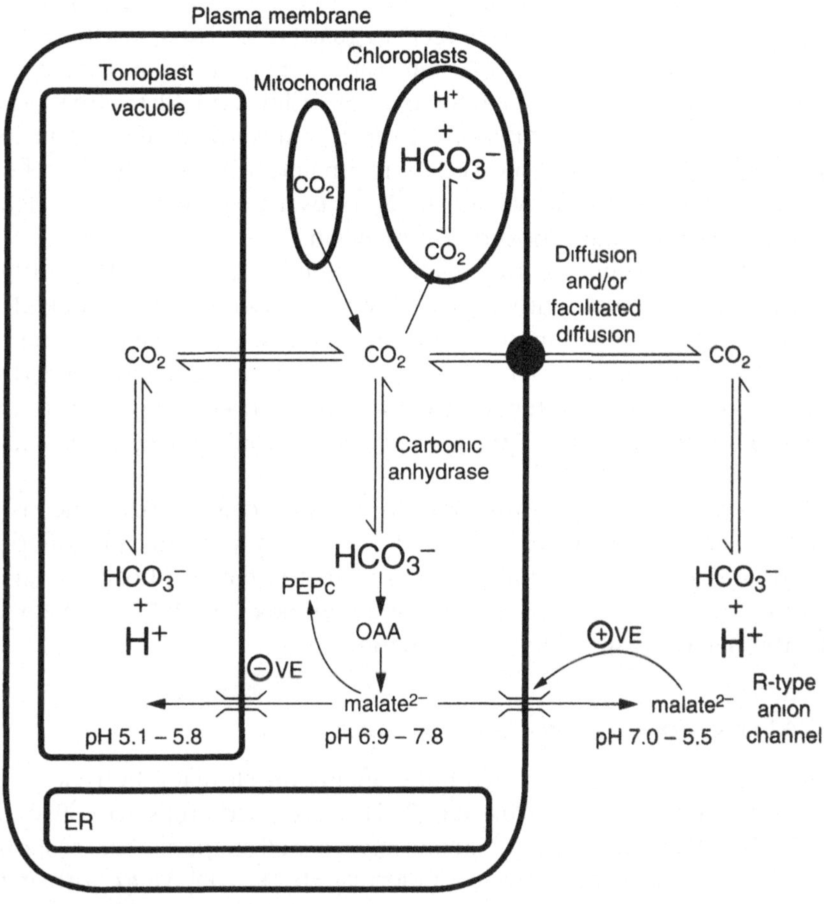

Figure 8.15 The predicted distribution of CO_2 and HCO_3 based on the prevailing pH gradients and metabolic activity in different compartments of a guard cell CO_2 may enter the guard cell by simple diffusion, although additional facilitated diffusion or active uptake may occur (Gimmler *et al* , 1990) HCO_3 is not thought to be taken up rapidly The possible link between perception of increased CO_2 and malate^{2-} activation of the plasma membrane R-type anion channel at the plasma membrane is also shown (Hedrich and Marten, 1993)

exactly opposite to the action of the other hydrolysis product, IP_3, which promotes closure (Gilroy *et al.*, 1990).

In contrast to diacylglycerols, fusicoccin was found to act synergistically with light at low concentrations, indicating the signalling pathways are separate (Assmann and Schwartz, 1992). Fusicoccin may activate the plasma membrane proton pump directly (e.g. Johansson *et al* , 1993) or indirectly via other proteins and fusicoccin was shown to stimulate currents through the H -ATPase in patch-clamp experiments (Serrano *et al*, 1988, Lohse and Hedrich, 1992, Assmann and Schwartz, 1992) However, fusicoccin may have additional effects in guard cells Thus, fusicoccin also caused a marked and irreversible stimulation of Rb^+ influx and a decrease in Rb efflux (Clint and MacRobbie, 1984, Clint and Blatt, 1989) (Fig 8 14), through a rapid inhibition of the K^+ outward rectifier (Blatt and Clint, 1989) In contrast, however, Assmann and Schwartz (1992) did not find any change in the activity of the K^+ outward rectifier in patch-clamp measurements during exposure to fusicoccin

Equally interesting are the results of Shimazaki *et al* (1992, 1993) who found that blue light stimulated stomatal opening was blocked by inhibitors of a protein kinase, called myosin light chain kinase in animal cells The properties of this enzyme remain to be characterized in plants, but in animal systems, myosin light chain kinase is activated by calcium and CaM, which are associated with stomatal closure in guard cells (see Section 8 6 1)

Reversible protein phosphorylation is a well characterized means to regulate enzyme activities, but so far the only proteins identified in guard cells that have actually been shown to change phosphorylation state is the chloroplast light-harvesting complex (LHCPII) (Kinoshita *et al*, 1993) and possibly PEPC (see Chapter 9)

8.5.4 Perception and transduction of auxin

The effects of natural and synthetic auxins on stomatal behaviour are difficult to evaluate (see Chapter 7) However, according to Lohse and Hedrich (1992), a range of auxins give a bell-shaped dose–response curve for stomatal opening in epidermal strips from *Vicia* with maximum stimulation at about a concentration of 5 μM concentration These auxin concentrations also caused a slow increase in H^+ pump currents in the whole-cell patch-clamp mode (Lohse and Hedrich, 1992) Similarly, Blatt and Thiel (1994) found increasing activation of the H^+-ATPase at auxin concentrations up to 10 μM and also observed activation of the K^+ inward rectifier in parallel At auxin concentrations higher than 10 μM, a reduction in the level of stimulation of opening was observed (Lohse and Hedrich, 1992) and Blatt and Thiel (1994) showed that the K^+ inward rectifier was inactivated at these concentrations while the K^+ outward rectifier and an anion 'leak' conductance were activated The increase in anion conductance may result from direct interaction of auxin with the R-type anion channel, causing a shift in activation potential to more negative values and time-dependent changes in current amplitude (Marten *et al* 1991)

Studies on auxin perception in other plant tissues have identified a number of auxin binding proteins which are potential candidates for auxin receptors (Palme, 1992). One such class of auxin binding proteins are soluble and predominantly located in the endoplasmic reticulum, but are thought to be secreted at the plasma membrane and mediate auxin effects through interaction with a putative 'docking' protein (see Fig. 8.16). In an attempt to characterize the interaction between the auxin binding protein and the 'docking' protein in guard cells, Thiel *et al.* (1993) monitored the effect of synthetic peptides corresponding to surface domains of the protein. A peptide spanning the C-terminus of the

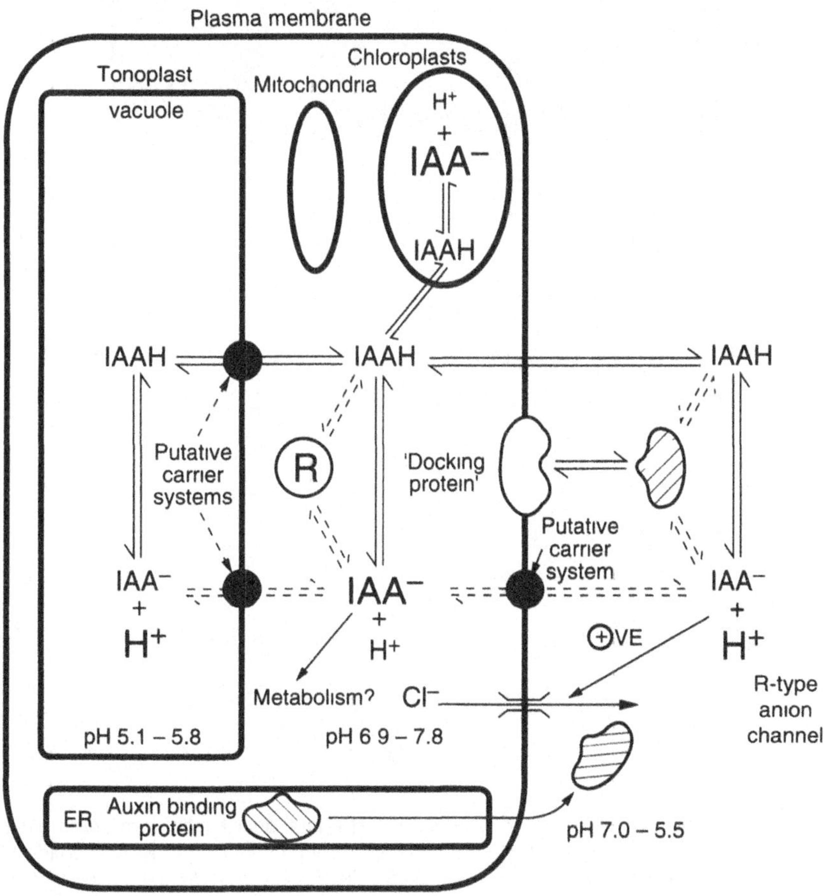

Figure 8.16 The predicted distribution of indole-3-acetic acid (IAA) based on the prevailing pH gradients in different compartments of a guard cell. The protonated form of IAA is membrane permeable, although there may be additional transport systems at the plasma membrane and/or tonoplast. The possible locations of putative auxin receptors are also indicated. Auxin may directly modulate the activity of the plasma membrane R-type anion channel (Marten *et al.*, 1991) or interact with a soluble auxin binding protein (Thiel *et al.*, 1993). The high permeability of guard cell membranes to auxin and presence of additional transport systems suggest an intracellular receptor (R) for auxin is also possible.

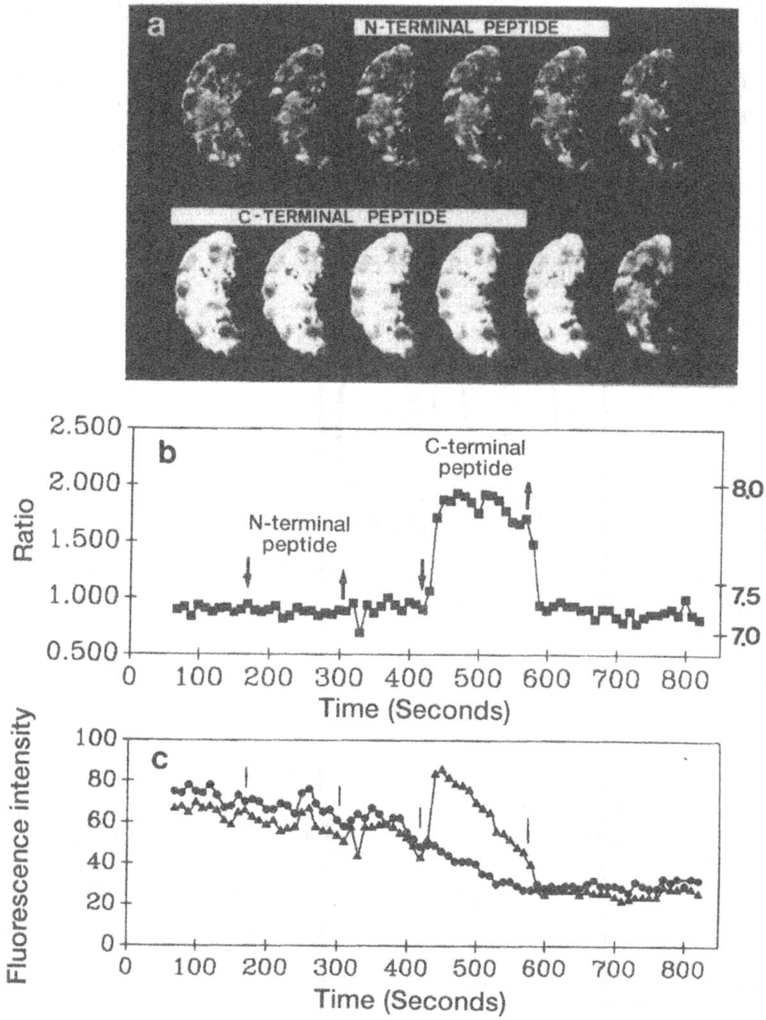

Figure 8.17 Changes in pH in the cytoplasm of stomatal guard cells from *Commelina communis* measured using confocal ratio imaging of the pH indicator, BCECF, loaded by microinjection into one guard cell of the pair A gallery of 12 ratio images is shown from a complete time course lasting 20 min (a) pH changes were quantified from a small region of cytoplasm near the nucleus, and are presented as the ratio values and the calibrated pH values determined at the end of the experiment (b) and the original fluorescent intensities at each wavelength (c) The circles are a measure of the dye concentration which slowly decreases during the experiment due to leakage or photobleaching The triangles represent the fluorescence at a wavelength that increases with alkalinization Note the dramatic, but reversible alkalinization in both the images (a) and the ratio trace (b) that occurs during exposure to a 12 amino acid synthetic peptide derived from the C-terminus of the auxin binding protein compared to the lack of activity of a peptide derived from near the N-terminus Data from Thiel *et al* (1993), with permission

auxin binding protein triggered rapid cytoplasmic alkalinization (see Fig 8 17) and subsequent inactivation of the K^+ inward rectifier and activation of the K^+ outward rectifier The predicted consequence of these changes would be inhibition of stomatal opening, i e the high concen tration auxin response The effects of the peptide on channel activities were abolished by clamping the cytoplasmic pH using butyrate as a per meant weak acid The action of 30 μM exogenous auxin on plasma membrane K^+ channel gating was also prevented by butyrate clamping (Blatt and Thiel, 1994), suggesting the pH changes induced by the peptide are related to the action of auxins *in vivo* Although these data show the C-terminal peptide of the auxin-binding protein has marked physiological effects in guard cells, further experiments are required to establish the sequence of events in auxin responses *in vivo*

References

Alberts, B , Bray, D , Lewis, J *et al* (1994) *Molecular Biology of the Cell*, 3rd edn, Garland Publishing, New York

Allan, A C , Fricker, M D , Ward, J L *et al* (1994) Two transduction pathways mediate rapid effects of abscisic acid in *Commelina* guard cells *Plant Cell* , **6**, 1319–1328

Allaway, WG and Hsaio, TC (1973) Preparation of rolled epidermis of *Vicia faba* L so that stomata are the only viable cells Analysis of guard cell potassium by flame photometry *Aust J Biol Sci* , **26**, 309–318

Allaway, WG and Milthorpe, FL (1976) Structure and functioning of stomata, in *Water Deficits and Plant Growth*, Vol 4, (ed TT Kozlowski), Academic Press, New York, pp 57–102

Allen, GJ and Sanders, D (1994) Two voltage-gated, calcium release channels coreside in the vacuolar membrane of broad bean guard cells *Plant Cell*, **6**, 685–694

Amodeo, G , Escobar, A and Zeiger, E (1994) A cationic channel in the guard cell tonoplast of *Allium cepa* *Plant Physiol* , **105**, 999–1006

Amodeo, G , Srivastava, A, and Zeiger, E (1992) Vanadate inhibits blue light-stimulated swelling of *Vicia* guard cell protoplasts *Plant Physiol* , **100**, 1567–1570

Anderson, BE , Ward, JM and Schroeder, JI (1994) Evidence for an extracellular reception site for abscisic acid in *Commelina* guard cells *Plant Physiol* , **104**, 1177–1183

Assmann, SA (1993) Signal transduction in guard cells *Ann Rev Cell Biol* , **9**, 345–375

Assmann, SA and Schwartz, A (1992) Synergistic effect of light and fusicoccin on stomatal opening *Plant Physiol* , **98**, 1349–1355

Assmann, SA , Simoncini, L and Schroeder, JI (1985) Blue light activates electrogenic ion pumping in guard cell protoplasts of *Vicia faba* *Nature*, **318**, 285–287

Atkinson, C J (1991) The flux and distribution of xylem sap calcium to adaxial and abaxial epidermal tissue in relation to stomatal behaviour *J Exp Bot*, **42**, 987–993

Atkinson, C J, Mansfield, T A and Davies, W J (1990) Does calcium in xylem sap regulate stomatal conductance? *New Phytol*, **116**, 19–27

Atkinson, C J, Mansfield, T A, Kean, A M and Davies, W J (1989) Control of stomatal aperture by calcium in isolated epidermal tissue and whole leaves of *Commelina communis* L *New Phytol*, **111**, 9–17

Atkinson, C J, Ruiz, L P and Mansfield, T A (1992) Calcium in xylem sap and the regulation of its delivery to the shoot *J Exp Bot*, **43**, 1315–1324

Baier, M and Hartung, W (1988) Movement of abscisic acid across the plasmalemma and the tonoplast of guard cells of *Valerianella locusta* *Botanica Acta*, **101**, 332–337

Bearce, B C and Kohl, H C, Jr (1970) Measuring osmotic pressure of sap within live cells by means of a visual melting point apparatus *Plant Physiol*, **46**, 515–519

Becker, D, Zeilinger, C Lohse, G *et al* (1993) Identification and biochemical characterization of the plasma-membrane proton ATPase in guard cells of *Vicia faba* L *Planta*, **190**, 44–50

Bertl, A, Blumwald, E Coronado, R *et al* (1992) Electrical measurements in endomembranes *Science*, **258**, 873–874

Bittisnich, D J, Entwisle L O and Neales, T F (1987) Acid-induced opening in *Vicia faba* and the role of guard cell wall elasticity *Plant Physiol*, **85**, 554–557

Blatt, M R (1985) Extracellular potassium activity in attached leaves and its relation to stomatal function *J Exp Bot*, **163**, 240–251

Blatt, M R (1987a) Electrical characteristics of stomatal guard cells the ionic basis of the membrane potential and the consequence of potassium chloride leakage from microelectrodes *Planta*, **170**, 272–287

Blatt, M R (1987b) Electrical characteristics of stomatal guard cells the contribution of ATP-dependent, electrogenic transport revealed by current–voltage and difference–current–voltage analysis *J Membr Biol*, **98**, 257–274

Blatt, M R (1988) Mechanisms of fusicoccin action a dominant role for secondary transport in a higher plant cell *Planta*, **174**, 187–200

Blatt, M R (1988) Potassium-dependent, biopolar gating of K channels in guard cells *J Membr Biol*, **102**, 235–246

Blatt, M R (1990) Potassium channel enhancement in intact stomatal guard cells rapid enhancement by abscisic acid *Planta*, **180**, 445–455

Blatt, M R (1991) Ion channel gating in plants physiological implica-

tions and integration from stomatal function *J Membr Biol*, **124**, 95–112
Blatt, M R (1992) K channels of stomatal guard cells Characteristics of the inward rectifier and its control by pH *J Gen Physiol*, **99**, 615–644
Blatt, M R and Armstrong, F (1993) K channels of stomatal guard cells abscisic-acid-evoked control of the outward rectifier mediated by cytoplasmic pH *Planta*, **191**, 330–341
Blatt, M R and Clint, G M (1989) Mechanisms of fusicoccin action kinetic modification and inactivation of K$^+$ channels in guard cells *Planta*, **178**, 509–523
Blatt, M R and Thiel, G (1993) Hormonal control of ion channel gating *Ann Rev Plant Physiol Plant Mol Biol*, **44**, 543–567
Blatt, M R, Thiel, G and Trentham, D R (1990) Reversible inactivation of K channels of *Vicia* stomatal guard cells following the photolysis of caged inositol, 1,4,5-trisphosphate *Nature*, **346**, 766–769
Blum, W, Key, G and Weiler, E W (1988) ATPase activity in plasmalemma-rich vesicles isolated by aqueous two-phase partitioning from *Vicia faba* mesophyll and epidermis characterization and influence of abscisic acid and fusicoccin *Physiol Plant*, **72**, 279–287
Bowling, D J F (1987) Measurement of the apoplastic activity of K$^+$ and Cl in the leaf epidermis of *Commelina communis* in relation to stomatal activity *J Exp Bot* **38**, 1351–1355
Bowling, D J F and Edwards, A (1984) pH gradients in the stomatal complex of *Tradescantia virginiana* *J Exp Bot*, **35**, 1641–1645
Bowling, D J F and Smith, G N (1990) Apoplastic transport in the leaf epidermis in relation to stomatal activity *Biochem Physiol Pflanz*, **186**, 309–316
Brindley, H M (1990a) Fluxes of ^{86}Rb in 'isolated' guard cells of *Vicia faba* L *Planta*, **181**, 432–439
Brindley, H M (1990b) Effects of light/dark and calcium-channel drugs on fluxes of ^{86}Rb in isolated guard cells of *Vicia faba* L *Planta*, **181**, 440–447
Clint, G M (1987) The effects of fusicoccin on anion fluxes in isolated guard cells of *Commelina communis* L *J Exp Bot*, **38**, 863–867
Clint, G M and Blatt M R (1989) Mechanisms of fusicoccin action evidence for concerted modulations of secondary K$^+$ transport in a higher plant cell *Planta*, **178**, 495–508
Clint, G M and MacRobbie, E A C (1984) Effects of fusicoccin in isolated guard cells of *Commelina communis* L *J Exp Bot*, **35**, 180–192
Cosgrove, D J and Hedrich, R (1991) Stretch-activated chloride, potassium, and calcium channels coexisting in plasma membranes of guard cells of *Vicia faba* L *Planta* **186**, 143–153
Crane, F L, Sun, I L Clarke, M G *et al* (1985) Transplasma-membrane

redox systems in growth and development *Biochim Biophys Acta*, **811**, 233–264

Curvetto, N and Delmastro, S (1990) A biochemical and physiological proposal for stomatal movement possible involvement of adenosine, 3′,5′-cyclic monophosphate *Plant Physiol Biochem* , **28**, 367–378

Curvetto, N , Darjania, L and Delmastro, S (1994) Effect of two cAMP analogues on stomatal opening in *Vicia faba* Possible relationship with cytosolic calcium concentration *Plant Physiol Biochem* , **32**, 365–372

Dayanandan, P and Kaufman, PB (1975) Stomatal movements associated with potassium fluxes *Am J Bot* , **62**, 221–231

De Silva, D L R , Cox, R C , Hetherington, A M and Mansfield, T A (1986) The role of abscisic acid and calcium in determining the behaviour of adaxial and abaxial stomata *New Phytol* , **104**, 41–51

De Silva, D L R , Hetherington, A M and Mansfield, T A (1985a) Synergism between calcium ions and abscisic acid in preventing stomatal opening *New Phytol* , **100**, 473–482

De Silva, D L R , Cox, R C , Hetherington, A M and Mansfield, T A (1985b) Suggested involvement of calcium and calmodulin in the responses of stomata to abscisic acid *New Phytol* , **101**, 555–563

Dittrich, P, Mayer, M and Meusel, M (1979) Proton-stimulated opening of stomata in relation to chloride uptake by guard cells *Planta*, **144**, 305–309

Donovan, N , Martin, S and Donkin, M E (1985) Calmodulin binding drugs trifluoperazine and compound, 48/80 modify stomatal responses of *Commelina communis* L *J Plant Physiol* , **118**, 165–176

Edwards, A and Bowling, D J F (1984) An electrophysiological study of the stomatal complex of *Tradescantia virginiana J Exp Bot* , **35**, 562–567

Edwards, A and Bowling, D J F (1985) Evidence for a CO_2 inhibited proton extrusion pump in the stomatal cells of *Tradescantia virginiana J Exp Bot* , **36**, 91–98

Edwards, M C and Bowling, D J F (1986) The growth of rust germ tubes towards stomata in relation to pH gradients *Physiol Mol Plant Pathol* , **29**, 185–196

Edwards, M C , Smith, G N and Bowling, D J F (1988) Guard cells extrude protons prior to stomatal opening – a study using fluorescence microscopy and pH microelectrodes *J Exp Bot* , **39**, 1541–1547

Eschel, A , Waisel, Y and Ramati, A (1974) The role of sodium in stomatal movements of a halophyte a study by X-ray microanalysis, in *Proc 7th Int Colloq on Plant Analysis and Fertilizer Problems*, (ed J Wehrmann), German Society for Plant Nutrition, Hannover

Fairley-Grenot, K A and Assmann, S M (1991) Evidence for G-protein regulation of inward K^+ channel current in guard cells of Fava bean

Plant Cell, **3**, 1037–1044
Fairley-Grenot, K A and Assmann, S A (1992a) Whole-cell K^+ current across the plasma membrane of guard cells from a grass *Zea mays* *Planta*, **186**, 282–293
Fairley-Grenot, K A and Assmann, S A (1992b) Permeation of Ca^+ through K^+ channels in the plasma membrane of *Vicia faba* guard cells *J Membr Biol*, **128**, 103–113
Fairley-Grenot, K A and Assmann, S M (1993) Comparison of K^+-channel activation and deactivation in guard cells from a dicotyledon (*Vicia faba* L) and a graminaceous monocotyledon (*Zea mays*) *Planta*, **189**, 410–419
Fischer, R A (1968a) Stomatal opening role of potassium uptake by guard cells *Science*, **160**, 784–785
Fischer, R A (1968b) Stomatal opening in isolated epidermal strips of *Vicia faba* I Responses to light and CO_2-free air *Plant Physiol*, **43**, 1947–1952
Fischer, R A (1971) Role of potassium in stomatal opening in the leaf of *Vicia faba* *Plant Physiol*, **47**, 555–558
Fischer, R A and Hsaio, T C (1968) Stomatal opening in isolated epidermal strips of *Vicia faba* II Responses to KCl concentration and the role of potassium absorption *Plant Physiol*, **43**, 1953–1958
Fitzsimons, P J and Weyers, J D B (1978) Separation and purification of protoplast types from *Commelina communis* L leaf epidermis *J Exp Bot*, **34**, 55–66
Fitzsimons, P J and Weyers, J D B (1986a) Volume changes of *Commelina communis* guard cell protoplasts in response to K^+, light and CO_2 *Physiol Plant*, **66**, 463–468
Fitzsimons, P J and Weyers, J D B (1986b) Potassium ion uptake by swelling *Commelina communis* guard cell protoplasts *Physiol Plant*, **66**, 469–475
Freudling, C, Starrach, N, Flach, D *et al* (1988) Cell walls as reservoirs of potassium ions for reversible volume changes of pulvinar motor cells during rhythmic leaf movements *Planta*, **175**, 193–203
Fricker, M D and Willmer, C M (1987) Vanadate sensitive ATPase and phosphatase activity in guard cell protoplasts of *Commelina* *J Exp Bot*, **38**, 642–648
Fricker, M D and Willmer, C M (1990a) Some properties of proton pumping ATPases at the plasma membrane and tonoplast of guard cells *Biochem Physiol Pflanz*, **186**, 301–308
Fricker, M D and Willmer, C M (1990b) Nitrate-sensitive ATPase activity and proton pumping in guard cell protoplasts of *Commelina* *J Exp Bot*, **41**, 193–198
Fujino, M (1967) Role of ATP and ATP-ase in stomatal movement *Sci Bull Fac Educ Nagasaki Univ*, **18**, 1–47

Garrec, J -P, Vavasseur, A , Michalowicz, G and Laffray, D (1983) Stomatal movements and repartition of the elements K, Cl, Na, P, Ca, Mg, and S in the stomatal complexes of *Vicia faba* and *Commelina communis* Electron probe studies *Z Pflanz* , **112**, 35–42

Gautier, H , Vavasseur, A , Lasceve, G and Boudet, A M (1992) Redox processes in the blue light response of guard cell protoplasts of *Commelina communis* L *Plant Physiol* , **98**, 34–38

Gepstein, S , Jacobs, M and Taiz, L (1982–83) Inhibition of stomatal opening in *Vicia faba* epidermal tissue by vanadate and abscisic acid *Plant Sci Lett* , **28**, 63–72

Gilroy, S , Bethke, PC and Jones, R L (1993) Calcium homeostasis in plants *J Cell Sci* , **106**, 453–462

Gilroy, S , Fricker, M D Read, N D and Trewavas, A J (1991) Role of calcium in signal transduction of *Commelina* guard cells *Plant Cell*, **3**, 333–344

Gilroy, S , Read, N D and Trewavas, A J (1990) Elevation of cytoplasmic calcium by caged calcium or caged inositol trisphosphate initiates stomatal closure *Nature*, **346**, 769–771

Gimmler, H , Weiss, C , Baier, M and Hartung, W (1990) The conductance of the plasmalemma for CO_2 *J Exp Bot* , **41**, 785–795

Gotow, K , Sakaki, T , Kondo N *et al* (1985) Light-induced alkalinisation of the suspending medium of guard cell protoplasts from *Vicia faba* L *Plant Physiol* , **79**, 825–828

Gradmann, D , Blatt, M R and Thiel, G (1994) Electrocoupling of ion transporters in plants *J Membr Biol* , **136**, 327–332

Gunar, I I , Zlotnikova, I F and Panichkin, L A (1975) Electrophysiological investigations of cells of the stomatal complex of spiderwort *Sov Plant Physiol* , **22**, 704–707

Hartung, W (1983) The site of action of abscisic acid at the guard cell plasmalemma of *Valerianella locusta* *Plant Cell Environ* , **6**, 427–429

Hartung, W and Slovik, S (1991) Physicochemical properties of plant growth regulators and plant tissues determine their distribution and redistribution stomatal regulation by abscisic acid in leaves *New Phytol* , **119**, 361–382

Hedrich, R , Barbier-Brygoo, H , Felle, H *et al* (1988) General mechanisms for solute transport across the tonoplast of plant vacuoles a patch-clamp survey of ion channels and proton pumps *Botanica Acta*, **101**, 7–13

Hedrich, R , Busch, H and Raschke, K (1990) Ca and nucleotide dependent regulation of voltage dependent anion channels in the plasma membrane of guard cells *EMBO J* , **9**, 3889–3892

Hedrich, R and Marten I (1993) Malate-induced feedback regulation of plasma membrane anion channels could provide a CO_2 sensor to guard cells *EMBO J*, **12**, 897–901

Hedrich, R and Neher, E (1987) Cytoplasmic calcium regulates voltage-dependent ion channels in plant vacuoles *Nature*, **329**, 833–836

Hedrich, R and Schroeder, J I (1989) The physiology of ion channels and electrogenic pumps in higher plants *Ann Rev Plant Physiol Plant Mol Biol*, **40**, 539–569

Hornberg, C and Weiler, E W (1984) High-affinity binding sites for abscisic acid on the plasmalemma of *Vicia faba* guard cells *Nature*, **310**, 321–324

Hosoi, S, Iino, M and Shimazaki, K-I (1988) Outward-rectifying K^+ channels in stomatal guard cell protoplasts *Plant Cell Physiol*, **29**, 907–911

Humble, G D and Hsiao, T C (1969) Specific requirement for potassium for light-activated opening of stomata in epidermal strips *Plant Physiol*, **44**, 230–234

Humble, G D and Hsaio, T C (1970) Light-dependent influx and efflux of potassium of guard cells during stomatal opening and closing *Plant Physiol*, **46** 483–487

Humble, G D and Raschke, K (1971) Stomatal opening quantitatively related to potassium transport Evidence from electron microprobe analysis *Plant Physiol*, **48**, 447–453

Iljin, WS (1957) Drought resistance in plants and physiological processes *Ann Rev Plant Physiol*, **8**, 257–346

Immamura, S (1943) Untersuchungen uber den mechanismus der turgorschwankung der Spaltoffnungschliezellen *Jap J Bot*, **12**, 251–346

Inoue, H and Katoh, Y (1987) Calcium inhibits ion-stimulated stomatal opening in epidermal strips of *Commelina communis* L *J Exp Bot*, **38**, 142–149

Inoue, H, Noguchi, M and Kubo, K (1985) Ion-stimulated stomatal opening induced by preillumination in epidermal strips of *Commelina communis Plant Physiol*, **79**, 389–393

Irving, H R, Gehring, C A and Parish, R W (1992) Changes in cytosolic pH and calcium of guard cells precede stomatal movements *Proc Natl Acad Sci USA*, **89**, 1790–1794

Ishikawa, H, Aizawa, H, Kishira H *et al* (1983) Light-induced changes of membrane potential in guard cells of *Vicia faba Plant Cell Physiol*, **24**, 769–772

Jamieson, A V and Willmer, C M (1984) Functional stomata in a variegated leaf chimera of *Pelargonium zonale* L without guard cell chloroplasts *J Exp Bot*, **35** 1053–1059

Jarvis, R G and Mansfield T A (1980) Reduced stomatal responses to light, carbon dioxide and abscisic acid in the presence of sodium ions *Plant Cell Environ* , **3**, 279–283

Johannes, E , Brosnan, J M and Sanders, D (1991) Calcium channels and signal transduction in plant cells *BioEssays*, **13**, 331–336

Johansson, F , Sommarin, M and Larsson, C (1993) Fusicoccin activates the plasma membrane H -ATPase by a mechanism involving the C-terminal inhibitory domain *Plant Cell*, **5**, 321–327

Karlsson, PE and Schwartz, A (1988) Characterization of the effects of metabolic inhibitors, ATPase inhibitors and a potassium-channel blocker on stomatal opening and closing in isolated epidermis of *Commelina communis* L *Plant Cell Environ* , **11**, 165–172

Kearns, E V and Assmann, S M (1993) The guard cell–environment connection *Plant Physiol* , **102**, 711–715

Keller, B U , Hedrich, R and Raschke, K (1989) Voltage-dependent anion channels in the plasma membrane of guard cells *Nature*, **341**, 450–452

Kinoshita, T , Shimazaki, K -I and Nishimura, M (1993) Phosphorylation and dephosphorylation of guard-cell proteins from *Vicia faba* L in response to light and dark *Plant Physiol* , **102**, 917–923

Kinoshita, T , Nishimura, M and Shimazaki, K -I (1995) Cytosolic Ca^{2+} regulates the plasma membrane H^+-ATPase in guard cells of Fava bean *Plant Cell*, in press

Lasceve, G , Couchat, P, Vavasseur, A and Bossy, J P (1987) Changes in K^+, Cl and P contents in stomata of *Zea mays* leaves exposed to different light and CO_2 levels *Physiol Plant* , **69**, 709–715

Lasceve, G , Gautier, H , Jappe, J and Vavasseur, A (1993) Modulation of the blue light response of stomata of *Commelina communis* by CO_2 *Physiol Plant* , **88**, 453–459

Laffray, D , Louget, P and Garrec, J P (1982) Microanalytical studies of potassium and chloride fluxes and stomatal movements of two species *Vicia faba* and *Pelargonium hortorum* *J Exp Bot* , **33**, 771–782

Laffray, D , Vavasseur, A , Garrec, J -P and Louget, P (1984) Effects of high carbon dioxide partial pressure on stomatal movements and related ion fluxes in *Pelargonium hortorum* and *Vicia faba* Electron probe studies *Physiol Veg* , **22**, 851–857

Lee, Y and Assmann, S M (1991) Diacylglycerols induce both ion pumping in patch-clamped guard-cell protoplasts and opening of intact stomata *Proc Natl Acad Sci USA*, **88**, 2127–2131

Lee, H J , Tucker, E B , Crain, R C and Lee, Y (1993) Stomatal opening is induced in epidermal peels of *Commelina communis* L by GTP analogs or pertussis toxin *Plant Physiol* , **102**, 95–100

Lemtiri-Chlieh, F and MacRobbie, E A C (1994) Role of calcium in the modulation of *Vicia* guard cell potassium channels by abscisic acid a patch clamp study *J Membr Biol*, **137**, 99–107

Leung, J, Bouvier-Durand, M, Morris, P-C *et al* (1994) *Arabidopsis* ABA response gene ABAI1 features of a calcium-modulated protein phosphatase *Science*, **264**, 1448–1452

Linder, B and Raschke, K (1992) A slow anion channel in guard cells, activating at large hyperpolarisation, may be principal for stomatal closing *FEBS Lett*, **313**, 27–30

Ling, V and Assmann, S M (1992) Cellular distribution of calmodulin and calmodulin-binding proteins in *Vicia faba* L *Plant Physiol*, **100**, 970–978

Lohse, G and Hedrich, R (1992) Characterization of the plasma membrane proton-ATPase from *Vicia faba* guard cells modulation by extracellular factors and seasonal changes *Planta*, **188**, 206–214

Losch, R (1985) Daily and seasonal courses of *Valerianella locusta* stomatal apertures and guard cell potassium contents field measurements and model predictions *Acta Hort*, **171**, 219–228

Luan, S, Li, W, Rusnak, F, Assmann, S M and Schreiber, S L (1993) Immunosuppressants implicate protein phosphatase regulation of K^+ channels in guard cells *Proc Natl Acad Sci USA*, **90**, 2202–2206

Macallum, A B (1905) On the distribution of potassium in animal and vegetable cells *J Physiol*, **32**, 95–128

MacRobbie, E A C (1980) Osmotic measurements on stomatal cells of *Commelina communis* L *J Membr Biol*, **53**, 189–198

MacRobbie E A C (1981) Effects of ABA in 'isolated' guard cells of *Commelina communis* L *J Exp Bot*, **32**, 563–572

MacRobbie E A C (1983) Effects of light/dark on cation fluxes in guard cells of *Commelina communis* L *J Exp Bot*, **34**, 1695–1710

MacRobbie, E A C (1984) Effects of light/dark on anion fluxes in isolated guard cells of *Commelina communis* L *J Exp Bot*, **35**, 707–726

MacRobbie, E A C (1987) Ionic relations of guard cells, in *Stomatal Function*, (eds E Zeiger, G D Farquhar and I R Cowan), Stanford University Press, Stanford, CA, pp 125–162

MacRobbie, E A C (1988) Control of ion fluxes in stomatal guard cells *Botanica Acta*, **101**, 140–148

MacRobbie, E A C (1989) Calcium influx at the plasmalemma of isolated guard cells of *Commelina communis* Effects of abscisic acid *Planta*, **178**, 231–241

MacRobbie, E A C (1990) Calcium-dependent and calcium-independent events in the initiation of stomatal closure by abscisic acid *Proc Roy Soc Lond*, *Ser B*, **241**, 214–219

MacRobbie, E A C (1992) Calcium and ABA-induced stomatal closure *Phil Trans Roy Soc Lond*, *Ser B*, **338**, 5–18

MacRobbie, E A C and Lettau, J (1980a) Ion content and aperture in 'isolated' guard cells of *Commelina communis* L *J Membr Biol*, **53**,

199–205

MacRobbie, E A C and Lettau, J (1980b) Potassium content and aperture in intact stomatal and epidermal cells of *Commelina communis* L *J Membr Biol*, **56**, 249–256

Mansfield, T A, Hetherington, A M and Atkinson, C J (1990) Some current aspects of stomatal physiology *Ann Rev Plant Physiol Plant Mol Biol*, **41**, 55–75

Marten, I, Busch, H, Raschke, K and Hedrich, R (1993) Modulation and block of the plasma membrane anion channel of guard cells by stilbene derivatives *Eur Biophys J*, **21**, 403–408

Marten, I, Lohse, G and Hedrich, R (1991) Plant growth hormones control voltage-dependent activity of anion channels in plasma membrane of guard cells *Nature*, **353**, 758–762

Marten, I, Zeilinger, C, Redhead, C *et al* (1992) Identification and modulation of a voltage-dependent anion channel in the plasma membrane of guard cells by high-affinity ligands *EMBO J*, **11**, 3569–3575

McAinsh, M R, Brownlee, C and Hetherington, A M (1990) Abscisic acid-induced elevation of guard cell cytosolic Ca^{+} precedes stomatal closure *Nature*, **343**, 186–188

McAinsh, M R, Brownlee, C and Hetherington, A M (1992) Visualising changes in cytosolic free Ca during the response of stomatal guard cells to abscisic acid *Plant Cell*, **4**, 1113–1122

Meidner, H and Edwards, M (1975) Direct measurements of turgor pressure potentials of guard cells, I *J Exp Bot*, **26**, 319–330

Meyer, K, Leube, M P and Grill, E (1994) A protein phosphatase, 2C involved in ABA signal transduction in *Arabidopsis thaliana Science*, **264**, 1452–1455

Moody, W and Zeiger, E (1978) Electrophysiological properties of onion guard cells *Planta*, **139**, 159–165

Morsucci, R, Curvetto N and Delmastro, S (1991) Involvement of cytokinins and adenosine, 3′,5′-cyclic monophosphate in stomatal movement in *Vicia faba Plant Physiol Biochem*, **29**, 537–547

Morsucci, R, Curvetto, N and Delmastro, S (1992) High concentrations of adenosine or kinetin riboside induces stomatal closure in *Vicia faba*, probably through inhibition of adenylate cyclase *Plant Physiol Biochem*, **30**, 383–388

Nejidat, A (1987) Effect of Ophiobolin A on stomatal movement role of calmodulin *Plant Cell Physiol*, **28**, 455–460

Nejidat, A, Itai, C and Roth-Bejerano, N (1989) Regulation of Mg,K-ATPase activity in guard cells of *Commelina communis* by phytochrome and ABA *Plant Cell Physiol*, **30**, 945–949

Nejidat, A, Roth-Bejerano, N and Itai, C (1986) K,Mg-ATPase activity in guard cells of *Commelina communis Physiol Plant*, **68**, 315–319

Nelson, S D and Mayo, J M (1977) Low K^+ in *Paphiopedilum leeanum* leaf epidermis implications for stomatal functioning *Can J Bot*, **55**, 489–495
Obermeyer, G, Armstrong, F and Blatt, M R (1994) Selective block by α-dendrotoxin of the K^+ inward rectifier at the *Vicia* guard cell plasma membrane *J Membr Biol*, **137**, 249–259
Outlaw, WH (1983) Current concepts on the role of potassium in stomatal movements *Physiol Plant*, **59**, 302–311
Outlaw, WJ and Manchester, J (1979) Guard cell starch concentration quantitatively related to stomatal aperture *Plant Physiol*, **64**, 79–82
Outlaw, WH, Manchester, J and Zenger, VE (1981) The relationship between protein content and dry weight of guard cells and other single cell samples of *Vicia faba* L *Histochem J*, **13**, 329–336
Outlaw, WJ, Manchester, J and Zenger, VE (1982) Potassium involvement not demonstrated in stomatal movements of *Paphiopedilum* Qualified confirmation of the Nelson–Mayo report *Can J Bot*, **60**, 240–244
Pallaghy, C K (1968) Electrophysiological studies in guard cells of tobacco *Planta*, **80**, 147–153
Pallaghy, C K (1971) Stomatal movement and potassium transport in epidermal strips of *Zea mays* the effect of CO_2 *Planta*, **101**, 287–295
Pallaghy, C K and Fischer, R A (1974) Metabolic aspects of stomatal opening and ion accumulation by guard cells in *Vicia faba* *Z Pflanz*, **71**, 332–344
Pallas, J E, Jr (1966) Mechanics of guard cell action *Quart Rev Biol*, **41**, 365–383
Palme, K (1992) Molecular analysis of plant signaling elements relevance of eukaryotic signal transduction models *Int Rev Cytol*, **132**, 223–283
Pantoja, O and Willmer, C M (1988) Redox activity and peroxidase activity associated with the plasma membrane of guard cell protoplasts *Planta*, **174**, 44–50
Pantoja, O and Willmer, C M (1991) Ferricyanide reduction by guard cell protoplasts *J Exp Bot*, **42**, 323–329
Parets-Soler, A, Pardo, J M and Serrano, R (1990) *Plant Physiol*, **93**, 1654–1658
Parmar, PN (1991) *D Phil Thesis*, University of Cambridge
Parmar, PN and Brearley, C A (1993) Identification of 3- and 4-phosphorylated phosphoinositides and inositol phosphates in stomatal guard cells *Plant J*, **4**, 255–263
Paterson, N W, Weyers, J D B and A'Brook, R (1988) The effect of pH on stomatal sensitivity to abscisic acid *Plant Cell Environ*, **11**, 83–89
Pekarek, J (1936) Uberdie aziditatsverhaltnisse in den epidermis und

schliesszellen bei *Rumex acetosa* im licht und im dunkeln *Planta*, **21**, 419–446

Penny, M G and Bowling, D J F (1974) A study of potassium gradients in the epidermis of intact leaves of *Commelina communis* L in relation to stomatal opening *Planta*, **119**, 17–25

Penny, M G and Bowling, D J F (1975) Direct determination of pH in the stomatal complex of *Commelina* *Planta*, **122**, 209–212

Penny, M G , Kelday, L S and Bowling, D J F (1976) Active chloride transport in the leaf epidermis of *Commelina communis* in relation to stomatal activity *Planta*, **130**, 291–294

Raghavendra, A S (1990) Blue light effects on stomata are mediated by the guard cell plasma membrane redox system distinct from the proton translocating ATPase *Plant Cell Environ* , **13**, 105–110

Raschke, K (1975) Stomatal action *Ann Rev Plant Physiol* , **26**, 309–340

Raschke, K (1979) Movements of stomata, in *Encyclopedia of Plant Physiology*, *Vol 7*, *Physiology of Movements*, (eds W Haupt and M E Feinleib), Springer, Berlin, pp 383–441

Raschke, K and Fellows, M P (1971) Stomatal movement in *Zea mays* shuttle of potassium and chloride between guard cells and subsidiary cells *Planta*, **110**, 296–316

Raschke, K , Hedrich, R , Reckmann, U and Schroeder, J I (1988) Exploring the biophysical and biochemical components of the osmotic motor that drives stomatal movement *Botanica Acta*, **101**, 283–294

Raschke, K and Humble, G D (1973) No uptake of anions required by opening stomata of *Vicia faba* guard cells release hydrogen ions *Planta*, **115**, 47–57

Raschke, K and Schnabl, H (1978) Availability of chloride affects the balance between potassium chloride and potassium malate in guard cells of *Vicia faba* L *Plant Physiol* , **62**, 84–87

Roth-Bejerano, N and Itai, C (1990) Regulation of ferricyanide reduction by epidermal tissue *Plant Cell Physiol* , **31**, 77–80

Roth-Bejerano, N , Nejidat, A , Rubenstein, B and Itai, C (1988) Effect of ferricyanide on potassium uptake by intact epidermal tissue and guard cell protoplasts *Plant Cell Physiol* , **29**, 677–682

Ruiz, L P and Mansfield, T A (1994) A postulated role for calcium oxalate in the regulation of calcium ions in the vicinity of stomatal guard cells *New Phytol* , **127**, 473–481

Ruiz, L P, Atkinson, C J and Mansfield, T A (1993) Calcium in the xylem and its influence on the behaviour of stomata *Phil Trans Roy Soc Lond* , *Ser B*, **341**, 67–74

Saftner, R A and Raschke, K (1981) Electrical potentials in stomatal complexes *Plant Physiol* , **67**, 1124–1132

Sanders, D (1984) Gradient-coupled chloride transport in plant cells, in *Chloride Transport Coupling in Cells and Epithelia*, (ed G A Gerencser), North-Holland, Amsterdam, pp 63–119

Sawhey, B L and Zelitch, I (1969) Direct determination of potassium ion accumulation in guard cells in relation to stomatal opening in light *Plant Physiol* , **44**, 1350–1354

Scarth, G W (1929) The influence of H-ion concentration on the turgor and movement of plant cells with special reference to stomatal behaviour *Proc Int Conf Plant Sci* , **2**, 1151–1162

Scarth, G W (1932) Mechanism of the action of light and other factors on stomatal movements *Plant Physiol* , **7**, 481–504

Schachtman, D P and Schroeder, J I (1994) Structure and transport mechanism of a high-affinity potassium uptake transporter from higher plants *Nature*, **370**, 655–658

Schauf, C L and Wilson, K J (1987) Effects of abscisic acid on K channels in *Vicia faba* guard cell protoplasts *Biochem Biophys Res Commun* , **145**, 284–290

Schnabl, H and Raschke, K (1980) Potassium chloride as stomatal osmoticum in *Allium cepa* L , a species devoid of starch in guard cells *Plant Physiol* , **65**, 88–93

Schroeder, J I (1989) Quantitative analysis of outward rectifying K channel currents in guard cell protoplasts from *Vicia faba* *J Membr Biol* , **107**, 229–235

Schroeder, J I (1988) K transport properties of K channels in the plasma membrane of *Vicia faba* guard cells *J Gen Physiol* **92**, 667–683

Schroeder, J I (1992) Plasma membrane ion channel regulation during abscisic acid-induced closing of stomata *Phil Trans Roy Soc Lond Ser B*, **338**, 1–112

Schroeder, J I and Fang, H H (1991) Inward-rectifying K channels in guard cells provide a mechanism for low-affinity K^+ uptake *Proc Natl Acad Sci USA*, **88**, 11583–11587

Schroeder, J I and Hagiwara, S (1989) Cytosolic calcium regulates ion channels in the plasma membrane of *Vicia faba* guard cells *Nature*, **338**, 427–430

Schroeder, J I and Hagiwara, S (1990) Repetitive increases in cytosolic Ca $^{+}$ of guard cells by abscisic acid activation of nonselective Ca permeable channels *Proc Natl Acad Sci USA*, **87**, 9305–9309

Schroeder, J I and Keller, B U (1992) Two types of anion channel cur rents in guard cells with distinct voltage regulation *Proc Natl Acad Sci USA*, **89**, 5025–5029

Schroeder, J I , Raschke, K and Neher, E (1987) Voltage dependence of K^+ channels in guard-cell protoplasts *Proc Natl Acad Sci USA* , **84**, 4108–4112

Schroeder, J I , Schmidt, C and Sheaffer, J (1993) Identification of high-affinity slow anion channel blockers and evidence for stomatal regulation by slow anion channels in guard cells *Plant Cell*, **5**, 1831–1841

Schwartz, A (1985) Role of Ca^{+} and EGTA on stomatal movements in *Commelina communis* L *Plant Physiol* , **79**, 1003–1005

Schwartz, A , Ilan, N and Assmann, S A (1991) Vanadate inhibition of stomatal opening in epidermal peels of *Commelina communis* chloride interferes with vanadate uptake *Planta*, **183**, 590–596

Schwartz, A , Ilan, N and Grantz, D A (1988) Calcium effects on stomatal movement in *Commelina communis* *Plant Physiol* , **87**, 583–587

Schwartz, A , Wu, WH , Tucker, E B and Assmann, S M (1994) Inhibition of inward K^+ channels and stomatal responses by abscisic acid an intracellular locus of phytohormone action *Proc Natl Acad Sci USA*, **91**, 4019–4023

Serrano, R (1989) Structure and function of plasma membrane ATPase *Ann Rev Plant Physiol Plant Mol Biol* , **40**, 61–94

Serrano, E E and Zeiger, E (1989) Sensory transduction and electrical signaling in guard cells *Plant Physiol* , **91**, 795–799

Serrano, E E , Zeiger, E and Hagiwara, S (1988) Red light stimulates an electrogenic proton pump in *Vicia* guard cell protoplasts *Proc Natl Acad Sci USA*, **85**, 436–440

Shimada, K , Ogawa, T and Shibata, K (1979) Isotachophoretic analysis of ions in guard cells of *Vicia faba* *Physiol Plant* , **47**, 173–176

Shimazaki, K , Iino, M and Zeiger, E (1986) Blue light-dependent proton extrusion by guard-cell protoplasts of *Vicia faba* *Nature*, **319**, 324–326

Shimazaki K I , Gotow, K , Sakaki, T and Kondo, N (1983) High respiratory rate of guard cell protoplasts from *Vicia faba* L *Plant Cell Physiol* , **24**, 1049–1056

Shimazaki, K -I , Kinoshita, T and Nishimura, M (1992) Involvement of calmodulin and calmodulin-dependent myosin light chain kinase in blue light-dependent H^+ pumping by guard cells protoplasts from *Vicia faba* L *Plant Physiol* , **99**, 1416–1423

Shimazaki, K and Kondo, N (1987) Plasma membrane H^+-ATPase in guard-cell protoplasts from *Vicia faba* L *Plant Cell Physiol* , **28**, 893–900

Shimazaki, K -I , Omasa, K , Kinoshita, T and Nishimura, M (1993) Properties of the signal transduction pathway in the blue light response of stomatal guard cells of *Vicia faba* and *Commelina benghalensis* *Plant Cell Physiol* , **34**, 1321–1327

Small, J and Maxwell, K M (1939) pH phenomena in relation to stomatal opening, I *Coffea arabica* and some other species *Protoplasma*, **32**, 272–288

Smith, G.N. and Willmer, C.M. (1988) Effects of calcium and abscisic acid on volume changes of guard cell protoplasts of *Commelina*. *J. Exp. Bot.*, **39**, 1529–1539.

Squire, G.R. and Mansfield, T.A. (1972) Studies on the mechanism of action of fusicoccin, the fungal toxin that induces wilting, and its interaction with abscisic acid. *Planta*, **105**, 71–78.

Squire, G.R. and Mansfield, T.A. (1974) The action of fusicoccin on stomatal guard cells and subsidiary cells. *New Phytol.*, **73**, 433–440.

Sze, H. (1985) H^+-translocating ATPases: advances using membrane vesicles. *Ann. Rev. Plant Physiol.*, **36**, 175–208.

Sze, H., Ward, J.M. and Lai, S. (1992) Vacuolar H^+-hydrolysing ATPases from plants: structure, function and isoforms. *J. Biol. Biomembr.*, **24**, 371–381.

Tester, M. (1990) Plant ion channels: whole-cell and single-channel studies. *New Phytol.*, 1**14**, 305–340.

Thiel, G. and Blatt, M.R. (1991) The mechanism of ion permeation through K^+channels of stomatal guard cells: voltage-dependent block by Na^+. *J. Plant Physiol.*, **138**, 326–334.

Thiel, G. and Blatt, M.R. (1994b) Phosphatase antagonist okadaic acid inhibits steady-state K^+ currents in guard cells of *Vicia faba*. *Plant J.*, **5**, 727–733.

Thiel, G., Blatt, M.R., Fricker, M.D., White, I.R. and Millner, P. (1993) Modulation of K^+channels in *Vicia* stomatal guard cells by peptide homologs to the auxin binding protein C-terminus. *Proc. Natl. Acad. Sci. USA*, **90**, 11493–11497.

Thiel, G., MacRobbie, E.A.C. and Blatt, M.R. (1992) Membrane transport in stomatal guard cells: the importance of voltage control. *J. Membr. Biol.*, **126**, 1–18.

Travis, A.J. and Mansfield, T.A. (1977) Studies of malate formation in 'isolated' guard cells. *New Phytol.*,**75**, 541–546.

Travis, A.J. and Mansfield, T.A. (1979a) Stomatal responses to light and CO_2 are dependent on KCl concentration. *Plant Cell Environ.*, **2**, 319–323.

Travis, A.J. and Mansfield, T.A. (1979b) Reversal of the CO_2^- responses of stomata by fusicoccin. *New Phytol.*, **83**, 607–614.

Turner, N.C. (1973) Action of fusicoccin on the potassium balance of guard cells of *Phaseolus vulgaris*. *Am. J. Bot.*, **60**, 717–725.

Ullrich, C.I., Köhler, K., Baier, M. *et al.* (1990) Neutral red as a redox dye induces K^+efflux and current-voltage changes in *Eremosphera*, *Lemna*, and guard cells. *Bot Acta*, **103**, 214–221.

Van Kirk, C.A. and Raschke, K. (1978a) Presence of chloride reduces malate production in epidermis during stomatal opening. *Plant Physiol.*, **61**, 361–364.

Van Kirk, C.A. and Raschke, K. (1978b) Release of malate from epidermal strips during stomatal closure *Plant Physiol*, **61**, 474–475

Vani, T and Raghavendra, A S (1992) Plasma membrane redox system in guard cell protoplasts of pea (*Pisum sativum* L) *J Exp Bot*, **43**, 291–297

Vani, T and Raghavendra, A S (1989) Tetrazolium reduction by guard cells in abaxial epidermis of *Vicia faba* blue light stimulation of a plasmalemma redox system *Plant Physiol*, **90**, 59–62

Vavasseur, A, Lasceve, G and Cousson, A (1994) Guard cell response to potassium ferricyanate *Physiol Plant*, **93**, 253–258

Villalba, J M, Luetzelschwab, M and Serrano, R (1991) Immunocytolocalization of plasma-membrane proton-ATPase in maize coleoptiles and enclosed leaves *Planta*, **185**, 458–461

Ward, J M and Schroeder J I (1994) Calcium-activated K channels and calcium-induced calcium release by slow vacuolar ion channels in guard cell vacuoles in the control of stomatal closure *Plant Cell*, **6**, 669–683

Weyers, J D B and Fitzsimons, PJ (1982) The non-osmotic volume of *Commelina* guard cells *Plant Cell Environ*, **5**, 417–421

Willmer, C M and Beattie L N (1978) Cellular osmotic phenomena during stomatal movements of *Commelina communis* I Limitations of the incipient plasmolysis technique for determining osmotic pressures *Protoplasma*, **95** 321–332

Willmer, C M and Pantoja, O (1992) The plasma membrane and tonoplast of guard cells In *Plant Membranes A Biophysical Approach to Structure, Development and Senescence*, (ed YY Leshem), Kluwer, Dordrecht, pp 221–238

Willmer, C M, Grammatikopoulos, G, Lasceve, G and Vavasseur, A (1995) Characterization of the vacuolar-type H ATPase from guard cell protoplasts of *Commelina J Exp Bot*, **46**, 383–389

Willmer, C M and Mansfield, T A (1969) A critical examination of the use of detached epidermis in studies of stomatal physiology *New Phytol*, **68**, 363–375

Willmer, C M and Pallas J E (1973) A survey of stomatal movements and associated potassium fluxes in the plant kingdom *Can J Bot*, **51**, 37–42

Willmer, C M and Pallas, J E (1974) Stomatal movements and ion fluxes within epidermis of *Commelina communis* L *Nature*, **252**, 126–127

Willmer, C M, Rutter, J C and Meidner, H (1983) Potassium involvement in stomatal movements of *Paphiopedilum J Exp Bot*, **142**, 507–513

Wu, W-H and Assmann, S M (1994) A membrane-delimited pathway of G-protein regulation of the guard-cell inward K^+ channel *Proc Natl Acad Sci USA*, **91**, 6310–6314

Yamashita, T (1952) Influences of potassium supply upon various properties and movement of guard cell. *Sieboldia, Acta Biol.*, **1**, 51–70.

Zeiger, E. (1983) The biology of stomatal guard cells. *Ann. Rev. Plant Physiol.*, **34**, 441–475.

Zeiger, E., Bloom, A.J. and Hepler, P.K. (1978) Ion transport in stomatal guard cells: a chemiosmotic hypothesis. *What's New in Plant Physiol.*, **9**, 29–32.

Zeiger, E., Moody, W., Hepler, P. and Varela, F. (1977) Light-sensitive membrane potentials in onion guard cells. *Nature*, **270**, 270–271.

9 The metabolism of guard cells

9.1 Introduction

Not all the K^+ accumulated by guard cells when stomata open is balanced by Cl^- uptake and malate anions produced in the guard cells normally brings about complete charge balance (but see Section 9.4.2). The biosynthesis of malate, with starch as at least part of the carbon source, also results in the production of protons which may be pumped out of the guard cells by ATPase activity to maintain a driving force for K^+ and Cl^- uptake. This chapter, therefore, concentrates on the carbon metabolism of guard cells paying particular attention to the pathway of carbon flow between starch and malate during stomatal movements and the regulation of this carbon flow. Several reviews are also to be found on this topic (e.g. Outlaw 1982, Robinson and Preiss, 1985). Additionally, this chapter will deal with aspects on the role of guard cell chloroplasts, on the energetics of stomatal movements and on the lipid metabolism of guard cells.

Major advances in our knowledge about the metabolism of guard cells have been made since it was discovered in the early 1970s that guard cells were able to fix and metabolize CO_2 in the same way as the initial steps of C4 photosynthesis. Willmer *et al.* (1973a) detected high activities of phosphoenolpyruvate (PEP) carboxylase and malate dehydrogenase (NAD and NADP specific, which are involved in malate synthesis) and NADP malic enzyme (which decarboxylates malate), in epidermal tissue of tulip and *Commelina communis* (they also detected high activities of RuBisCo but this was possibly due to mesophyll cell contamination). Since the lower epidermis was found to have nearly four times more PEPc activity than the upper epidermis and also has about four times more stomata than the upper epidermis it was concluded the enzyme was primarily located in the guard cells (Willmer *et al.*, 1973b). Willmer and Dittrich (1974) also found that when epidermal tissue of *Commelina* and tulip were exposed to $^{14}CO_2$ the turnover pattern of labelled compounds was similar to that found in C4 plants. Subsequently, other studies detected considerable activity of enzymes involved in malate metabolism in epidermal tissue and specifically in guard cells of a variety of species.

Although early biochemical work was carried out with epidermal tissue, and very useful information was obtained, such work can be criticized for a number of reasons (Willmer *et al.*, 1987; Outlaw, 1982). First, about 95% of the epidermal tissue consists of cell types other than guard cells and there is usually some mesophyll contamination of the epidermis which is difficult to eradicate. Furthermore, epidermal tissue often contains potent inhibitors of various enzyme activities which are difficult to neutralise. By the early 1980s, however, methods were developed for isolating relatively large quantities of purified guard cell protoplasts (GCPs) allowing more unequivocal biochemical results to be obtained (see Box 9.1). For these reasons the biochemical aspects presented here will concentrate on results obtained with GCPs or freeze-dried guard cells dissected out from the leaf and treated in a very special way using techniques adapted by Outlaw (1980).

Box 9.1 Preparation of guard cell protoplasts of *Commelina communis*

Introduction

In order to study the metabolism of guard cells and their molecular and cell biology it is usually necessary to work with guard cells rather than epidermal strips. The only way to obtain large numbers of pure guard cells is to make protoplasts.

Procedure

The method described below of preparing guard cell protoplasts (GCPs) of *Commelina communis* is based on that used by Fitzsimons and Weyers (1983).

1. Abaxial epidermis is peeled from the three youngest, fully expanded leaves of the main axis of the plant and floated, cuticle uppermost, in Petri dishes containing 270–300 mol m^{-3} mannitol and 10 mM MES–KOH buffer, pH 6.2, until enough has been collected (seven to eight 5 cm diameter Petri dishes should produce about 1 million purified GCPs). Some investigators suggest that a 30 min preplasmolysis treatment in 150 mM mannitol is beneficial and prevents the protoplasts from breaking into smaller 'subprotoplasts' and that 20 μM Ca^{2+} be added to the medium to give better recovery (Clint, 1985). However, the latter may not be necessary; indeed the epidermis of some species (including *C. communis*) contains an abundance of Ca^{2+} and the medium quickly accumulates millimolar concentrations of the ion.

   There is a knack to peeling epidermal strips from leaves and transferring them with a seeker or some similar object from dish to dish and some practice may be needed before adequate amounts can be collected. The method of obtaining epidermal strips described by Weyers and Travis (1981) is also worth trying.

2. Siphon or pipette off the plasmolysing medium and replace with 5 cm^3 per dish of the digest medium consisting of 300 mM mannitol, 2% (w/v) cellulysin (Calbiochem-Behring, La Jolla, CA, USA), 0.05% (w/v) pectolyase (Seishin, Pharmaceuticals, Tokyo, Japan), 0.5% bovine serum albumin (BSA) and 10 mM MES–KOH buffer, pH 5.5. Incubate at 30°C with illumination (an angle poise lamp with a tungsten, 150 W bulb about 0.5 m away from the Petri dishes is satisfactory), and with very slow agitation of the dishes if possible.
3. After about 1.5 h adhering mesophyll cell protoplasts and some epidermal cell protoplasts are released. These are separated from the digesting epidermal strips by collecting the medium from the dishes with a Pasteur pipette and centrifuging it for 5 min at 400 *g* in a bench centrifuge. The pellet is discarded and the supernatant replaced into the dishes. This procedure helps to ensure that the Percoll gradient (see step 5) is not over-loaded and to produce more than a 99% purity of the GCPs.
4. Separation of the GCPs occurs usually after 3–4 h incubation. This is checked by gently swirling a dish and then observing the dish contents under the low power (×10 objective) of a microscope. When the GCPs are released they are collected with a pipette with a widened, fired tip, and centrifuged in the plasmolysing medium, i.e. the medium minus the digestive enzymes and BSA, for 5 min at 100 *g*. The partially digested epidermis remaining in the Petri dishes is washed three times with the plasmolysing medium and centrifuged as before. The pellets, containing GCPs and epidermal and subsidiary cell protoplasts as contaminants, are pooled and washed three times in the plasmolysing medium to ensure that the cellulysin and pectolyase are washed out of the protoplast sample.
5. The GCPs are purified on a stepped Percoll gradient consisting of 1 cm^3 of a bottom layer of 90% Percoll and 3 cm^3 of an upper layer of 45% Percoll. The washed protoplasts, in a volume of about 1–2 cm^3, are layered on top and centrifuged at 100 *g* for 5 min in a bench centrifuge. The Percoll solutions contained 300 mM mannitol and 10 mM MES buffer (the solid MES and mannitol are added to the 90 or 45% Percoll which results in pH values of 7.3 and 6.8 without addition of KOH and this is satisfactory). (Although dialysis of the Percoll against several changes of distilled water in a cold-room overnight does eradicate the sodium impurity present – and possibly other contaminants – normally such treatment is unnecessary.) GCPs collect at the 90/45% interface, epidermal and subsidiary cell protoplasts remain at the top of the gradient while intact and broken mesophyll cell protoplasts and other debris collects at the bottom of the 90% Percoll. The GCPs are collected with a pipette with a widened, flamed tip and washed three times with 10 cm^3 aliquots of a solution containing 300 mM mannitol and 10 mM MES–KOH

buffer, pH 6 2 (or other appropriate medium dependent on the experiment) The K^+ concentration in the medium resulting from adjustment of the buffer is equivalent to *c* 7 mM Occasionally the GCPs do not appear spherical and may have a crimped appearance In such cases a decrease in the mannitol concentration from 300 mM to 270 mM often results in the formation of spherical, viable protoplasts

6 Rather than measure the protein or chlorophyll content of an aliquot of the GCPs suspension it is generally more convenient to measure the protoplast population with a haemocytometer slide (an improved Neubauer haemocytometer slide is ideal) and a microscope at ×100 magnification Chlorophyll and protein contents of guard cells of *C communis* have been determined (*see Table 3 1*) and so the protein or chlorophyll contents of a known volume of the suspension of GCP can be calculated

Figure 9 B1 shows stages in the isolation of GCPs from epidermal strips of *C communis*

It is important to appreciate that the ease with which GCPs can be prepared and their general condition is dependent on the condition of the plant material Plants should be grown ideally in a high photon flux density (*C communis* is a tropical and subtropical plant) and well watered under humid conditions Under such growing conditions leaves are large and the lower epidermis is loosely attached to the mesophyll

Guard cell protoplasts are very strong relative to protoplasts of other cell types and it is not easy to burst them Indeed, one problem encountered when using them in biochemical studies is how to break them open to create an homogenous mixture of cell content Even high concentrations of surfactants such as Triton X-100 do not adequately disrupt the protoplasts and forcing them many times through a fine gauge needle (e g 25 gauge) is necessary to obtain satisfactory results

References

Clint, G M (1985) The investigation of stomatal ionic relations using guard cell protoplasts I Methodology *J Exp Bot*, **36**, 1726–1738

Fitzsimons, P J and Weyers, J D B (1983) Separation and purification of protoplast types from *Commelina communis* L leaf epidermis *J Exp Bot*, **34**, 55–66

Weyers, J D B and Travis, A J (1981) Selection and preparation of leaf epidermis for experiments on stomatal physiology *J Exp Bot*, **32**, 837–850

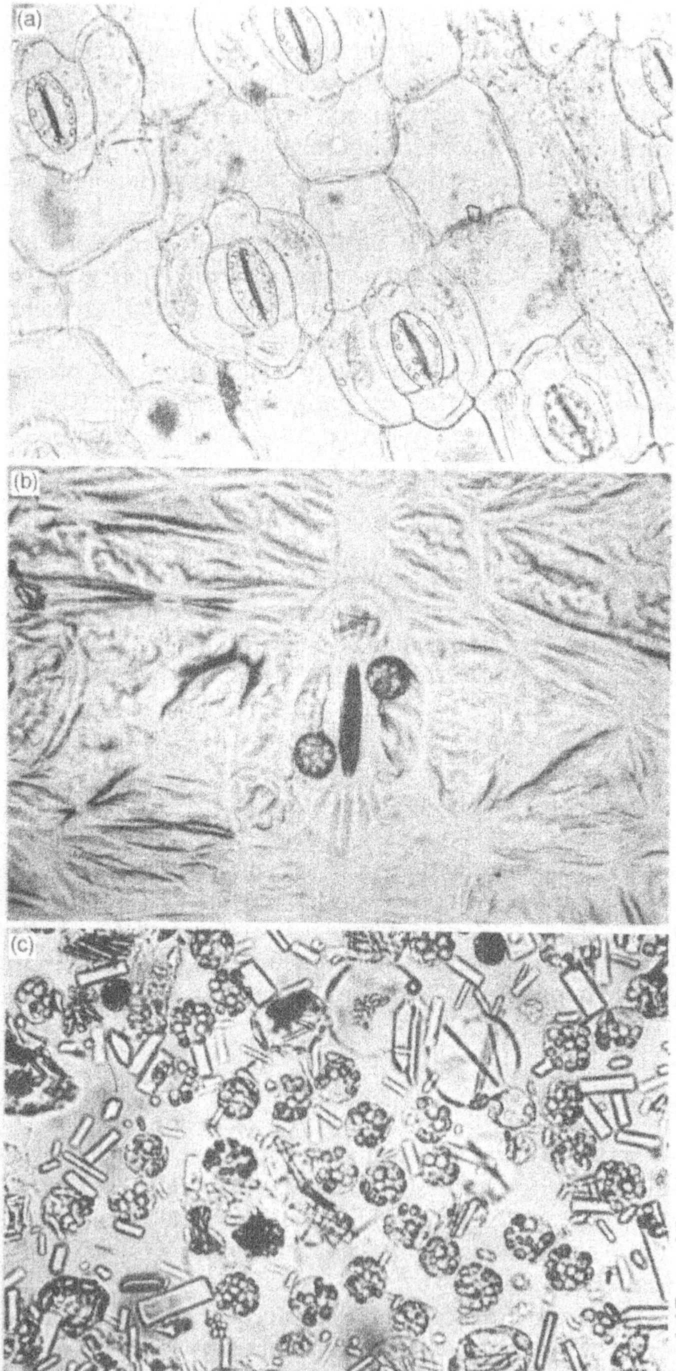
(a)
(b)
(c)

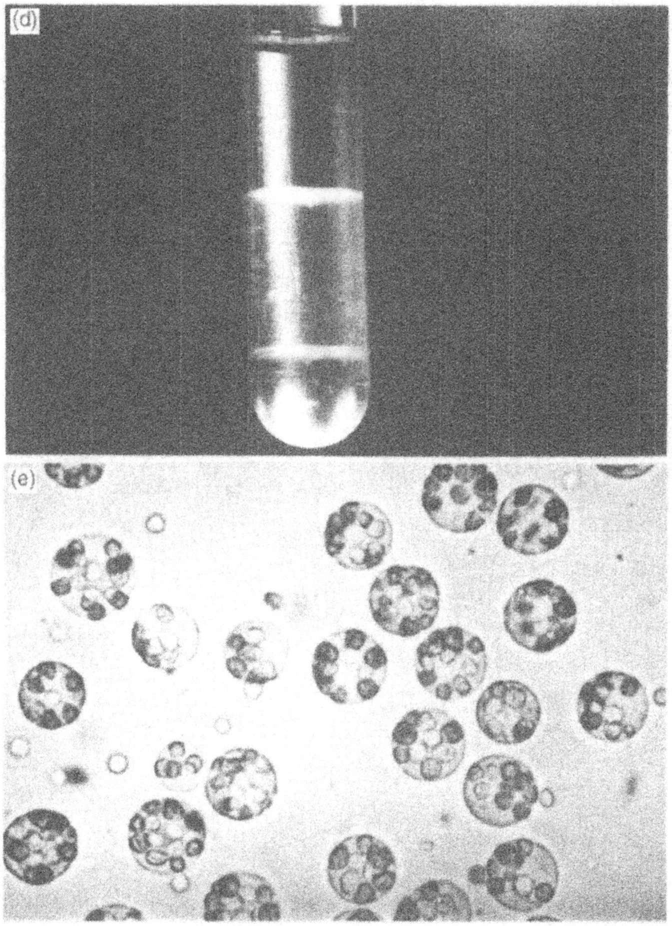

Figure 9.B1 Typical stages in the isolation of guard cell protoplasts from *Commelina communis* (a) Epidermis before digestion with enzymes, (b) Guard cell protoplasts about to be released from the retaining cell walls, (c) unpurified suspension of guard cell protoplasts in 270 mM mannitol, (d) guard cell protoplasts separated on a stepped 90/45% Percoll gradient, (d) purified sample of guard cell protoplasts taken from the 90/45% interface

9.2 The flow of carbon from starch to malate during stomatal opening

Figure 9.1 shows the possible major routes of carbon flow during stomatal opening and concomitant malate synthesis. Support for the existence of these pathways comes from labelling studies of guard cells and from detection of enzyme activities in guard cells and a study of their kinetics.

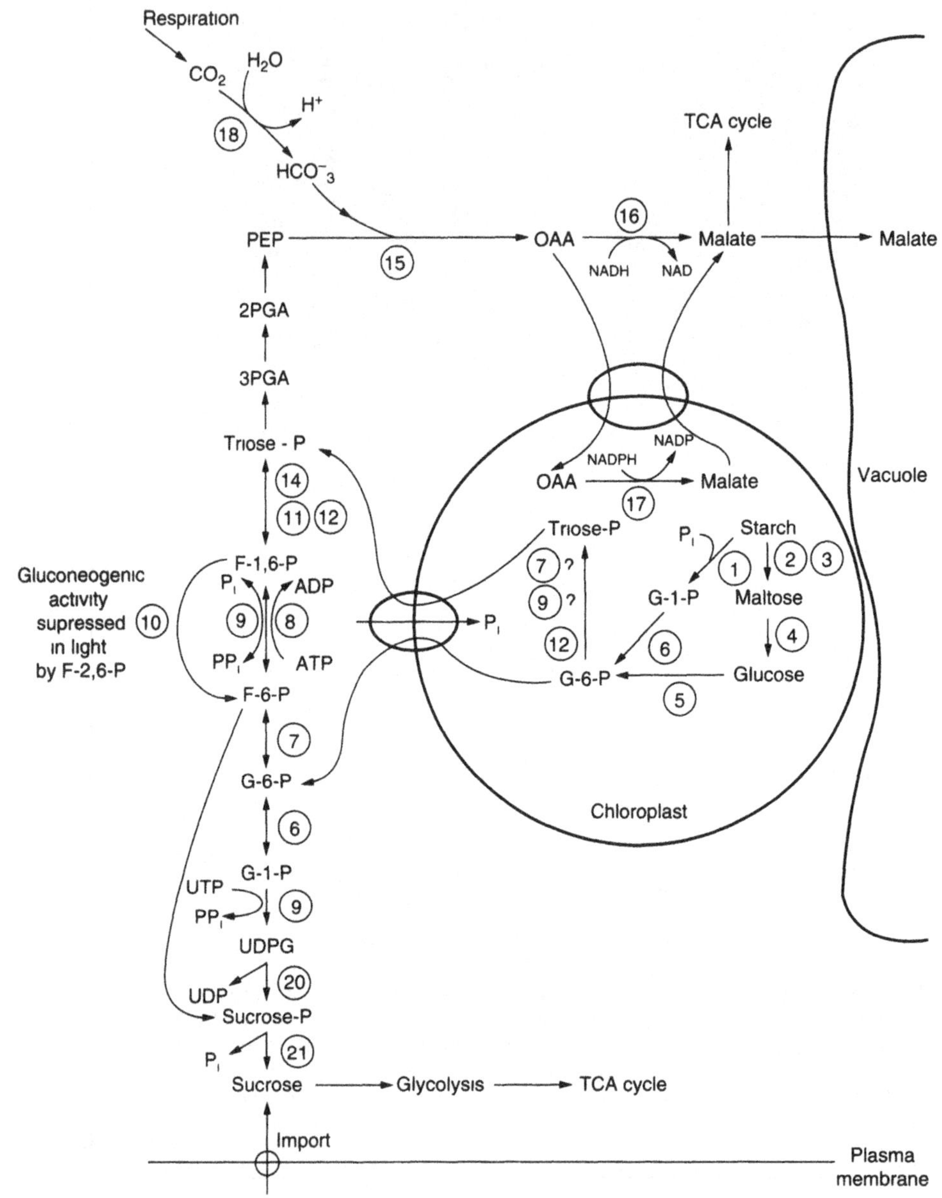

Figure 9.1 Possible pathways which may operate in starch-containing guard cells during stomatal opening in the light Key to enzymes 1, phosphorylase, 2, α-amylase and β-amylase, 3, R-enzyme (starch debranching enzyme), 4, maltase, 5, hexokinase, 6, phosphoglucomutase, 7, Phosphohexoisomerase, 8, ATP-phosphofructokinase, 9, PP_i-phosphofructokinase, 10, fructose-1,6-bisphosphatase, 11, aldolase, 12, phosphotriose isomerase, 13, NADP glyceraldehyde phosphodehydrogenase, 14, NAD glyceraldehyde phosphodehydrogenase, 15, phosphoenolpyruvate carboxylase, 16, NAD malate dehydrogenase, 17, NADP malate dehydrogenase, 18, carbonic anhydrase, 19, UDP-glucose pyrophosphorylase, 20, sucrose phosphate synthetase, 21, sucrose phosphate phosphatase

The various steps in the pathways and the regulation of key enzymes will now be discussed. In order to appreciate the relative rates and importance of enzyme activities in guard cells a comparison of activities in guard cells and other leaf cell types is presented in Table 9.1 (see also Outlaw, 1982). Additionally, Table 9.2 shows the distribution of some of the enzyme activities in various cell compartments of the guard cell.

There is much evidence indicating that starch supplies carbon skeletons for malate production during stomatal opening (except in starch-free guard cells where fructans may substitute for starch, see Section 9.4). Moreover, starch is an ideal substance from which malate (a source of anions and osmoticum) can be made since it is non-ionic and essentially insoluble and therefore has little osmotic activity.

Most observations indicate that guard cell starch levels decrease when stomata open and increase when stomata close, although some investigators do not observe this inverse correlation. However, as Raschke (1979) and Allaway (1981) point out, if there is simultaneous consumption within the guard cell and replenishment of carbohydrate from outside then strict synchrony may not occur. Also malate levels generally increase and decrease in guard cells in an inverse manner to that of the starch levels.

Within the chloroplasts of guard cells some enzymes concerned in the breakdown of starch and many of the glycolytic enzymes have been detected. Curiously, however, Robinson and Preiss (1987) found considerable activity of starch degrading enzymes (amylase, phosphorylase and R-enzyme) in the cytoplasm of guard cells of *C. communis*, although total cell activities were relatively low (see Table 9.2). Nevertheless, presumably starch degradation occurs primarily through phosphorylase activity since its activity was almost twice that of amylase (Robinson and Preiss, 1987). Moreover, they found very low R-enzyme (starch debranching enzyme) activities and could not detect hexokinase activity. Therefore, starch breakdown is primarily through phosphorylase activity to produce glucose-1-phosphate (G-1-P). Glucose-1-phosphate can then be converted to glucose-6-phosphate (G-6-P) and hence to other hexose phosphates and 3-C compounds within the chloroplast. Against this view is the finding of Robinson and Preiss (1987) that most of the phosphoglucomutase activity, which catalyses the conversion of G-1-P to G-6-P, is outside the chloroplast (Table 9.2).

A phosphate translocator has been detected in the envelope of guard cell chloroplasts which has a higher affinity for G-6-P than triose phosphate or 3-PGA (Overlach *et al.*, 1993) and hence G-6-P may be the major substance transported across the chloroplast envelope. Nevertheless, some evidence suggests that triose phosphate may be transported out of the chloroplasts in the light (Raschke, personal communication). However, studies indicate that some enzymes involved in

Table 9.1 Enzyme activities in guard cells and other cell types within a leaf of *Vicia faba* and *Commelina communis*

Enzyme	Cell type	Activity			Species	Reference
		µmol mg $^{-1}$ chl h $^{-1}$	µmol mg $^{-1}$ protein h $^{-1}$	pmol cell $^{-1}$ h $^{-1}$		
Ribulose-1,5-bisphosphate carboxylase (EC 4 1 1 39)	guard	52	1 48†	–	*V faba*	a
	mesophyll	130	4 33†	–		
	guard	<4	<0 1	–	*V faba*	b
	epidermal	–	16 2	–		
	spongy mesophyll	202	5 0	–		
	palisade	110	5 5	–		
	guard	3 5	0 1	0 01	*V faba*	c
	mesophyll			142 50	*V faba*	d
	guard	ND	ND	ND	*C communis*	e
	mesophyll	62 3	5 6	14 3		
	guard	5 5–11 5*	0 11–0 23	–	*C communis*	f
	mesophyll	173*	14 4	–		
	guard	140	4 0†		*V faba*	g
	mesophyll	460	15 3†			
Ribulose-5 phosphate kinase (EC 2 7 1 19)	guard	175	5 0	0 75	*V faba*	b
	epidermal	–	2 4	–		
	spongy mesophyll	1632	82 0	–		
	palisade	1994	50 0	–		
NADP Glyceraldehyde -3-phosphate dehydrogenase (EC 1 2 1 13)	guard	<18	<0 5	<0 2	*V faba*	b
	epidermal	–	<0 4	–		
	spongy mesophyll	115	5 8	–		
	palisade	130	3 2	–		
	guard	ND	ND	ND	*V faba*	h
	mesophyll	73	0 9	–		
	guard	2670	33 0†	–	*V faba*	a
	mesophyll	630	21 0†	–		
	guard	–	–	2 95	*V faba*	i
	guard	414*	8 3	–	*C communis*	j
Carbonic anhydrase‡ (EC 4 2 1 1)	guard	17214	314	–	*C communis*	e
	mesophyll	10944	960	–		
Starch synthase (EC 2 4 1 21)	guard	23 0*	0 46	–	*C communis*	f
	guard	6 5*	0 13	–	*C communis*	j
ADPG pyrophosphorylase (EC 2 7 7 27)	guard	50*	1 0	–	*C communis*	f
	mesophyll	275*	5 5	–		
	guard	137*	2 7	–	*C communis*	j
	guard	98	2 8	–	*V faba*	k
	epidermal	–	0 6	–		
	spongy mesophyll	112	2 8	–		
	palisade	44	2 2	–		

Enzyme	Cell type	Activity			Species	Reference
		µmol mg^{-1} chl h^{-1}	µmol mg^{-1} protein h^{-1}	pmol cell^{-1} h^{-1}		
Branching enzyme (max activity with 1 mg primer) (EC 2 4 1 18)	guard	1208*	24 7	–	*C communis*	j
Amylase (α-amylase EC 3 2 1 1, β-amylase EC 3 2 1 2)	guard	137*	2 75	–	*C communis*	j
Starch phosphorylase (EC 2 4 1 1)	guard	227*	4 55	–	*C communis*	j
R enzyme (EC 3 2 1 41)	guard	85*	1 70	–	*C communis*	k
NADP Malate dehydrogenase[§] (EC 1 1 1 82)	guard	642	8 03[†]	–	*V faba*	l
	mesophyll	66	2 2[†]	–		
	guard	600	17 1	–	*V faba*	m
	guard	–	–	0 58	*V faba*	i
	guard	200	5 7	–	*C communis*	m
Pyruvate phosphate dikinase (EC 2 7 9 1)	guard	<14	<0 4	<0 05	*V faba*	n
	epidermal	ND	ND	ND		
	guard	21	0 6	0 1	*V faba*	c
	guard	70	1 2	–	*C communis*	e
Cytochrome *c* reductase (EC 1 6 99 3)	guard	723	9 04	–	*V faba*	h
	mesophyll	20 6	0 69	–		
NAD Malic enzyme (EC 1 1 1 82)	guard	1416	40 0	6 0	*V faba*	n
	epidermal	–	16 2	–		
	spongy mesophyll	283	7 1	–		
	palisade	100	5 0	–		
Fumarase (EC 4 2 1 12)	guard	576	16 5	2 6	*V faba*	o
	spongy mesophyll	186	4 6	11 0		
	palisade	80	4 0	9 9		
	guard	–	–	0 6	*V faba*	d
	mesophyll	–	–	1 2		
Cytochrome *c* oxidase (EC 1 9 3 1)	guard	505	14 4[†]	–	*V faba*	h
	mesophyll	18 8	9 04[†]	–		
ATP-phosphofructokinase	guard	–	–	0 56	*V faba*	d
	guard	–	–	6 3	*V faba*	i

Enzyme	Cell type	Activity: μmol mg^{-1} chl h^{-1}	Activity: μmol mg^{-1} protein h^{-1}	Activity: pmol cell^{-1} h^{-1}	Species	Reference
(EC 2 7 1 11)	guard	150†	4 3	–	*V faba*	d
	palisade	7 2	0 4	–		
	guard	27*	0 5	–	*C communis*	j
Adolase (EC 4 1 2 13)	guard	108 5*	2 17	–	*C communis*	j
Phospho-glucose isomerase (EC 5 3 1 9)	guard	987*	19 7	–	*C communis*	j
Fructose-1 6-bisphosphate phosphatase (EC 3 1 3 11)	guard	160	2 00†	–	*V faba*	a
	mesophyll	170	5 67†	–		
	guard	54*	1 08	–	*C communis*	j
Pyrophosphate phosphofructo-kinase (EC 2 7 1 90)	guard	285*	0 57	–	*C communis*	j
	guard	–	–	10 2	*V faba*	i
	guard	322†	9 2	–	*V faba*	p
	palisade	8†	0 4	–		
Triose-phosphate isomerase (EC 5 3 1 1)	guard	32 4*	647 8	–	*C communis*	j
	guard	1072	30 6†	4 8	*V faba*	o
	spongy mesophyll	522	13 1†	32 0		
	palisade	199	10 0†	24 0		
NAD Glyceraldehyde-3-phosphate dehydrogenase (EC 1 2 1 12)	guard	1874*	37 5	–	*C communis*	j
Sucrose phosphate synthase (EC 2 4 1 14)	guard	168†	4 8	–	*V faba*	q
	spongy mesophyll	88†	2 2	–		
	palisade	50†	2 5	–		
	epidermal	–	0 5	–		
Sucrose synthase (EC 2 4 1 13)	guard	38 5†	1 1	–	*V faba*	q
	spongy mesophyll	3 0†	0 1	–		
	palisade	3 0†	0 1	–		
	epidermal	–	1 4	–		
Phosphoenol pyruvate carboxylase (EC 4 1 1 31)	guard			8 0–9 0	*C communis*	see Table 9 3
	guard			1 6–10	*V faba*	see Table 9 3
NADP Malic enzyme (EC 1 1 1 40)	guard	357	10 2	1 5	*V faba*	n
	palisade	10	0 5	–		
	spongy mesophyll	113	2 8	–		
	epidermal	–	12 6	–		

Enzyme	Cell type	Activity			Species	Reference
		μmol mg^{-1} chl h^{-1}	μmol mg^{-1} protein h^{-1}	pmol cell^{-1} h^{-1}		
	guard	–	–	0 6	*V faba*	i
	guard	2550	73 0	11 0	*V faba*	c
	guard	110	2 0	–	*C communis*	e
Phosphoenol-pyruvate carboxykinase (EC 4 1 1 49)	guard	47	1 3	0 2	*V faba*	p
	guard	ND	ND	ND	*V faba*	c
NAD Malate dehydrogenase (EC 1 1 1 37)	guard	580	16 5	2 5	*V faba*	c
	guard	30 000	375 0[†]	–	*V faba*	l
	guard	7 272	132 0	–	*C communis*	e
	mesophyll	1 622	142 0	–		
6-phospho-gluconate dehydrogenase (EC 1 1 1 44)	guard	160	4 6[†]	0 7	*V faba*	o
	spongy mesophyll	46	1 2[†]	2 8		
	palisade	15	0 8[†]	1 0		
Glucose-6-phosphate dehydrogenase (EC 1 1 1 49)	guard	–	–	2 4	*V faba*	d
Hydroxy-pyruvate reductase (EC 1 1 1 81)	guard	ND	ND	ND	*V faba*	o
	spongy mesophyll	138	3 45[†]	8 4		
	palisade	64	3 20[†]	9 0		
Glycollate oxidase (EC 4 2 1 1)	guard	ND	ND	ND	*C communis*	e
	mesophyll	1756	32	–		
Acid phosphatase (EC 3 1 3 2)	guard	1700	34	9 03	*C communis*	r
	guard	–	–	1 18	*V faba*	d
	mesophyll	–	–	6 83		
Mannosidase (α-mannosidase EC 3 2 1 24, β-mannosidase EC 3 2 1 25)	guard	–	–	0 15	*V faba*	d
	mesophyll	–	–	0 19		

* Calculated assuming a protein chlorophyll ratio of 50 and 12 in a guard cell and a mesophyll cell, respectively of *C communis* (see Table 3 1)

[†] Calculated assuming a protein chlorophyll ratio of 35, 20 and 40 in a guard cell, palisade cell and spongy mesophyll cell, respectively (ratio assumed to be 30 if palisade and spongy mesophyll are not differentiated), of *V faba* (see Table 3 1)

[‡] Activities are expressed as 'enzyme units mg^{-1} protein, mg^{-1} chlorophyll or protoplast^{-1} (see Reyss and Prioul, 1975)

[§] Maximum light activation

[a] Shimazaki *et al*, 1989, [b] Outlaw *et al*, 1979a, [c] Schnabl, 1981, [d] Schnabl and Kottmeier, 1984b, [e] Birkenhead and Willmer 1986, [f] Robinson *et al*, 1984, [g] Shimazaki, 1989, [h] Shimazaki *et al*, 1983, [i] Hedrich *et al*, 1985, [j] Robinson and Preiss, 1987, [k] Outlaw and Tarczynski, 1984, [l] Gotow *et al*, 1985, [m] Willmer, unpublished, [n] Outlaw *et al* 1981b, [o] Hampp *et al*, 1982, [p] Hite *et al*, 1992, [q] Hite *et al*, 1993, [r] Fricker and Willmer 1987

Table 9.2 Compartmentalization of enzymes within guard cells percentage distribution of the enzyme between chloroplasts, mitochondria and cytosol or between chloroplasts and the remainder of the cell

Enzyme	Distribution (%) chloroplast	cytosol	cytosol + mitochondrion	mitochondrion	Species	Reference
Phosphoenolpyruvate carboxylase	8 5		91 5		*V faba*	d
	10		90		*C communis*	c
	1		99		*V faba*	b
	5	90		5	*V faba*	a
	1		99		*V faba*	b
	4	89		7	*C communis*	e
Aldolase	28		72		*C communis*	c
Fructose-1,6-bisphosphatase	58		4 2		*C communis*	c
	0 5		99 5		*V faba*	b
	47		53		*V faba*	d
Phosphoglucose isomerase	18		82		*C communis*	c
Phosphoglucomutase	26		74		*C communis*	c
ATP-phosphofructokinase (ATP-PFK)	0		100		*C communis*	c
	27		73		*V faba*	b
PP_i-phosphofructokinase (PP_i-PFK)	0		100		*C communis*	c
	2		98		*V faba*	b
3-Phosphoglycerate (kinase)	30		70		*V faba*	d
Glyceraldehyde-3-phosphate dehydrogenase (NADP)	65		35		*C communis*	c
	62		38		*V faba*	d
	65		35		*V faba*	b
Glyceraldehyde-3-phosphate dehydrogenase (NAD)	15		85		*C communis*	c
	15		85		*V faba*	d

Enzyme	Distribution (%)				Species	Reference
	chloroplast	cytosol		mitochondrion		
Carbonic anhydrase	53	47		0	*C communis*	e
Triose-phosphate isomerase	36		65		*V faba*	d
	25		75		*C communis*	c
NAD malate dehydrogenase	6	43		51	*C communis*	e
	14	62		24	*V faba*	a
NADP malic enzyme	0 2		99 8		*V faba*	b
Pyruvate kinase	20	80		0	*V faba*	a
ADPG pyrophosphorylase	73		27		*C communis*	c
Starch synthase	77		23		*C communis*	c
Branching enzyme	76		24		*C communis*	c
Amylase	35		65		*C communis*	c
Phosphorylase	43		57		*C communis*	c
R enzyme	25		75		*C communis*	c
NADP malate dehydrogenase	75		25		*V faba*	d
	72		28		*V faba*	b
	76	17		7	*V faba*	a
Cytochrome c oxidase	13	0		87	*V faba*	a

[a] Gotow *et al* , 1985, [b] Hedrich *et al* , 1985, [c] Robinson and Preiss, 1987, [d] Shimazaki *et al* , 1989, [e] Willmer and Birkenhead, unpublished

the conversion of G-6-P to triose phosphate (ATP-phosphofructokinase, aldolase and phosphoglucoisomerase) are of relatively low activity or non-detectable in guard cell chloroplasts and at much higher levels in the cytoplasm (see Table 9.2) and hence the transport of hexose phosphate seems more favourable.

During stomatal opening it has been suggested that a compatible cytoplasmic osmoticum, probably sucrose, is synthesized as well as malate being synthesized and accumulated in the vacuole (see Section 9.5). Thus, once out of the chloroplasts into the cytosol the 3-C compounds or hexose phosphate can be readily converted to PEP via a section of the glycolytic sequence and hence to malate or they can also be readily channelled into sucrose synthesis. From G-6-P the pathway of sucrose synthesis can be achieved as indicated in Fig. 9.1. High levels of the enzyme sucrose phosphate synthetase, an enzyme involved in the pathway, have been detected in guard cells of *Vicia faba* (Hite *et al.*, 1993).

Glucose-6-phosphate can also be readily converted to fructose-6-phosphate (F-6-P) via hexisomerase found mainly in the cytoplasm (Robinson and Preiss, 1987). At this point two cytoplasmically located enzymes could potentially convert F-6-P to fructose-1,6-phosphate, i.e. pyrophosphate fructose-6-phosphate phosphotransferase (PP_i-PFK) and ATP-PFK. Hite *et al.* (1992) speculate that the ATP-dependent PFK activity provides carbon skeletons while the PP_i-dependent PFK activity responds to energy demands.

A method of regulating carbon flow between hexose phosphates and triose phosphates has been proposed by Hedrich *et al.* (1985) which stems from their observation that fructose-2,6-bisphosphate levels greatly increase upon illumination of GCPs of *V. faba*. Fructose-2,6-phosphate appears to function as a regulator metabolite in glycolysis and gluconeogenesis of photosynthetic tissue by enhancing glycolysis through activation of PP_i-PFK and suppressing gluconeogenesis through an inhibition of cytoplasmic fructose 1,6-bisphosphatase. If this is also the case in guard cells then increased fructose-2,6-phosphate levels would stimulate the flow of carbon from starch to malate during stomatal opening in the light. The decline of fructose-2,6-phosphate in darkness would enable fructose-1,6-bisphosphatase to become active favouring gluconeogenesis and carbon flow from malate to starch during stomatal closure.

Fructose 1,6 bisphosphate is then probably converted to triose phosphate, 3-PGA and hence to PEP within the cytoplasm (see Hite *et al.*, 1992). Detection of enzyme activities involved in this sequence (aldolase, phosphotriose isomerase and NAD glyceraldehyde phosphodehydrogenase) have been mainly localized in the cytoplasm (Robinson and Preiss, 1987). If a phosphate/triose phosphate translocator is dominant in chloroplasts of guard cells then triose phosphate could be channelled into either sucrose or PEP production.

In the cytoplasm PEP carboxylase converts HCO_3 and PEP to OAA which, in turn is rapidly converted to malate via malate dehydrogenase activity

Bicarbonate is the substrate utilized by PEP carboxylase rather than CO_2, and carbonic anhydrase reversibly catalyses the hydration of CO_2 There is little information on this enzyme in guard cells but Birkenhead and Willmer (unpublished data) found that the enzyme is very active in guard cells of *Commelina* with the activity being about equal in the chloroplastic and cytosolic compartments

PEP carboxylase is a key enzyme involved in malate synthesis and therefore a knowledge of its regulation in guard cells is of considerable importance in understanding how stomata function The enzyme occurs universally in the cytoplasm of all cell types in higher plants and catalyses the irreversible carboxylation of PEP Different isoforms exist in C3, C4 and CAM plants, and Schulze *et al* (1992) also found different isoforms in the mesophyll, epidermal and guard cells of *V faba* Reviews by Willmer (1983) and Outlaw (1990) document the widely different values of PEP carboxylase activity and other characteristics of the enzyme purported to be from guard cells which have been obtained by various investigators It is believed that, although part of the variation may be due to species differences, it is also due to different assay procedures and different sources of the enzyme being employed For example, enzyme characteristics were studied using freeze-dried guard cell pairs, epidermal tissue and GCPs Also microanalytical assay procedures of Outlaw and his co-workers (using guard cell pairs dissected from freeze-dried leaf tissue) and standard macro assay techniques (using $^{14}CO_2$ incorporation or NADH oxidation) were used and assays conducted with crude, semi-purified or purified extracts As indicated earlier, it is also generally conceded that biochemical data obtained from using epidermal strips is dubious because of mesophyll contamination of the epidermis and that epidermal tissue consists of less than 5% guard cells (Outlaw, 1982) Additionally, epidermal tissue contains enzyme inhibitors, some of which are not eradicated by passage of extracts through Sephadex columns (Willmer *et al* , 1987)

Because of criticisms concerning the use of epidermal tissue, data presented in Table 9 3 on guard cell PEP carboxylase are restricted to those obtained with GCPs, freeze-dried guard cell pairs and a relatively purified PEP carboxylase isoform isolated from *V faba* epidermis PEP carboxylase from guard cells of *V faba* has a native molecular mass of 467 kDa, two different subunits exist of 110 and 112 kDa indicating that the enzyme is a tetramer (Denecke *et al* , 1993) In general, guard cell PEP carboxylase is similar to PEP carboxylase from leaves of C4 plants although there are considerable differences in the maximum activities of guard cell PEP carboxylase, rates are high and comparable to those

Table 9.3 Characteristics of PEP carboxylase from guard cells

Variable	Sample	Observation	Reference
Stability	*V faba*, freeze-dried guard cells 0 02% BSA, 25°C	linear activity for up to at least 60 min	a
	V faba GCP 0–5°C, crude and desalted extracts	50% activity loss after 30 and 60 min with crude and desalted extracts, respectively	b
	V faba, partially purified guard cell enzyme from epidermis	cold pretreatment inactivates enzyme (but no discontinuity in Arrhenius plots)	j
pH	*V faba* GCP	broad peak of 8 0–8 6, suggestion of two peaks within this range	b
	V faba, partially purified guard cell enzyme from epidermis	8 0–8 2 optimum but a combination of malate and G-6-P shifted optimum to pH 7 6	j
Temperature	*V faba* freeze-dried guard cells	maximum activity 25°C (initial rate 35% greater at 40°C than 25°C but instability above 35°C)	c
	V faba, partially purified guard cell enzyme from epidermis	maximum activity 35°C (activation energy 51 kJ mol–1)	k
V_{max} (pmol guard cell $^{-1}$ h $^{-1}$)	*V faba* freeze-dried guard cells	3 9	a
	V faba freeze-dried guard cells	6 9, unaffected by stomatal aperture, pH 7 0–8 5, malate (0–15 mM), G-6-P (0–5 mM)	j
	V faba guard cells	10	d
	V faba guard cells desalted, pH 8 3	1 6–2 2	b
	V faba guard cells	3 7*	e
	V faba, partially purified guard cell enzyme from epidermis	9 1	i
	C communis guard cells desalted, pH 7 0–7 4	8 3†	f
	C communis guard cells crude extracts, pH 8 3	8 5	g
$K_{M,\ PEP\ Mg}$	*V faba*, freeze-dried guard cells	0 06 mM (pH 8 7, 0 5 mM HCO_3^-, 1 mM Mg^{2+})	a
	V faba, freeze-dried guard cells	0 2–1 5 mM (pH 7 7, 0 5 mM HCO_3^-, 1 mM Mg^{2+})	c
	V faba GCP	0 21 mM (pH 8 3, 2 mM HCO_3^-, 5 mM Mg^{2+}), PEP stimulates up to 0 75 mM and inhibits at higher concentrations	b
	V faba, partially purified guard cell enzyme from epidermis	0 06 mM (pH 8 3) 0 35 mM (pH 7 0)	k
	V faba freeze-dried guard cells	0 06 mM (pH 8 0 + G-6-P), 3 43 mM (pH 7 0 + malate)	l
	V faba, GCP	0 01–0 2 mM (swelling GCP decrease $K_{m\ PEP}$)	m

Variable	Sample	Observation	Reference
Malate	*V faba* GCP	malate inhibits activity, increasing malate increases $K_{m\ PEP}$ and decreases V_{max} (at pH 8 3 0 7 mM malate inhibits by 50% at 1 mM PEP)	2
	V faba, partially purified guard cell enzyme from epidermis	malate inhibits at pH 8 3 but not at pH 7 0	11
	C communis GCP	malate inhibits activity (3 mM malate inhibits by 50% [$K_{i\ mal}$=0 4 mM] at pH 7 4 and 2 mM PEP)	6
G-6-P	*V faba* freeze-dried guard cells	no effect on V_{max}, decreases $K_{m\ PEP}$ (0 1 mM at pH 8 5, 6 2 mM at pH 7 0)	10
	V faba, partially purified guard cell enzyme from epidermis	stimulates activity at pH 7 0 but not at pH 8 0	11
	C communis GCP	reverses malate inhibition	6
K^+ and osmotic stress	*V faba* GCP desalted extract 10 min in K^+	12 mM KCl maximizes activity and nullifies effects of 1 mM malate on V_{max}	2
	V faba freeze-dried guard cells	50% inhibition by 200 mM NaCl or Na acetate	3
Native molecular mass and subunits	*V faba* partially purified guard cell enzyme from epidermis	467000, tetramer with subunits of M_r 110 and 112 kDa	11

* Calculated from original values on the basis of 2 5 pg chlorophyll per guard cell
† Calculated from original values on the basis of 200 pg protein per guard cell
See Table 3 1 for conversion values

[a] Outlaw and Kennedy 1978, [b] Schnabl and Kottmeier, 1984a, [c] Outlaw *et al* , 1979, [d] Schnabl, 1981, [e] Gotow *et al* , 1985, [f] Donovan *et al* , 1985, [g] Birkenhead and Willmer, 1986, [h] Robinson and Preiss, 1987, [i] Hedrich *et al* , 1985, [j] Tarczynski and Outlaw, 1993, [k] Denecke *et al* , 1993, [l] Wang *et al* , 1994

found in mesophyll cells of C4 plants, activity is dependent on Mg^{2+} ions, the substrate being a $PEPMg^{2+}$ chelate, the enzyme has a pH optimum around 8 0 although this is reported to be shifted to lower pH values in the presence of malate and G-6-P Like PEP carboxylase from other sources, activity is inhibited by malate and stimulated by G-6-P in a pH sensitive manner Malate inhibition is reversed by G-6-P Reported values of $K_{m\,PEP}$ vary between 0 06 and 3 43 mM depending on pH, malate and G-6-P concentrations and other factors (see Table 9 3) Higher pH values decrease the $K_{m\,PEP}$ while increasing malate levels generally decrease V_{max} and increase the K_m Moreover, Schnabl and Kottmeier (1984a) found that PEP stimulated activity upto 0 75 mM while higher concentrations inhibited activity

Another possible explanation of the wide variation in $K_{m\,PEP}$ values and V_{max} may be explained by the findings of Kottmeier and Schnabl (1986) that, as GCP of *V faba* swelled the K_m decreased Thus, an increase in GCP swelling of 50% resulted in the K_m being lowered from 0 2 mM PEP to 0 01 mM PEP Additionally, they observed that the $K_{m\,PEP}$ decreased as the pH of the *incubating medium* for the GCP was increased (over a range of 7 5–8 5) and the *assay medium* kept at a constant pH 8 3 (in the absence of K^+), the K_m also decreased as the pH of the *assay medium* was increased over a range of 5 6–8 5 The significance of these findings is unclear

According to Tarczynski and Outlaw (1993) and Wang *et al* (1994) V_{max} is not affected significantly by the extent of stomatal aperture, pH (7 0–8 5), malate concentration or G-6-P concentration However, they found that the $K_{m\,PEP}$ was greatly affected by physiological concentrations of G-6-P (guard cell concentration, 0 6–1 2 mM) and malate (cytoplasmic concentrations in cells are typically about 8 0 mM) such that they consider PEP carboxylase is regulated by its immediate cytosolic environment, such as a relatively low pH and limiting substrate concentrations Further work by Zhang *et al* (1994), however, found that there was a post-translational modification to PEP carboxylase during stomatal opening

Although *in vivo* light activation of PEP carboxylase in GCPs of *V faba* (Gotow *et al* , 1985, Wang *et al* , 1994) or *C communis* (Willmer *et al* , 1990) was not detected, with the finding that light regulation of phosphorylation and dephosphorylation of PEP carboxylase occurs in C4 (e g Jiao and Chollet, 1988) and CAM (e g Nimmo *et al* , 1986) plants, with the phosphorylated form of the enzyme being more active than the dephosphorylated form, a search for such a system was made in guard cells Thus, Nelson *et al* (1993) and Nimmo (personal communication) found evidence of the phosphorylation of PEP carboxylase from guard cells of *C communis* which was correlated with stomatal opening or swelling of GCPs However, Schnabl *et al* (1992) were unable to detect a light-induced phosphorylation of PEP carboxylase from GCPs of *V faba* In many ways guard cells are more like CAM cells rather than mesophyll cells of C4 plants (which contain the PEP carboxylase) because they exhibit large fluctuations in malate content and possibly possess some Calvin cycle activity In CAM plants phosphorylation and increased activity of PEP carboxylase occurs in the dark with dephosphorylation and loss of activity occurring in the light (e g Nimmo *et al* , 1986) However, it should be appreciated that in guard cells, when stomata are normally open in the light, malate levels are high while in the dark, when stomata are normally closed, malate levels are low This is the opposite to what happens in CAM cells when malate levels in the light and dark are low and high, respectively

Guard cells also possess high levels of activity of NAD specific malate dehydrogenase (cytoplasmic and mitochondrial) and considerable activity of a chloroplastic NADP specific malate dehydrogenase which catalyses the reduction of OAA to malate In those species so far investigated NADP malate dehydrogenase from guard cells is light activated *in vivo* (Gotow *et al* , 1985, Scheibe *et al* , 1990, Willmer, unpublished) In GCPs of *V faba* (Gotow *et al* , 1985, Willmer, unpublished data, see Fig 9 2) and *C communis* (Willmer, unpublished) the enzyme is fully activated within 5 min of illumination of the protoplasts, undergoing a 4- to 6-fold increase in activity The enzyme is inactivated to minimum levels within 5 min of the protoplasts being placed in the dark (Gotow *et al* , 1985, Willmer, unpublished) Maximum activities in GCP of *C communis* are about 200 μmol mg^{-1} chlorophyll h^{-1} (equivalent to 4 μmol mg^{-1} protein h^{-1}) and 600–700 μmol mg^{-1} chlorophyll h^{-1} (equivalent to 12–14 μmol mg^{-1} protein h^{-1}) in GCPs of *V faba* Such activities are higher than values quoted for CAM tissues but generally lower than activities found in C4 plants Gotow *et al* (1985) also found that the *in vivo* light activation of the enzyme was inhibited by DCMU This indicates that the enzyme is activated by reductants from linear electron flow within the chloroplast (as in other cell types NADP malate dehydrogenase is located within the chloroplasts of guard cells) In C4 plants it is believed that NADP malate dehydrogenase is activated upon reduction of disulphide groups by reduced thioredoxin-m The source of the reducing power is considered to be ferredoxin and thus there will be competition between thioredoxin and NADP for the reduced ferredoxin The NADP/NADPH ratio in the chloroplast, and probably the cell in general, will therefore probably influence activity of the enzyme

Certainly the activation and inactivation of NADP malate dehydrogenase is rapid enough to be involved in the regulation of stomatal movements which themselves can be relatively rapid There is also evidence of an OAA translocator being present in the chloroplast envelope of guard cells (Raschke, personal communication) The high activity of light-stimulated NADP malate dehydrogenase and the presence of an OAA/malate shuttle in the chloroplast envelope means that there are two potentially

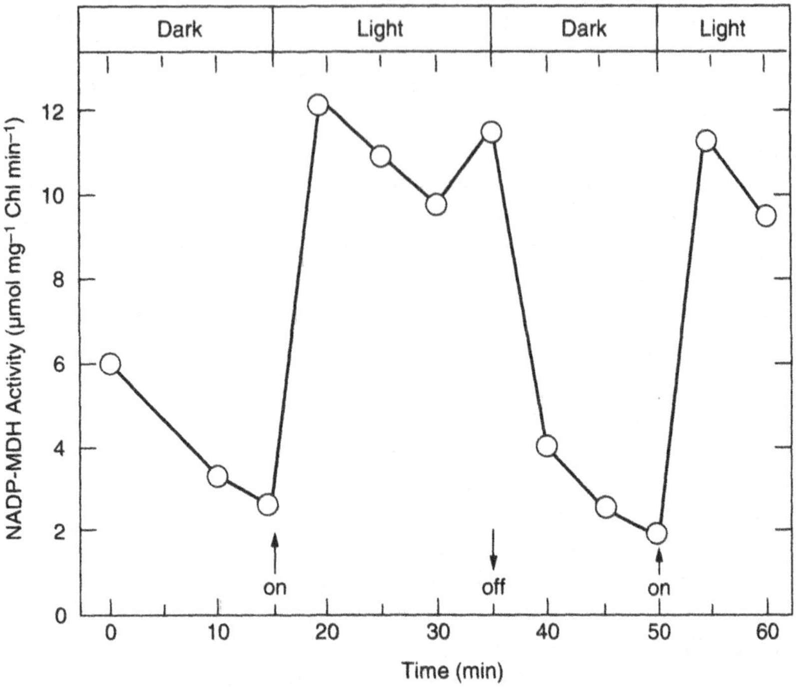

Figure 9.2 Light activation and dark inactivation of NADP-MDH in guard cell protoplasts (GCPs) of *Vicia faba*. Protoplasts were incubated in 1.0 ml of 0.4 M mannitol, 10 mM KCl, 1 mM $CaCl_2$ and 5 mM MES–Tris (pH 6.5) at 25°C. The activity at time 0 was that in GCPs kept on ice after isolation until experiment. After Gotow *et al.* (1985).

major routes of malate formation. The significance of these two alternative routes of malate formation is unclear but it is a possible site of regulation of carbon flow between starch and malate. It is also apparent that in darkness, when stomatal opening does occur in CAM species, for example, the NADP specific malate dehydrogenase will be relatively inactive and most of the malate will arise from NAD specific malate dehydrogenase activity. Scheibe *et al.* (1990), also conclude that the activity of the chloroplastic NADP malate dehydrogenase in guard cells of *Pisum sativum* barely suffices to meet the malate requirement for stomatal opening even in the light.

9.3 The fate of malate during stomatal closure

Figure 9.3 presents some possible pathways of carbon flow during stomatal closure. When stomata close, generally malate levels in guard cells decrease while starch levels increase though the stoichiometry of the malate decrease/starch increase is not necessarily 1:1. Three fates of malate have been suggested:

1. Malate is released from the guard cells.
2. Malate is metabolized in the mitochondria of the guard cells.
3. Malate can be channelled back into starch synthesis within the guard cell.

All three possibilities probably occur in guard cells although the extent to which they occur is not known.

Since starch synthesis is too slow to remove all the malate formed during opening and, energetically, it is expensive to convert malate to starch the exit of at least some malate from the guard cells is a likely alternative. Thus, Van Kirk and Raschke (1978) found that a substantial proportion of malate was released from guard cells to the apoplast when stomata closed. Also, Dittrich and Raschke (1977) found that malate was the only labelled compound released from epidermal tissue of *C. communis* with closing stomata which was exposed to $^{14}CO_2$. However, some studies have shown that numerous labelled compounds are released from epidermal tissue of *C. cyanea* at all stages of stomatal movements (see Milthorpe *et al.*, 1979; Willmer 1981) while Freer-Smith and Willmer (1981) found that malate was the chiefly labelled compound (about 90% of the total ^{14}C) released from epidermal strips of *C. communis* whether stomata were opening or closing.

Thus, some caution is needed in the interpretation of such studies; the results may, for example, only reflect the extent to which cell damage with accompanying cell 'leakiness' is occurring.

Malate can also readily cross mitochondrial membranes and be used as a source of energy for active ion transport systems occurring in the guard cells or for general cell metabolism. The presence of labelled acids of the TCA cycle when epidermal tissue is fed $^{14}CO_2$ or [^{14}C]malate indicates this to be the case. Furthermore, as Table 9.1 illustrates, mitochondrial enzyme activity is greater in guard cells than in mesophyll cells and, like mitochondria from mesophyll cells, cyanide resistant respiration also occurs in guard cell mitochondria (Shimazaki *et al.*, 1983; Pantoja and Willmer, 1988; Vani and Raghavendra, 1994).

There is also evidence from ^{14}C-labelling studies that part of the carbon from malate is channelled into starch synthesis via gluconeogenesis (Dittrich and Raschke 1977; see also Milthorpe *et al.*, 1979; Willmer 1981). High levels of NADP specific malic enzyme (e.g. Outlaw *et al.*, 1981; Birkenhead and Willmer, 1986) and NAD specific malic enzyme (Outlaw *et al.*, 1981b) activity have been detected in guard cells and, therefore, decarboxylation of malate can readily occur, producing pyruvate and CO_2. According to Hedrich *et al.* (1985) NADP malic enzyme of guard cells is cytoplasmic, the same as in CAM plants, though in C4 plants it is a chloroplastic enzyme. The released CO_2 could be refixed via whatever Calvin cycle exists in guard cells though the decarboxylation

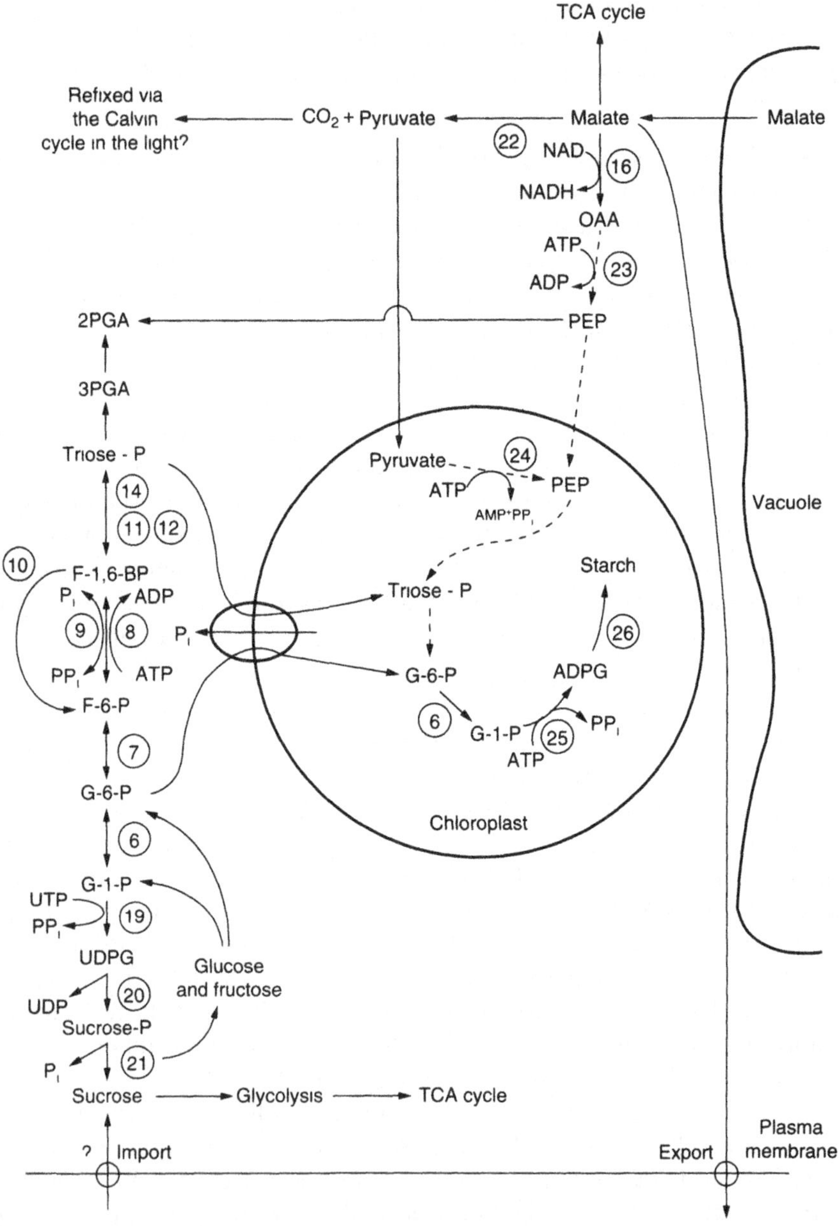

Figure 9.3 Possible pathways which may operate in starch-containing guard cells during stomatal closing in the dark Key to enzymes (the same numbers apply to particular enzymes as used in Fig 9 1) 6, phosphoglucomutase, 7, phosphohexoisomerase, 8, ATP-phosphofructokinase, 9, PP_i-phosphofructokinase, 10, fructose-1,6-bisphosphatase, 11, aldolase, 12, phosphotriose isomerase, 13, NADP glyceraldehyde phosphodehydrogenase, 14, NAD phosphodehydrogenase, 16, NAD malate dehydrogenase, 19, UDP-glucose pyrophosphonylase, 20, sucrose phosphate synthetase, 21, sucrose phosphate phosphatase, 22, NADP malic enzyme, 23, phosphoenolpyruvate carboxykinase, 24, pyruvate phosphate dikinase, 25, ADPG pyrophosphorylase, 26, starch synthetase The dashed arrows indicate very tentative steps

step would usually be expected to occur in darkness when stomata are closing, conditions which would inactivate the Calvin cycle.

The pyruvate formed from the decarboxylation step must first be converted to PEP before it can be channelled into gluconeogenesis. However, the glycolytic step, PEP to pyruvate, catalysed by pyruvate kinase is essentially irreversible and investigators have detected no activity or only low activities of two potential enzymes which synthesize PEP, i.e. pyruvate P_i dikinase and PEP carboxykinase, in guard cells. Thus, this part of the pathway of carbon flow from malate to starch remains problematical. Once PEP is available, triose phosphate or PGA can be formed and transported into the guard cell chloroplast via the phosphate translocator as suggested by Robinson and Preiss (1985). However, this creates a potential problem in that the presence of fructose-1,6-bisphosphatase is required in order that 3-C compounds be converted to starch, but according to Hedrich *et al.* (1985) the activity of this enzyme in guard cell chloroplasts of *V. faba* is negligible while considerable activity is found in the cytoplasm. This aspect is controversial, however. Shimazaki *et al.* (1989) and Robinson and Preiss (1987) detected significant activity of fructose-1,6-bisphosphatase in GCPs of *V. faba* and *C. communis* (see Table 9.2), respectively, and in *V. faba* GCPs is was light activated (Shimazaki, personal communication). However, fructose-1,6-bisphosphate activity has not been detected in GCPs of *Pisum sativum* (Raschke, personal communication). Whatever the final conclusion, carbon can be channelled in to the chloroplasts as G-6-P, PGA or triose phosphate since an appropriate translocator has been detected in guard cell chloroplasts of *Pisum sativum* (Overlach *et al.*, 1993). This translocator preferentially imports G-6-P, but it also has affinities for PGA and triose phosphate at similar levels to those found in mesophyll cell chloroplasts. Thus, as Fig. 9.3 shows it is possible that 3-C compounds or G-6-P could be imported into the chloroplasts and converted to starch. During darkness, when stomata are normally closing, fructose-2,6-phosphate levels decline (Hedrich *et al.*, 1985) enabling cytoplasmic fructose-1,6-bisphosphophosphatase activity to increase, favouring gluconeogenesis and starch formation.

The need for compatible osmotica in the cytoplasm will decrease as concentrations of osmotica decrease in the vacuole. Thus the channelling of carbon into sucrose and the need to import sucrose will be less important. Presumably, however, carbon from sucrose could be channelled into hexose phosphate and imported into the chloroplast as hexose phosphate or 3-C compounds.

In mesophyll cells starch synthesis is controlled at the level of adenosine diphosphoglucose pyrophosphorylase (ADPG pyrophosphorylase) activity (see, e.g. Preiss, 1982). The enzyme is located in plastids and catalyses the conversion of G-1-P to ADPG utilizing ATP and liberating pyrophosphate (PP_i) in the process. ADPG pyrophosphorylase is regulated by the ratio of 3-PGA concentration (the activator) to inorganic phosphate (P_i) concentration (the inhibitor). Thus a high and low ratio promotes and inhibits starch synthesis, respectively. 3-PGA also increases V_{max} and relieves the P_i inhibition of ADPG pyrophosphorylase. Within chloroplasts of mesophyll cells, the concentration of PGA and P_i depends on the activity of RuBP carboxylase and ATP formation, respectively; during photosynthesis PGA is present at a high concentration while the P_i concentration is low, thereby promoting starch synthesis. In the dark the opposite occurs.

In guard cells, however, not only is the presence of RuBP carboxylase probably low (but see Section 9.4), but starch levels decrease in the light when stomata normally open and increase in the dark when stomata normally close. The regulation of ADPG pyrophosphorylase activity in guard cells is therefore of considerable interest and two independent studies have been made, i.e. those of Outlaw and Tarczynski (1984) and Robinson *et al.* (1984). In the former study the microanalysis enzyme amplification system (see Outlaw, 1980) of freeze-dried stomata of *V. faba* was used while in the latter study partially digested epidermal strips of *C. communis* with 'mesophyll and epidermal cells ... removed', were used. Outlaw and Tarczynski (1984) found similar specific activities in guard cells and mesophyll cells while Robinson *et al.* (1984) found activities to be about three times higher in mesophyll cells (on a protein basis). In both studies 3-PGA was an effective activator and P_i a strong inhibitor of guard cell ADPG pyrophosphorylase activity. In the study of Outlaw and Tarczynski (1984), 3-PGA and phosphate levels were considerably higher in illuminated palisade cells than in guard cells and illumination did not effect these levels in guard cells.

Thus guard cell ADPG pyrophosphorylase responds to, and may be regulated by, PGA:P_i ratios in the same manner as the enzyme from mesophyll cells, but if PGA and P_i levels stay constant in guard cells (Outlaw and Tarczynski, 1984) then a mechanism for the regulation of starch synthesis by this enzyme remains to be established. However, the site of ADPG pyrophosphorylase activity is the chloroplast and the P_i and 3-PGA levels therefore need to be measured in these organelles before it can be concluded that the PGA:P_i ratio is not involved in the regulation of ADPG pyrophosphorylase in guard cells.

The final steps in the biosynthesis of starch from ADPG involve starch synthase and the branching enzyme. Although Robinson and Preiss (1987) found considerable activity of the latter enzyme in GCPs of *C. communis,* activities of the former were very low and would appear to be a rate-limiting step.

Some enzymes of the pentose phosphate pathway, i.e. 6-phosphogluconate and glucose-6-phosphate dehydrogenase, have also been detected in guard cells (see Table 9.1).

9.4 The role of chloroplasts in guard cells

Chloroplasts are a regular feature of guard cells (see Chapter 2) with the exception of that in the orchid genus, *Paphiopedilum*, and some leaves in a variegated chimera of *Pelargonium zonale*. However, their role in the functioning of stomata remains a controversial aspect.

9.4.1 Is the Calvin cycle present in GCPs?

Although there is a general view that the Calvin cycle is of low activity or possibly absent in the guard cell chloroplasts, such that it cannot operate at levels sufficient to meet the solute requirements for stomatal opening, there is strong support for the existence of substantial activity by some investigators.

There is much disagreement about the level of activity of RuBisCo in guard cells. In the early work of Willmer *et al.* (1973a) high levels of RuBP carboxylase activity were detected in the epidermis of *C. communis* and tulip although ^{14}C-labelling experiments with the same tissues indicated that the Calvin cycle was not very active (Willmer and Dittrich, 1974) and it was considered that the high carboxylase activity was due, at least in part, to mesophyll tissue contamination. In *C. communis* epidermis substantial work concluded that the Calvin cycle was not active perhaps due to the absence of phosphoribulokinase (Raschke and Dittrich 1977). Work by Outlaw *et al.* (1979a), using single guard cell pairs dissected from freeze-dried leaves, were also unable to detect the presence of the Calvin cycle in guard cells of *V. faba* and tobacco since they found no evidence of RuBP carboxylase activity and insignificant levels of phosphoribulokinase and NADP specific glyceraldehyde phosphodehydrogenase activity. More recently, and in contrast, Shimazaki (1989) and Shimazaki *et al.* (1989) detected considerable activity of RuBisCo and other Calvin cycle enzymes in guard cells of *Vicia* (see Tables 9.1 and 9.2).

Immunological studies are also conflicting. Many reports indicate RuBisCo is absent or in very low levels (less than 1% relative to mesophyll cells) in guard cells of many different species (Outlaw *et al.*, 1979a, 1982; Vaughn, 1987; Reckmann *et al.*, 1990). Immunofluorescence techniques, however, detected considerable amounts of RuBP carboxylase in guard cells of CAM plants, small amounts in some C3 plants but none in C4 plants (Soundararajan and Smith, 1982). Zemel and Gepstein (1985), however, observed that positive results for RuBisCo only occurred if the walls of *V. faba* guard cells were partially digested (possibly facilitating better penetration of antibodies into cells). Tarczynski *et al.* (1989), however, are convinced from their electrophoretic studies that RuBisCo is virtually absent from *V. faba* guard cells.

Nevertheless, other information suggests that the Calvin cycle is present in guard cells. For example, Wu and Assmann (1993) observed that chloroplasts isolated from *V. faba* guard cells exhibited CO_2-dependent O_2 evolution in light while Cardon and Berry (1992) observed that chlorophyll *a* fluorescence in intact guard cells of *Tradescantia albiflora* was sensitive to O_2 and CO_2 in a manner suggesting oxygenase and carboxylase activities of RuBisCo. Also, the groups of Zeiger and Tallman consider that light quality and availability of K^+ can modulate Calvin cycle activity in guard cells. Poffenroth *et al.* (1992) found that production of soluble sugars by guard cells of *V. faba* (epidermis was used with only intact guard cells) exposed to red light in the presence of KCl was completely inhibited by DCMU, an inhibitor of non-cyclic electron transport. Further studies by Tallman and Zeiger (1988) and Talbott and Zeiger (1993) confirmed that sucrose accumulation, in the absence of potassium, occurred in guard cells of *V. faba* irradiated with red light when there was no measurable starch degradation. They concluded that light quality modulates alternative mechanisms of osmotic accumulation in guard cells, including potassium uptake, photosynthetic sugar production and starch breakdown.

The crucial question, however, is whether Calvin cycle activity is sufficient to contribute significantly to the 'osmotic motor' of guard cells. According to Poffenroth *et al.* (1992) sugar production via the Calvin cycle contributes significantly to the osmotica necessary for stomatal opening. Furthermore, in very recent work it has been concluded that there is a major contribution of sucrose to guard cell osmotic pressures of onion and *V. faba* during periods of the afternoon although the source of this sucrose was not clearly established (Amodeo *et al.*, 1994; Talbott *et al.*, 1994). Reckmann *et al.* (1990), however, concluded that only about 2% of the osmotica (assuming it to be glucose) needed for stomatal opening comes from the Calvin cycle. Additionally it has been observed that considerable amounts of sucrose accumulate in the guard cell wall possibly being swept along in the transpiration stream from mesophyll cells (Outlaw, personal communication).

The data of Cardon and Berry (1992) also suggest that the oxygenase activity of RuBisCo occurs in guard cells and therefore photorespiration. However, glycollate oxidase and hydroxypyruvate reductase activity (peroxisomal enzymes involved in photorespiration) were not detected in guard cells of *C. communis* and *V. faba*, respectively (see Table 9.1). Nevertheless, occasionally microbodies fitting the description of peroxisomes (see Chapter 3) are observed in guard cells.

It is pertinent to note, however, that stomata can function in the absence of the Calvin cycle in guard cells; it happens in orchids of the *Paphiopedilum* genus and a variegated cultivar of *Pelargonium zonale* their guard cells of which do not contain chloroplasts. Moreover, stomata open and close in CAM plants in the dark when the Calvin cycle is not functioning

9.4.2 The photosystems of GCPs

Initially it was thought that guard cell chloroplasts did not possess photosystem (PS) II but it is now well established that both PS I and II are present

Chlorophyll *a/b* ratios for guard cells have been used to indicate whether their chloroplasts are enriched in PS I values approximately between 2 and 3 (typical of C3 plants) indicate that considerable amounts of both photosystems are present while values above 3 (as in C4 plants) indicate that PS I dominates Many species have been investigated for their chlorophyll *a/b* ratios by a number of independent investigators and, although a large variation has been obtained spanning C3 and C4 values, the majority of values are between 2 and 3 Earlier measurements using epidermal tissue are suspect because of possible contamination from underlying mesophyll cells and the very low content of chlorophyll in guard cells making accurate measurements difficult However, Yemm and Willis (1954) used microspectrophotometric analysis of single plastids of guard cells in *Rumex patientia*, *Chrysanthemum maximum* and *V faba* and found values of 2–3 More recently GCPs and 'isolated guard cells' have been used in conjunction with modern techniques to determine which light reactions are present in guard cells (see review by Zeiger *et al* , 1986) It has now been established that light harvesting pigments for the light harvesting centre (LHC), PS I and PS II, oxygen evolution, DCMU-sensitive electron transport and fluorescent transients associated with the formation of the high energy state are present in guard cells (Outlaw *et al* 1981a, Zeiger *et al* , 1981, 1984, Melis and Zeiger, 1982, Ogawa *et al* , 1982, Shimazaki *et al* , 1982, Vaughn and Outlaw, 1983, Hipkins *et al* , 1983, Martin *et al* , 1984, Mawson *et al* , 1984) Immunocytochemical studies have also detected structural components of the PS II complex in guard cell chloroplasts of *V faba* (Zemel *et al* , 1988) Additionally guard cell chloroplasts of *V faba* also exhibit State I to State II transitions indicating that energy transfer between PS II and I occurs (Mawson and Cummins, 1986) Observations of fluorescent transients and the 518 nm electrochromic shift observed in isolated chloroplasts from *V faba* guard cells also indicated their capacity for cyclic and non-cyclic photophosphorylation (Shimazaki and Zeiger, 1985) Low temperature emission spectra and variable fluorescence induction kinetics of white tissue from leaves of *Chlorophyllum comosum* (in which guard cells were the only cells to contain chloroplasts) also indicated that both PS I and II were present and that PS II was probably larger in guard cell chloroplasts than in mesophyll chloroplasts

(Zeiger *et al.* 1981). This study also concluded that guard cell chloroplasts oxidize water and have a functional PS II connected through an intermediate plastoquinone pool to PS I resulting in the generation of reduced NADP, i.e. a linear (non-cyclic) electron transport mechanism was present. Low temperature fluorescence emission spectra of guard cells of *V. faba*, were also made by Outlaw *et al.* (1981a) enabling PS I/PS II ratios to be evaluated.

Guard cell protoplasts of *V. faba* contain the same types of carotenoids as those in the leaf mesophyll and include neoxanthin, violaxanthin, antheraxanthin, lutein, zeaxanthin and β-carotene (Karlsson *et al.*, 1992; Masamoto *et al.* 1993; Srivastava and Zeiger, 1993). The xanthophyll cycle has also been detected in guard cells (Masamoto *et al.*, 1993). In mesophyll cell chloroplasts the cycle is thought to protect the photosynthetic apparatus from photoinhibition. Thus in dim light or darkness zeaxanthin is converted to violaxanthin via antheraxanthin in an epoxidation reaction which destroys excited O_2. In bright light the reaction is reversed in a de-epoxidation reaction (see Lawlor 1993). It has been suggested that, as well as serving as protectors against photoinhibition, carotenoids and specifically zeaxanthin is a blue light photoreceptor in guard cells (Srivastava and Zeiger, 1993; Quinones *et al.*, 1993).

Some quantitative differences between mesophyll and guard cell chloroplasts have been observed, however: guard cell chloroplasts have a reduced sensitivity to low PFD in terms of *in vitro* photophosphorylation (Shimazaki and Zeiger 1985) and fluorescence transients (Mawson *et al.* 1984), an apparent larger PS II unit size (Melis and Zeiger, 1982) and significant differences in the sensitivity and magnitude of State I to State II transitions (Mawson and Cummins, 1986).

Observations of stomatal movements in epidermal strips of a number of species in the presence of inhibitors of PS I or PS II or in red or far-red light also suggested that PS I and the accompanying cyclic photophosphorylation was necessary for substantial opening to occur.

Activities of PS I and PS II in guard cells will be small on a cell basis relative to that occurring in chloroplasts of mesophyll cells (nevertheless, on a chlorophyll basis cyclic and non-cyclic photophosphorylation in guard cells approach 80% of those in mesophyll cells, Shimazaki and Zeiger 1985) and therefore, may not be of great significance. Nevertheless, if the Calvin cycle is absent or of low activity in guard cells then this raises the question of what purpose the NADPH produced in linear electron flow serves. (In other green cells of higher plants which possess PS I and II the step, PGA to phosphoglyceraldehyde of the Calvin cycle, utilizes NADPH produced in linear electron flow). Undoubtedly it will be utilized by NADP specific malate dehydrogenase in the reduction of OAA to malate and it could also be used in N and S reduction though the latter processes have not been associated with guard cells The ATP and reducing equivalents produced in the chloroplasts of guard cells can be transported out, possibly via a phosphoglycerate/dihydroxyacetone phosphate shuttle (Shimazaki *et al* , 1989), and used in innumerable reactions including ion transport processes at the plasmalemma and tonoplast The ATP and reducing power may also be used in starch/sugar interconversion within the chloroplast

9.4.3 Polysaccharide pools in guard cells

Except for a few genera, most notable being the *Allium* genus, green and non-green plastids of guard cells contain an abundance of starch The plastids, therefore, have often been considered as amyloplasts acting as a sink and source of carbon for malate synthesis and breakdown during stomatal opening and closing

In the starch-free chloroplasts a search was made for other soluble polysaccharides which could supply part of the carbon skeleton for malate synthesis in guard cells It has been concluded that the mucilaginous polysaccharides detectable in *Allium cepa* epidermis did not serve as a carbon source for malate synthesis and that fructans or other soluble polysaccharides which could substitute for starch were not detectable (Schnabl, 1977, Schnabl and Ziegler, 1977) Darbyshire and Allaway (in Allaway, 1981), however, found considerable amounts of fructans (polymerized to a degree of about 8 or 9 hexose units) in dark-treated epidermis of *Allium cepa* (in which guard cells were the only intact cells), most of which disappeared when the epidermis was given a light-treatment Thus these investigators believe that fructans could be mobilized to supply carbon for malate synthesis in guard cells of *A cepa* Moreover, epidermal tissue of onion contains high levels of PEP carboxylase activity and malate is the chiefly labelled product when $^{14}CO_2$ is supplied to the tissue (Willmer, unpublished)

9.5 Metabolite transport into guard cells and sucrose metabolism

On the assumption that the stores of guard cell starch are not sufficient to satisfy the life-time demands for carbon to make malate and as a source of energy in the TCA cycle, and if the Calvin cycle is absent or of low activity in guard cells, then import of carbon by the cells must occur (The Calvin cycle is an autocatalytic process being able to regenerate substrate to perpetuate the cycle and also allow carbon to be syphoned off for sucrose synthesis)

In some species (e.g. *Phyllitis scolopendrium* and other ferns) epidermal cells possess an abundance of chloroplasts and there is some evidence indicating that they can photosynthesize and may, therefore, supply guard cells with carbon. However, epidermal cells are often free of chloroplasts and in such cases the guard and epidermal cells must receive carbon from mesophyll cells. There is some experimental evidence to suggest this is the case (see Milthorpe *et al.*, 1979, Willmer 1981).

For example, Outlaw and Fisher (1975) showed that following a pulse of $^{14}CO_2$ to a leaflet of *V. faba*, label eventually reached the epidermal layers. Also, the observation that exogenously supplied sugars, and particularly G-1-P, to epidermal tissue of numerous species resulted in starch formation within the guard cells is a further indication that metabolite transport could occur between mesophyll cells and guard cells (Konagamitsu and Ono, 1959, Reddy and Rama Das, 1986). Moreover, sucrose accumulates in the apoplast of guard cells (Outlaw, personal communication) and at certain periods of the afternoon Amodeo *et al.* (1994) and Talbott *et al.* (1994) consider it to be a major osmoticum of guard cells of onion and *V. faba* (also see Section 9.4.1).

There are other reasons to involve sugars in guard cell reactions during stomatal movements. The water potentials of the cytoplasm and vacuole maintain equilibrium as K^+, Cl^- and malate accumulate in the vacuole and some other compensatory solute must accumulate in the cytoplasm as this happens (if substantial K^+ or Cl^- remained in the cytoplasm it is possible that they would interfere with many enzyme reactions). Under such circumstances pressure potentials in the two compartments will remain the same though the matrix potential of the cytoplasm may decrease. Moreover, MacRobbie (e.g. MacRobbie and Lettau, 1980a,b) concludes that potassium salts alone could not account for the observed guard cell osmotic potentials of open stomata and there must be some other osmoticum or osmotica involved. There is evidence that sucrose is the candidate. Outlaw and Manchester (1979) and Poffenroth *et al.* (1992) have recorded increases in guard cell sucrose concentrations (130 mM in guard cells of open stomata, Outlaw and Manchester, 1979) when stomata open. Furthermore, Hite *et al.* (1993) discovered high levels of sucrose phosphate synthase (the sucrose synthetic enzyme) and sucrose synthase (the sucrose degradative enzyme) activity in *V. faba* guard cells relative to mesophyll cells (see Table 9.1). They also measured acid invertase activity in guard cells which is a vacuolar enzyme degrading sucrose. Such high levels of enzyme activities for degrading and synthesizing sucrose would be required if sucrose, delivered from the apoplast, is utilized by the guard cells in a variety of ways including being stored in the vacuole, used as a compatible osmoticum in the cytoplasm and used as a carbohydrate intermediate for starch or malate synthesis.

9.6 The energy relations of stomatal movements

Because of the complexity of guard cell metabolism and because their metabolism is less well characterized than other cell types little has been discussed on the energetics of guard cells. However, a substantial attempt has been made by Assmann and Zeiger (1987). Major points of interest are the energy costs of opening and closing movements and the energy sources during the movements.

Both opening and closing movements have been described as active, energy-consuming steps. The opening movement can be more readily conceived of as being active since guard cells are at their lowest pressure (in a 'relaxed' state) when stomata are closed and increase their pressure as stomata open. Moreover, a number of studies have shown that metabolic inhibitors (including uncouplers of phosphorylation activity and inhibitors of electron flow) (e.g. Willmer and Mansfield, 1970; Pallaghy and Fischer, 1974) and an oxygen-free atmosphere (e.g. Akita and Moss, 1973) prevent stomatal opening. However, some studies have also shown that metabolic inhibitors, such as sodium azide, which is an inhibitor of respiration, low temperatures and oxygen-free air (Akita and Moss, 1973) prevent stomatal closure. Also CCCP, a H^+ ionophore which breaks down H^+ gradients thereby preventing ATP formation in chloroplasts and mitochondria and hyperpolarization of the plasma membrane, prevents closure of open stomata (Karlsson and Schwartz, 1988). This suggests that closing is an active process.

Guard cells contain an abundance of mitochondria while, in most plants, chloroplasts are poorly developed which suggests that oxidative phosphorylation is the major source of ATP for 'active' stomatal movements. The dark respiratory rate of guard cells is also high relative to that of mesophyll cells (e.g. Shimazaki *et al.*, 1983; Birkenhead *et al.*, 1985; Vani and Raghavandra, 1994). Moreover stomata can open in the dark (and do as a normal course of events in CAM plants) and stomatal closing, which may be active, is brought about in darkness, conditions under which photophosphorylation will not be occurring. Under these situations the energy source must be from oxidative phosphorylation or a system other than photophosphorylation.

According to Assmann and Zeiger (1987) ATP synthesis by either guard cell chloroplasts or mitochondria could accommodate the energetic requirements for stomatal opening. Earlier, Schwartz and Zeiger (1984) had concluded that stomatal opening in darkness depended on oxidative phosphorylation while opening in the light appeared to require photophosphorylation and the blue light specific opening could rely on oxidative phosphorylation or a membrane-bound electron transport carrier as an energy source for guard cell H^+ extrusion. Efforts have also been made to quantify the proportion of energy coming from the chloroplasts and mitochondria within guard cells during stomatal movements (e g Dahse *et al* , 1990, Mawson, 1993) However, studies are difficult because when one source of energy is inhibited other sources tend to compensate so that the real value is obscured Furthermore, processes of guard cell respiration and photosynthesis are closely coupled (Mawson, 1993)

Some early inhibitor studies have also shown that, in the light, stomatal opening is dependent to some extent on cyclic photophosphorylation (Willmer and Mansfield, 1970, Raghavendra and Das, 1972), although such results need to be confirmed with modern methods

The energy relations of guard cells at the biochemical level are complex and there must be regulatory interactions to keep ATP levels within certain limits and to maintain acceptable NAD(P)H NAD(P) ratios during stomatal activity It is known that many key regulatory enzymes in both anabolic and catabolic pathways can be allosterically controlled by the individual levels of cellular ATP, ADP and AMP (see Dawes, 1986) In general high levels of ADP or AMP activate regulatory enzymes involved in catabolic pathways whereas ATP inhibits them (thus ATP is regenerated from substrates only when necessary) and ADP or AMP inhibit many regulatory enzymes involved in biosynthetic pathways which are ATP dependent Other enzymes respond to the relative concentrations of ATP, ADP and AMP rather than the concentrations of the individual nucleotides Thus the ratio of ATP/ADP and ATP/AMP are important in regulating these enzymes and it is considered that the ratios of the adenine nucleotides should be buffered from wide fluctuations in order to maintain efficient cell functioning In fact adenine nucleotides are in equilibrium with each other due to the universal enzyme, adenylate kinase, present at high levels in mitochondria and chloroplasts The enzyme catalyses the reversible reaction, ATP + AMP 2ADP From this equation Atkinson (1968) derived a term he called the adenylate energy charge (AEC) which is a measure of the metabolic energy stored in the adenylate system (zero when all cell adenylate is AMP and 1 when it is all in the form of ATP) Thus, processes which utilize ATP, such as H^+ pumping, will tend to reduce AEC And indeed there is some evidence that enzymes involved in generating ATP are more active at low rather than high AEC while enzymes utilizing ATP are less active at low AEC values than high ones

A more sensitive indicator of the energy status of a cell is the phosphorylation potential ΔG_p, or the Gibbs free energy required to synthesize ATP from ADP and P_i (Nicholls, 1982)

Thus

$$\Delta G° + 2\ 303 RT \log\ (\mathrm{ATP}) / (\mathrm{ADP})(\mathrm{P_i}) \qquad (9\ 1)$$

where ΔG° is the Gibbs energy change when ATP is hydrolysed to ADP, R is the gas constant and T is the absolute temperature

Upon swelling of GCPs, ATP is utilized to pump protons out of the cell Thus ATP levels within GCPs might be expected to fall resulting in a decrease of AEC, ATP/ADP ratios and ΔG_p However, most cell types appear to regulate AEC between 0 8 and 0 9, and the decrease of ATP with increased ADP and AMP should stimulate increased activity from enzymes in ATP synthesizing pathways (e g oxidative phosphorylation, glycolysis, oxidative pentose phosphate pathway and possibly photophosphorylation) thereby maintaining relative constancy of ATP levels, AEC and ATP/ADP ratios There may also be a conservation of ATP in the cell by reduction of rates in biosynthetic pathways

Schnabl (1980) found in short term experiments enzyme activities changed as GCP swelled For example, ATP-phosphofructokinase (involved in glycolysis) and PEPc increased activity at the onset of protoplast swelling but PEPc subsequently decreased after 1 min In contrast, fructose bisphosphatase (involved in gluconeogenesis) and malic enzyme decreased on swelling Glucose-6-phosphate dehydrogenase (of the oxidative pentose phosphate pathway) also decreased at the onset of swelling but rapidly regained activity after 3 min before decreasing again As some of these enzymes are known to be regulated by AEC (e g ATP-phosphofructokinase) or by levels of reduced and oxidized pyridine nucleotides (e g G-6 P dehydrogenase), it might be expected that changes in enzyme activity would reflect adjustments in these parameters

Schnabl (1985) also investigated ATP/ADP ratios in GCPs as they swelled The ratio fluctuated widely (from 2 to 15) although an initial decrease in the ATP/ADP ratio in the first 5 min of swelling (from 9 to 6) was attributed to utilization of ATP for active transport processes However, most published values (e g Heber *et al* , 1982, Hampp, 1985) for other cell types are much lower and do not show the wide fluctuations obtained by Schnabl (1985) For nearly all active cells investigated the AEC values are between 0 8 and 0 95 O Donnell and Willmer (unpublished) also found values for GCPs of *C communis* in this range at all stages of swelling whether in the light or dark Apparently AEC values are tightly regulated by, for example, phosphofructokinase and fructose bisphosphatase activities and values in leaves remain constant when the leaves are switched from light to dark

Phosphorylation potentials ranging from –9 to –11 were obtained for GCPs of *C communis* by O Donnell and Willmer (unpublished) which are similar to other published values for other cell types However, values for guard cells were computed using whole cell metabolite concentrations when in fact adenine nucleotides, adenosine phosphates and phosphate are compartmentalized For example, Stitt *et al* (1982) found that in wheat mesophyll cells 47% of the adenine nucleotides resided in the chloroplasts, 44% in the cytosol and 9% in mitochondria

9.7 Lipid metabolism in guard cells

Relatively little work has investigated the lipid metabolism of leaf epidermis and even less on that in guard cells However, what has been done suggests that the epidermal and the guard cell lipid metabolism is distinctly different from that occurring in the mesophyll which is not surprising considering the epidermis is covered by a cuticle and waxes (see Section 3 1), which are presumably secreted by the cells of the epidermal layer

The epidermis rather than the mesophyll is the site of the synthesis of very long chain fatty acids (Kolattukudy, 1968, Kolattukudy and Buchner, 1972, Lessire *et al* , 1982, Lessire and Stumpf, 1983, Sato, 1985) Sato (1985) made a study of lipid biosynthesis in different protoplast types from leaves of *V faba* and found considerable differences between the mesophyll cell protoplasts and the guard and epidermal cell protoplasts (Tables 9 4 and 9 5) When [^{14}C]acetate was fed to the protoplasts, after 60 min the dominantly labelled fatty acid in mesophyll cell protoplasts was oleic acid (18 carbons long with one double bond in the chain) while in guard and epidermal cell protoplasts palmitic acid (16 carbons long, fully saturated), oleic acid and to a lesser extent arachidic and behenic acid (20 and 22 carbons long, respectively, both saturated) were the chiefly labelled fatty acids Distribution of label in the different lipid classes were similar except there was much more label in triacyl glycerol and an unidentified group of lipids in the guard and epidermal cell protoplasts than the mesophyll protoplasts

Table 9 4 Distribution of radioactivity in fatty acids of lipids after labelling with [2–^{14}C] acetate for 60 min

Fatty acid	Distribution of radioactivity (%)			
	MCP	GCP	heavy ECP	light ECP
Palmitic (16 0)	19	42	38	46
Margaric (17 0)	0	0	2	3
Stearic (18 0)	Tr*	5	5	8
Oleic (18 1)	81	37	37	27
Arachidic (20 0)	0	6	8	10
Behenic (22 0)	0	10	10	6

* Tr trace less than 2% Data from Sato (1985)

Table 9.5 Distribution of radioactivity in lipid classes after labelling with [2-^{14}C] acetate for 60 min

Lipid class	Distribution of radioactivity (%)			
	MCP	GCP	heavy ECP	light ECP
Phosphatidylcholine	34 (65)	44	37	34
Phosphatidylethanolamine	5 (21)	6	7	15
Phosphatidylinositol	1	5	4	1
Phosphatidylglycerol	14 (56)	14	8	9
Phosphatidic acid	2	0	0	0
Sterol glycoside	0	2	2	2
Monogalactosyl diacylglycerol	2 (276)	1	1	1
Digalactosyl diacylglycerol +sulfolipid	0 (254)	0	0	0
Fatty acids	32 (23)	2	2	4
Triacylglycerol	9 (67)	17	25	24
Unidentified	1	9	14	10

Values in parentheses for the MCP fraction represent mass contents expressed in terms of pmol fatty acids cell. Data from Sato (1985).

References

Akita, S. and Moss, D.N. (1973) The effect of oxygen free atmosphere on net photosynthesis and transpiration of barley (*Hordeum vulgare* L.) and wheat (*Triticum aestivum* L.) leaves. *Plant Physiol.*, **52**, 601–603.

Allaway, W.G. (1981) Anions in stomatal operation, in *Stomatal Physiology*, (eds P.G. Jarvis and T.A. Mansfield), Cambridge University Press, Cambridge.

Amodeo, G., Talbot, L.D., Nouhi, S. and Zeiger, E. (1994) Potassium efflux from guard cells does not necessarily imply stomatal closure. *Plant Physiol. Suppl.*, **105**, 100.

Assmann, S.M. and Zeiger, E. (1987) Guard cell bioenergetics, in *Stomatal Function*, (eds G.D. Farquhar, I.R. Cowan and E. Zeiger), Stanford University Press, Stanford, CA. pp. 163–193.

Atkinson, D.E. (1968) The energy charge of the adenylate pool as a regulatory parameter. Interaction with feed back modifiers. *Biochemistry*, **7**, 4030–4034.

Birkenhead, K. and Willmer, C.M. (1986) Some biochemical characteristics of guard cell and mesophyll protoplasts from *Commelina communis* L *J Exp Bot*, **37**, 119–128

Birkenhead, K, Laybourn-Parry, J and Willmer, C M (1985) Measurement of guard cell respiration rates using a Cartesian-diver technique *Planta*, **163**, 214–217

Cardon, Z G and Berry, J (1992) Effects of O_2 and CO_2 concentration on the steady-state fluorescence yield of single guard cell pairs in intact leaf discs of *Tradescantia albiflora* *Plant Physiol*, **99**, 1238–1244

Dahse, I, Willmer, C M and Meidner, H (1990) Tentoxin suppresses stomatal opening by inhibiting photophosphorylation *J Exp Bot*, **41**, 1109–1113

Dawes, E A (1986) *Microbial Energetics*, Blackie & Son, London

Denecke, M, Schulz, M, Fischer, C and Schnabl, H (1993) Partial purification and characterization of stomatal phosphoenolpyruvate carboxylase from *Vicia faba* *Physiol Plant*, **87**, 96–102

Dittrich, P and Raschke, K (1977) Malate metabolism in isolated epidermis of *Commelina communis* L in relation to stomatal functioning *Planta*, **134**, 77–81

Donovan, N, Gibb, E, Donkin, M E and Martin, E S (1985) A comparison of the kinetic properties of phosphoenolpyruvate carboxylase from guard-cell and mesophyll-cell protoplasts of *Commelina communis* *Planta*, **164**, 115–120

Freer-Smith, P and Willmer, C M (1981) Guard cell metabolism in epidermis of *Commelina communis* L during stomatal opening and closing, *J Exp Bot*, **32**, 535–43

Fricker, M D and Willmer, C M (1987) Vanadate sensitive ATPase activity and phosphatase activity in guard cell protoplasts of *Commelina* *J Exp Bot*, **38**, 642–648

Gotow, K, Tanaka, K, Kondo, N, Kobayashi, K and Sÿono, K (1985) Light activation of NADP-malate dehydrogenase in guard cell protoplasts from *Vicia faba* L *Plant Physiol*, **79**, 829–832

Hampp, R, Outlaw, W H, Jr and Tarczynski, M C (1982) Profile of basic carbon pathways in guard cells and other leaf cells of *Vicia faba* L *Plant Physiol*, **70**, 1582–1585

Hedrich, R, Raschke, K and Stitt M (1985) A role of fructose-2,6-bisphosphate in regulating carbohydrate metabolism in guard cells *Plant Physiol*, **79**, 977–982

Hipkins, M F, Fitzsimons, P J and Weyers, J D B (1983) The primary processes of photosystem II in purified guard-cell and mesophyll-cell protoplasts from *Commelina communis* L *Planta*, **159**, 554–560

Hite, R C, Bodson, M J and Outlaw, W H, Jr (1992) Enzymic potential for fructose 6-phosphate phosphorylation by guard cells and by palisade cells in leaves of the broad bean *Vicia faba* L *Histochem J*, **24**, 368–374

Hite, D R C , Outlaw, WH , Jr and Tarczynski, M C (1993) Elevated levels of both sucrose-phosphate synthase and sucrose synthase in *Vicia* guard cells indicate cell-specific carbohydrate interconversions *Plant Physiol* , **101**, 1217–1221

Jiao, J -A and Chollet R (1988) Light/dark regulation of maize leaf phosphoenolpyruvate carboxylase by *in vivo* phosphorylation *Arch Biochem Biophys* , **261**, 409–417

Karlsson, PE and Schwartz, A (1988) Characterization of the effects of metabolic inhibitors, ATPase inhibitors and potassium channel blockers on stomatal opening and closing in isolated epidermis of *Commelina communis* L *Plant Cell Environ* , **11**, 165–172

Karlsson, PE , Bogomolni, R A and Zeiger, E (1992) High performance liquid chromatography of pigments from guard cell protoplasts and mesophyll tissue of *Vicia faba* L *Photochem Photobiol* , **55**, 605–610

Kolattukudy, PE (1968) Further evidence for an elongation-decarboxylation mechanism in the biosynthesis of paraffins in leaves *Plant Physiol* , **43**, 375–383

Kolattukudy, PE and Buchner, J S (1972) Chain elongation of fatty acids by cell-free extracts of epidermis from pea leaves *Biochem Biophys Res Commun* , **46**, 801–807

Konagamitsu, Y and Ono, H (1959) Starch formation in albino part of variegated leaves *Sieboldia*, **2**, 137–142

Kottmeier, C and Schnabl, H (1986) The K_m-value of phosphoenolpyruvate carboxylase as an indicator of the swelling state of guard cell protoplasts *Plant Sci* , **43**, 213–217

Lawlor, D W (1993) *Photosynthesis*, Longman, London, pp 43–47

Lessire, R and Stumpf, PK (1983) Nature of fatty acid synthase systems in parenchymal and epidermal cells of *Allium porrum* L leaves *Plant Physiol* , **73**, 614–618

Lessire, R , Abdul-Karim, T and Cassagne, C (1982) Origin of the wax very long chain fatty acids in leek, *Allium porrum* L , leaves a plausible model, in *The Plant Cuticle*, (eds D F Cutler, K L Alvin and C E Prive), Academic Press, London, pp 167–180

MacRobbie, E A C and Lettau, J (1980a) Ion content and aperture in 'isolated guard cells of *Commelina communis* L *J Membr Biol* , **53**, 199–205

MacRobbie, E A C and Lettau, J (1980b) Potassium content and aperture in 'intact stomatal and epidermal cells of *Commelina communis* L *J Membr Biol* , **56**, 249–256

Martin, G E , Outlaw, WH , Jr, Anderson, L C and Jackson, S G (1984) Photosynthetic electron transport in guard cells of diverse species *Plant Physiol* , **75**, 336–337

Masamoto, K , Kinoshita, T and Shimazaki, K (1993) Light-induced de-epoxidation of violaxanthin in guard cell protoplasts of *Vicia faba*

Plant Cell Physiol, **94**, 935–938

Mawson, B T (1993) Modulation of photosynthesis and respiration in guard and mesophyll cell protoplasts by oxygen concentration *Plant Cell Environ*, **16**, 207–214

Mawson, B T and Cummins (1986) The kinetics of *in vivo* state transitions in mesophyll and guard cell chloroplasts monitored by 77 K fluorescence emission spectra *Plant Physiol*, **82**, 873–879

Mawson B T, Franklin A Filion, WG and Cummins, WR (1984) Comparative studies of fluorescence from mesophyll and guard cell chloroplasts in *Saxifraga cernia* analysis of fluorescence kinetics as a function of excitation intensity *Plant Physiol*, **74**, 481–486

Melis, A and Zeiger, E (1982) Chlorophyll a fluorescence transients in mesophyll and guard cells Modulation of guard cell photophosphorylation by CO_2 *Plant Physiol*, **69**, 642–647

Milthorpe F I, Thorpe N and Willmer C M (1979) Stomatal metabolism – a current assessment of its features in *Commelina*, in *Structure, Function and Ecology of Stomata*, (ed D N Sen), Bishen Singh and Mahendra Pal Singh, Dehra Dun, India

Nelson, J PS, Fewson, C A and Nimmo, H G (1993) Control of guard cell phosphoenolpyruvate carboxylase *J Exp Bot Suppl*, **44**, 3

Nicholls, D G (1982) *Bioenergetics*, Academic Press, London

Nimmo, G A, Nimmo H G Hamilton, I D *et al* (1986) Purification of the phosphorylated night form and dephosphorylated day form of phosphoenolpyruvate carboxylase from *Bryophyllum fedtschenkoi* *Biochem J*, **239**, 213–220

Ogawa, T Grantz, D Boyer, J and Govindjee (1982) Effects of cations and abscisic acid on chlorophyll a fluorescence in guard cells of *Vicia faba* *Plant Physiol* **69**, 1140–1144

Outlaw, W H, Jr (1980) A descriptive evaluation of quantitative histochemical methods based on pyridine nucleotides *Ann Rev Plant Physiol*, **31**, 299–311

Outlaw, WH, Jr (1982) Carbon metabolism in guard cells, in *Cellular and Subcellular Localization in Plant Metabolism*, (eds L L Cresy and G Hrazdina), Plenum, New York, pp 185–222

Outlaw, WH, Jr (1990) Kinetic properties of guard-cell phosphoenolpyruvate carboxylase *Biochem Physiol Pflanz*, **186**, 317–325

Outlaw, WH Jr and Fisher, D B (1975) Compartmentation in *Vicia faba* leaves 1 Kinetics of C in the tissues following pulse labelling *Plant Physiol*, **55**, 699–703

Outlaw, WH, Jr and Kennedy, J (1978) Enzymic and substrate basis for the anaplerotic step in guard cells *Plant Physiol*, **62**, 648–652

Outlaw, WH, Jr and Manchester, J (1979) Guard cell starch concentration quantitatively related to stomatal aperture *Plant Physiol*, **64**, 79–82

Outlaw, H W, Jr and Tarczynski, M C (1984) Guard cell starch biosynthesis regulated by effectors of ADP glucose pyrophosphorylase *Plant Physiol*, **74**, 424–429

Outlaw, WH, Jr, Manchester J, Di Camelli, C A *et al* (1979a) Photosynthetic carbon reduction pathway is absent in chloroplasts of *Vicia* faba guard cells *Proc Natl Acad Sci USA*, **76**, 6371–5

Outlaw, WH, Jr Manchester J and Di Camelli, C A (1979b) Histochemical properties of *Vicia faba* guard cell phosphoenolpyruvate carboxylase *Plant Physiol*, **64**, 269–272

Outlaw, W H, Jr Mayne, B C Zenger, V E and Manchester, J (1981a) Presence of both photosystems in guard cells of *Vicia faba* L Implications for environmental signal processing *Plant Physiol*, **67**, 12–16

Outlaw, WH, Jr, Manchester, J and Brown, PH (1981b) High levels of malic enzyme activities in *Vicia faba* L epidermal tissue *Plant Physiol*, **68**, 1047–1051

Outlaw, WH, Jr Tarczynski, M C and Anderson, L C (1982) Taxonomic survey for the presence of ribulose-1,5-bisphosphate carboxylase activity in guard cells *Plant Physiol*, **70**, 1218–1220

Overlach, S, Diekmann, W and Raschke, K (1993) Phosphate translocator of isolated guard-cell chloroplasts from *Pisum sativum* L transports glucose 6-phosphate *Plant Physiol*, **101**, 1201–1207

Pallaghy, C K and Fischer R A (1974) Metabolic aspects of stomatal opening and ion accumulation by guard cells of *Vicia faba Z Pflanz*, **71**, 332–344

Pantoja, O and Willmer, C M (1988) Redox activity and peroxidase activity associated with the plasma membrane of guard cell protoplasts *Planta*, **174**, 44–50

Poffenroth, M, Green, D B and Tallman, G (1992) Sugar concentrations in guard cells of *Vicia faba* illuminated with red or blue light analysis by high performance liquid chromatography *Plant Physiol*, **98**, 1460–1471

Preiss, J (1982) Regulation of the biosynthesis and degradation of starch *Ann Rev Plant Physiol*, **33**, 431–454

Quinones, M A, Lu, Z and Zeiger E (1993) Zeaxanthin concentrations co-segregate with the magnitude of the blue light response of adaxial guard cells and leaf stomatal conductances in an F_2 population of Pima cotton *Plant Physiol Suppl*, **102**, 15

Raghavendra, A S and Das, VS R (1972) Control of stomatal opening by cyclic photophosphorylation *Curr Sci*, **41**, 150–151

Raschke, K (1975) Stomatal action *Ann Rev Plant Physiol*, **26**, 309–40

Raschke, K (1979) Movements of stomata, in *Encyclopedia of Plant Physiology, Vol 7, Physiology of Movements*, (eds W Haupt and M Feinleib, Springer Berlin

Raschke, K and Dittrich, P (1977) [^{14}C]carbon-dioxide fixation by isolated leaf epidermis with stomata closed or open *Planta*, **134**, 69–75

Reckmann, U , Scheibe, R and Raschke, K (1990) Rubisco activity in guard cells compared with the solute requirement for stomatal opening *Plant Physiol* , **92**, 246–253

Reddy, A R and Rama Das, VS (1986) Stomatal movements and sucrose uptake by guard cell protoplasts of *Commelina beghalensis* L *Plant Cell Physiol* , **27**, 1565–1570

Reyss, A and Prioul, J L (1975) Carbonic anhydrase and carboxylating activities from plants (*Lolium multiflorum*) adapted to different light regimes *Plant Sci Lett* , **5**, 189–195

Robinson, N L and Preiss, J (1985) Biochemical phenomena associated with stomatal function *Physiol Plant* **64**, 141–146

Robinson, N L and Preiss, J (1987) Localization of carbohydrate metabolizing enzymes in guard cells of *Commelina communis* *Plant Physiol* , **85**, 360–364

Robinson, N L , Zeiger, E and Preiss, J (1984) Regulation of ADPglucose synthesis in guard cells of *Commelina communis* *Plant Physiol* , **73**, 862–864

Sato, N (1985) Lipid biosynthesis in epidermal, guard and mesophyll protoplasts from leaves of *Vicia faba* L *Plant Cell Physiol* , **26**, 805–811

Scheibe, R , Reckmann, U , Hedrich, R and Raschke, K (1990) Malate dehydrogenases in guard cells of *Pisum sativum* *Plant Physiol* , **93**, 1358–1364

Schnabl, H (1977) Isolation and identification of soluble polysaccharides in epidermal tissue of *Allium cepa* *Planta*, **135**, 307–311

Schnabl, H (1980) CO_2 and malate metabolism in starch-containing and starch-lacking guard-cell protoplasts *Planta*, **149**, 52–58

Schnabl, H (1981) The compartmentation of carboxylating and decarboxylating enzymes in guard cell protoplasts *Planta*, **152**, 307–313

Schnabl, H (1985) Regulation of volume changes in guard cell protoplasts, in *The Physiological Properties of Plant Protoplasts*, (ed PE Pilet), Springer, Berlin

Schnabl, H and Kottmeier, C (1984a) Properties of phosphoenolpyruvate carboxylase in desalted extracts from isolated guard-cell protoplasts *Planta*, **162**, 220–225

Schnabl, H and Kottmeier, C (1984b) Determination of malate levels during the swelling of vacuoles isolated from guard-cell protoplasts *Planta*, **161**, 27–31

Schnabl, H , Denecke, M and Schulz, M (1992) *In vitro* and *in vivo* phosphorylation of stomatal phosphoenolpyruvate carboxylase from *Vicia faba* L *Botanica Acta*, **105**, 367–369

Schnabl, H and Ziegler, H (1977) The mechanism of stomatal movement in *Allium cepa* L., *Planta*, **136**, 37–43

Schwartz, A. and Zeiger, E. (1984) Metabolic energy for stomatal opening. Role of photophosphorylation and oxidative phosphorylation. *Planta*, **161**, 129–136

Schulz, M., Hunte, C. and Schnabl, H. (1992) Multiple forms of phosphoenolpyruvate carboxylase in mesophyll, epidermal and guard cells of *Vicia faba* L. *Physiol. Plant.*, **86**, 315–321

Shimazaki, K. I. (1989) Ribulosebisphosphate carboxylase activity and photosynthetic O_2 evolution rate in *Vicia* guard cell protoplasts. *Plant Physiol.*, **91**, 459–463

Shimazaki, K. I. and Zeiger, E. (1985) Cyclic and non-cyclic photophosphorylation in isolated guard cell chloroplasts from *Vicia faba* L. *Plant Physiol.*, **78**, 211–214

Shimazaki, K.-I., Gotow, K. Sakaki, T. and Kondo, N. (1983) High respiratory activity of guard cell protoplasts from *Vicia Faba* L. *Plant Cell Physiol.*, **24**, 1049–1056

Shimazaki, K. I., Gotow, K. and Kondo N. (1982) Photosynthetic properties of guard cell protoplasts from *Vicia faba* L. *Plant Cell Physiol.*, **23**, 871–879

Shimazaki, K.-I., Terada, J. Tanaka, K. and Kondo, N. (1989) Calvin–Benson cycle enzymes in guard cell protoplasts from *Vicia faba* I. implications for the greater utilization of phosphoglycerate/dihydroxvacetone phosphate shuttle between chloroplasts and the cytosol. *Plant Physiol.*, **90**, 1057–1064

Soundarajan, M. and Smith, B. N. (1982) Localization of ribulose bisphosphate carboxvlase in the guard cells by an indirect, immunofluorescence technique. *Plant Physiol.*, **69**, 273–277

Srivastava, A. and Zeiger, E. (1993) A role of zeaxanthin in blue light photoreception of guard cells. *Plant Physiol. Suppl.*, **102**, 15

Stitt, M. McLilley, R. and Held. H. W. (1982) Adenine nucleotide levels in the cytosol, chloroplasts and mitochondria of wheat leaf protoplasts. *Plant Physiol.*, **70**, 971–977

Talbott, L. D. and Zeiger, E. (1993) Sugar and organic acid accumulation in guard cells of *Vicia faba* in response to red and blue light. *Plant Physiol.*, **102**, 1163–1169

Talbott, L. D., Huang, C. and Zeiger, E. (1994) Guard cells of intact *Vicia* leaves can use K^+ and sucrose as their main osmotica. *Plant Physiol. Suppl.*, **105** 100

Tallman, G. and Zeiger, E. (1988) Light quality and osmoregulation in *Vicia* guard cells. evidence for involvement of three metabolic pathways. *Plant Physiol.*, **88**, 887–895

Tarczynski, M. C. and Outlaw, W. H., Jr. (1993) The interactive effects of pH, L-malate, and glucose-6-phosphate on guard cell phosphoenolpyruvate carboxvlase. *Plant Physiol.*, **103**, 1189–1194

Tarczynski, M C , Outlaw, WH , Jr, Arold, N *et al* (1989) Electrophoretic assay for ribulose 1,5-bisphosphate carboxylase/oxygenase in guard cells and other leaf cells of *Vicia faba* L *Plant Physiol* , **89**, 1088–1093

Vani, T and Raghavendra, A S (1994) High mitochondrial activity but incomplete engagement of the cyanide-resistant alternative pathway in guard cell protoplasts of pea *Plant Physiol* , **105**, 1263–1265

Van Kirk, C A and Raschke, K (1978) Release of malate from epidermal strips during stomatal closure *Plant Physiol* , **53**, 360–5

Vaughn, K C (1987) Two immunological approaches to the detection of ribulose 1,5-bisphosphate carboxylase in guard cell chloroplasts *Plant Physiol* , **84**, 185–196

Vaughn, K C and Outlaw, WH , Jr (1983) Cytochemical and cytofluorometric evidence for guard cell photosystems *Plant Physiol* , **71**, 420–424

Wang, X -C , Outlaw, WH , Jr, De Bedout, J A and Du, Z (1994) Kinetic characterization of phosphoenolpyruvate carboxylase extracted from whole-leaf and from guard-cell protoplasts of *Vicia faba* L (C_3 plant) with respect to tissue pre-illumination *Histochem J* , **26**, 152–160

Willmer, C M (1981) Guard cell metabolism, in *Stomatal Physiology*, (ed PG Jarvis and TA Mansfield), Cambridge University Press, Cambridge

Willmer, C M (1983) Phosphoenolpyruvate carboxylase activity and stomatal operation *Physiol Veg* , **21**, 943–953

Willmer, C M and Mansfield, T A (1970) Effects of some metabolic inhibitors and temperature on ion-stimulated stomatal opening in detached epidermis *New Phytol* , **69**, 983–992

Willmer, C M and Dittrich, P (1974) Carbon dioxide fixation by epidermal and mesophyll tissue of *Tulipa* and *Commelina* *Planta*, **117**, 123–132

Willmer, C M , Pallas, J E , Jr and Black, C C , Jr (1973a) Carbon dioxide metabolism of leaf epidermal tissue *Plant Physiol* , **52**, 448–52

Willmer, C M , Kanai, R , Pallas, J E and Black, C C (1973b) Detection of high levels of phosphoenolpyruvate carboxylase in leaf epidermal tissue and its possible significance in stomatal movements *Life Sci* , **12**, 151–155

Willmer, C M , Jamieson, A and Birkenhead, K (1987) Leaf epidermal tissue is unsuitable to use for studying biochemical aspects of stomatal functioning *Plant Sci* , **52**, 105–110

Willmer, C M , Petropoulos, Y and Manetas, Y (1990) No light activation and high malate sensitivity of phosphoenolpyruvate carboxylase in guard cell protoplasts of *Commelina communis* L *J Exp Bot* , **41**, 1103–1107

Wu, W and Assmann, S M (1993) Photosynthesis by guard cell chloro-

plasts of *Vicia faba* L.: effects of factors associated with stomatal movements. *Plant Cell Physiol.*, **34**, 1015–1022.
Yemm, E.W. and Willis, A.J. (1954) Chlorophyll and photosynthesis in stomatal guard cells. *Nature*, **173**, 726.
Zeiger, E., Armond, P. and Melis, A. (1981) Fluorescence properties of guard cell chloroplasts. Evidence for linear electron transport and light-harvesting pigments of photosystem I and II. *Plant Physiol.*, **67**, 17–20.
Zeiger, E., Gotow, K., Mawson, B. and Taylor, S. (1986) The guard cell chloroplast: properties and function, in *Proc Int. Photosynthesis Congress*, (ed. J. Biggins, Nijhoff/Junk), The Hague.
Zemel, E., Leizerovich, I. and Gepstein, S. (1988) Photosystem II in guard cells of *Vicia faba*: immunological detection. *Plant Physiol.*, **88**, 518–521.
Zemel, E. and Gepstein, S. (1985) Immunological evidence for the presence of ribulose bisphosphate carboxylase in guard cell chloroplasts. *Plant Physiol.*, **78**, 586–590.
Zhang, S.Q., Outlaw, W.H., Jr and Chollet, R. (1994) Lessened inhibition of guard cell phosphoenolpyruvate carboxylase velocity during stomatal opening. *FEBS Lett.*, **352**, 45–48.

10 Recent developments and future research

10.1 Introduction

The previous nine chapters exemplify the wealth of new information on the function and functioning of stomata that has accumulated over the last decade. It should be apparent by now that guard cells are highly specialized cells and they have adopted an array of cellular and metabolic processes which enable stomata to be exquisitely tuned to the environment and the demands within the plant. Stomata can be viewed as one of the major adaptations that allow plants to survive in virtually all terrestrial environments on our planet. However, there still remains much to be elucidated. Progress in some areas of research, such as 'patchy' stomatal responses, has challenged previously accepted views, whilst other areas of research await development of simplifying concepts to harmonize the apparent profusion of conflicting data. Equally, there are key topics, such as the molecular basis of stomatal responses to CO_2, where there has been relatively little research. In addition, new fields of research are just opening up with the application of molecular biology techniques to stomatal research.

For the majority of the preceding text we have attempted to thread a path through the existing literature that presents a concensus view where possible or presents both sides of the argument if it is unresolved. In this chapter we speculate on areas where we anticipate future research will develop which should lead to a fuller understanding of stomatal function and how these developments will reflect on plant biology in general.

10.2 Stomatal distribution and patterning

Stomatal frequency is influenced by a wide number of factors, but the reported changes under different CO_2 regimes have excited considerable attention from both a paleological perspective and in the context of projected increases in the atmospheric CO_2 levels. A number of studies have investigated the genetic selection of stomatal frequency in crop plants and, although it appears that this feature is highly heritable (Jones, 1987), there is no concensus about how many genes are involved. Suh *et al.* (1976) crossed varieties of Sorghum and concluded that there was overdominance for stomatal frequency (heritability estimate was 51%) and one gene or linked block of genes was involved. However, in maize and wheat, multigene regulation of stomatal frequency appears to be the case (Heichel, 1971; Dylenok *et al.*, 1981).

Attempts are being made to elucidate the genetic regulation of stomatal frequency and patterning in leaves at the molecular level but there are surprisingly few reported mutants of these characteristics. In part this may arise from lack of a rapid screening system for isolating such mutants which require laborious microscopical observations and measurements to detect them. In recent work Yang and Sack (1993) have identified a mutant in *Arabidopsis* with altered stomatal characteristics giving rise to clusters of adjoining stomata, and hence termed *tmm* for 'too many mouths'.

10.3 Cell and tissue differentiation

Immunofluorescence techniques have enabled visualization of the complex and highly co-ordinated changes in cytoskeletal arrays during stomatal differentiation. The predictable anatomical geometry of mature stomatal complexes have made stomata one of the most useful models to examine polarized and asymmetric cell divisions in plants. At the moment there are clear descriptions of the morphological changes taking place but little information on the mechanisms underlying the spatial organization and temporal co-ordination of these processes across the stomatal complex. In particular, the symmetrical positioning and stability of microtubule organizing centres on the ventral cell walls of both guard and subsidiary cells is intriguing. It will be interesting to determine whether the positioning of the microtubule organizing centre reflects associations between the cytoskeleton and the cell wall, analogous to the connections between the cytoskeleton and extracellular matrix in animal cells. Localization of the protein $p34^{cdc2}$ to the preprophase band in developing stomatal complexes (Colasanti *et al.*, 1993) also suggests an exciting link between the universal $p34^{cdc2}$/cyclin control system, believed to regulate the cell cycle in all eukaryotic cells, and specification of the plane of cell division in plant cells.

10.4 Mechanical relations of the epidermis

A burst of activity in the seventies and eighties resulted in a number of models being developed which analysed the mechanical deformations of guard cells during stomatal movements. Since then there has been little additional work in this area. However, with the arrival of newer and different technologies it is possible that some of the remaining questions may be answered. For example, three-dimensional images of intact living guard cells can be rapidly acquired using confocal microscopy and these may provide an accurate morphological description of the shape changes occurring during stomatal movements (see Fig. 10.1). With confocal microscopy quantitative data on organelle numbers, volumes and surface areas within guard cells can also be determined, prerequisites for a number of models and calculations concerning cellular osmotic and turgor changes.

It has also been suggested that changes of guard cell volume or turgor are themselves sensed and stretch-activated ion channels may have a role in such a system.

10.5 Diffusion of gases and the role of stomata

Attention has now moved away from considerations of the diffusion of gases through stomatal pores, the theory of which is well understood, and has focused on less tractable problems such as analysis of boundary layers and description of diffusion regimes within plant communities. These developments are paralleling improvements in instrumentation and methodology, such as that needed in thermal imaging of heat transfer to map boundary layer thickness, as well as the availability of increased computational power to resolve complex models.

10.6 Environmental effects

Although much is known about how stomata respond to environmental cues there are still aspects which attract particular debate. For example, there is continuing controversy about whether guard cells respond to VPD, relative humidity or transpiration rate. Additionally, the concept of 'feedforward' control in humidity responses is still not universally accepted.

Another topic which is receiving much attention is concerned with how the functioning of a leaf might be adapted to the increasing atmos-

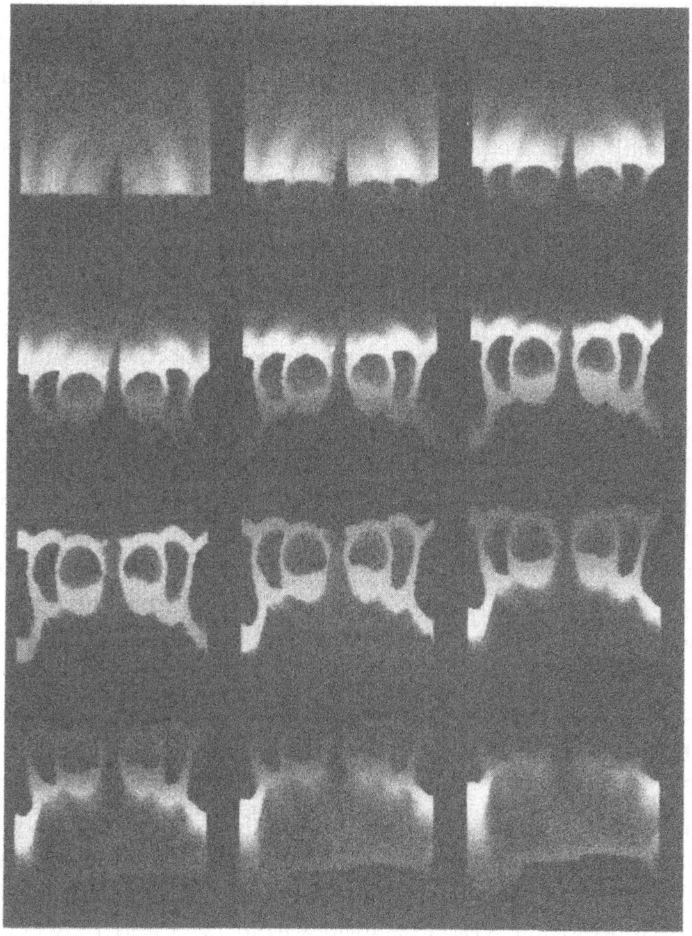

Figure 10.1 Visualization of the three-dimensional wall morphology of a living stomatal complex from *Commelina communis* from a series of confocal optical sections. The cell wall of an intact epidermal strip was stained with primulin and excited at 442 nm with fluorescence emission collected between 470 and 630 nm. A series of 40 optical sections were collected at 0.7 μm intervals through the stomatal complex. In the views shown, a small portion of the image containing about half the guard cell complex has been digitally extracted, and intensity thresholded to reveal the fluorescence from the wall. The views are taken from a tilt series calculated using an algorithm that codes the position of each thresholded voxel as an intensity value, with white nearest the viewer and black furthest away. The cross-sectional shape of the guard cell lumen and the relationship between the guard cells and the subsidiary cells is clearly visible in these projections. As the tissue remains intact and is still living, changes in the three-dimensional organization can be followed over time and parameters, such as cell volume and surface area, readily measured. Courtesy of Wood *et al.* (unpublished).

pheric CO_2 concentrations; stomata could specifically adapt to the higher CO_2 concentrations or the mesophyll could adapt in some way while the stomatal responses result from their close coupling to mesophyll activity via intracellular CO_2 levels or via other diffusable substances. There are also large gaps in our knowledge about the mechanism by which the CO_2 signal is perceived and transduced by guard cells, although it is clear that a multitude of factors can modulate CO_2 responses at the whole plant level.

A major objective continues to be the development of robust models which describe and predict stomatal behaviour at the leaf or whole plant level and allow scaling up to analyse stomatal behaviour in communities growing in different environments. The discovery of 'patchy' stomatal behaviour has already had an impact in this area but the extent to which 'patchy' responses occur in plants in their natural environment is still controversial.

The interrelationships of the various environmental inputs to produce a net stomatal response also remains a source of interest and possible explanations of stomatal behaviour via control theory are steadily increasing in complexity. It may eventually be possible, however, to proportion the extent of the various inputs and to ascertain the level of the direct and indirect effects on stomata.

10.7 Hormone action

Research over the last decade has primarily focused on the role of abscisic acid (ABA) in regulation of stomatal movements. Even with this one hormone there is considerable debate about its site of biosynthesis, its distribution and concentration within cells and between cells, and its site and mode of action. In addition to the rapid stomatal responses to ABA, there appear to be longer-term effects and studies into changes of stomatal sensitivity to ABA and ABA-induced protein expression are in progress. Preliminary experiments, for example, indicate that ABA can influence the expression of protein synthesis in guard cell protoplasts of *Commelina communis* (see Fig. 10.2). Similar studies have been undertaken by Outlaw and co-workers (personal communication). If these provisional results are substantiated it suggests there may be yet another level at which ABA regulates stomatal movements over longer-term periods. Indeed, Taylor *et al.* (1995) have shown that guard cells have all the requisite machinery to drive reporter gene expression from some ABA responsive promoters, but not others.

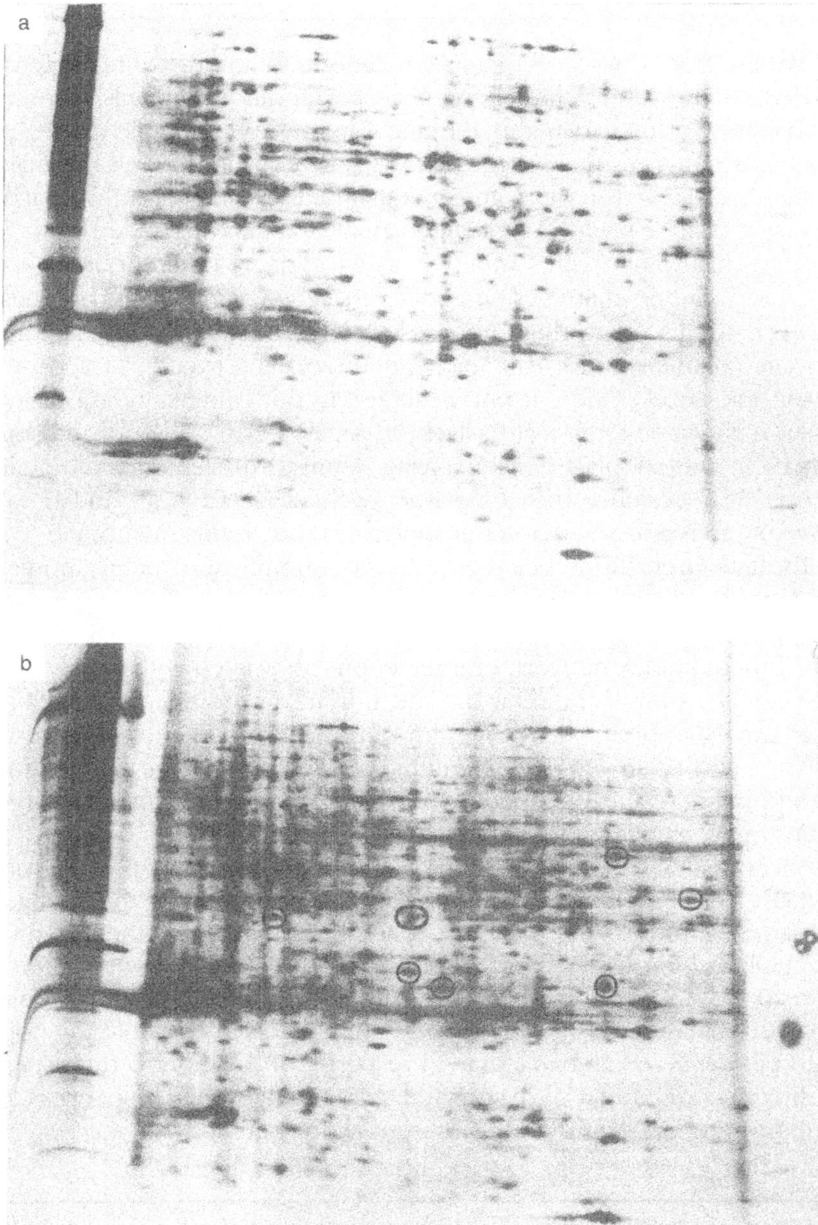

Figure 10 2 Two dimensional electrophoresis of polypeptides from guard cell protoplasts of *Commelina communis* incubated in the presence or absence of ABA (10 mM) for 6 h with ^{35}S methionine for the final 2 h An X ray film was placed over the gel to visualize the radioactive polypeptides Some polypeptides that were absent in the control treatment a or increased their expression are circled in b the ABA treated protoplasts (Unpublished data of Willmer Zhanguo Xin and Li)

10.8 Ion transport and signalling

It is evident from Chapter 8 that a large amount of information has accumulated on the various membrane transporters in guard cells, particularly on those at the plasma membrane. There is a reasonable degree of agreement between most of the reports on the properties of the main ion channels. However there are an equal number of apparently isolated reports of channels with marginally or widely different properties. It is expected that some of this variety is a reflection of different assay systems used rather than genuine differences in transport properties. However, it is also likely that future research will uncover additional roles for some channels in, for example, subtle signalling events, as opposed to bulk ion fluxes. One current weakness in our understanding of ion transport *in vivo* stems from a lack of detailed knowledge on the ion changes in the apoplast accompanying stomatal movements, although there is clear evidence that apoplastic levels of K^+, Cl^-, Ca^{2+} and H^+ all contribute to regulation of ion transporters at the plasma membrane.

Although interactions between different stimuli have been appreciated for a considerable number of years, there is still no unifying model to describe these interactions at a molecular level. Figure 10.3 depicts part of the complex network of interactions associated with perception of several key stimuli and their transduction to various effectors, such as ion channels and pumps. The wealth of possible connections shown in Fig. 10.3, and their different strengths, may be part of the explanation for the varying results obtained in factorial experiments examining interactions between parameters.

Conceptually there is also an interesting debate on whether individual guard cells continuously adjust their aperture to match the strength of a given stimulus, i.e. exhibit a graded response, or whether they essentially exhibit a threshold response. In the latter case each stoma will respond once a given threshold is exceeded, but within the population of stomata there is variation in the threshold value so that the population as a whole exhibits a graded response. In relation to this aspect Gradmann *et al.* (1994) have shown that apparent graded responses can result from repeated switching between opening and closing modes.

10.9 Intermediary metabolism and its regulation

A number of conjectural points remain regarding the carbon metabolism of guard cells. Debate and argument will continue about the importance of the Calvin cycle in guard cells; interest into the apoplastic movement of sucrose from the mesophyll to guard cells is continuing. The pathway of carbon flow from malate to starch during stomatal closing remains

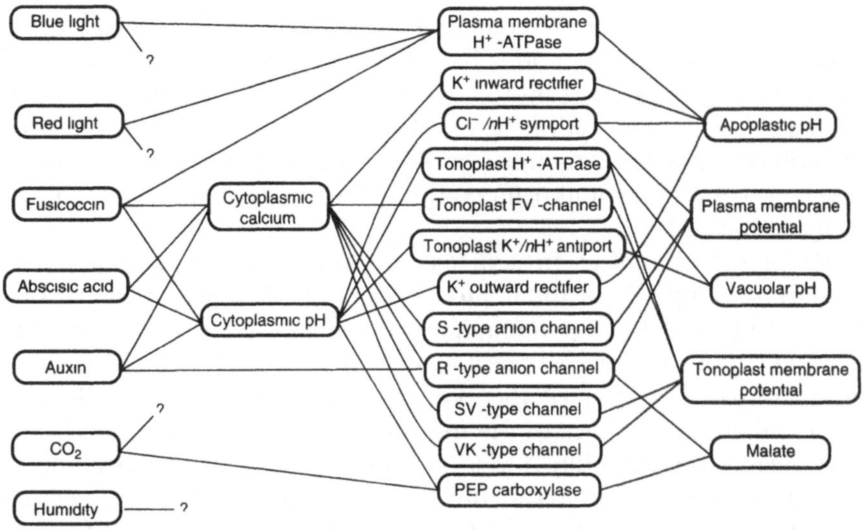

Figure 10.3 The complex signal transduction network in stomatal guard cells Links are shown where an interaction has been demonstrated between key stimuli, signalling intermediates and major effectors

problematical and research into this aspect appears to be at an impasse. Currently the regulation of various enzymes and proteins is of interest, particularly those that become phosphorylated. For example, the regulation of PEP carboxylase and whether it involves phosphorylation for its increased activity is an intense area of interest at the moment. Thus the regulation of protein kinases and phosphatases will be an important source of interest.

10.10 Molecular biology of stomata

Although guard cells are highly specialized cells, there have been few attempts until recently to characterize their protein complement that, in turn, might give rise to some of their unique morphology and behaviour. Even single-dimension gels indicate major differences in protein patterns between guard cells and other cell types within a leaf (Ohya and Shimazaki, 1989; Becker *et al* , 1993; see Fig. 10 4). There are also major differences in protein patterns between guard cell protoplasts isolated from closely related species. For example, in Fig. 10.4 extracts from guard cell protoplasts of *C communis* are compared with extracts from a morphologically similar but as yet unidentified *Commelina* species (Cotelle and Vavasseur, unpublished) and some considerable differences in the polypeptide patterns are observed. Also, Key and Weiler (1988) have raised a monoclonal antibody to a particular antigen only expressed on the plasma membrane of guard cells from *Vicia faba*. Another group of polypeptides have been identified in guard cells and which are enriched, on a protein basis, in these cells relative to that in other cell types (Webb *et al.*, 1993). These investigators also found that the polypeptides were immunologically related to a number of ABA-regulated proteins present in plant embryos, and termed late embryo abundant proteins (LEA).

Research that utilizes protein extracts from guard cells will always tend to be hampered by the limited amount of material that can be prepared. Molecular biology techniques are more suited to dealing with small quantities of material because they allow massive amplification of products with the use of the polymerase chain reaction (PCR) and cloning methods. Studies of the molecular biology of guard cells is still in its infancy but is rapidly developing and undoubtedly will become an important research area in the future. For example, Neill and Hey (1993) are attempting to construct a cDNA library from guard cells of *Pisum argenteum* using a PCR-based strategy and they are also analysing epidermal and guard cell extracts for wilt- and ABA-induced mRNAs and proteins. A number of relatively specific guard cell promoters have also been identified. The *rhal* gene in *Arabidopsis thaliana* encodes a small GTP-binding protein: fusions of the *rhal* promoter with a reporter gene gave high levels of expression in developing guard cells of transformed *Arabidopsis* (Terryn *et al.*, 1993).

Another gene, encoding a plant acidic chitinase, is reportedly expressed only in guard cells within the epidermal layer of leaves (Samac and Shah, 1991). The authors considered that this 'plant defence' gene is highly expressed in guard cells because stomata are the entry ports into plants for many pathogens and therefore these cells might be specialized in defence against pathogens.

Taylor *et al.* (1995) have used a slightly different strategy to probe ABA-induced gene expression in guard cells. They made transgenic plants of *Arabidopsis* and *Nicotiana* containing promoter regions of ABA regulated genes from *Craterostigma plantagineum* fused to a reporter gene. One such construct was expressed in guard cells of *Arabidopsis thaliana* and *Nicotiana tabacum* (Taylor *et al.*, 1995). Additionally they found that not all ABA-regulated gene promoters were expressed in guard cells of both species in response to drought stress or exogenous ABA.

One of the most exciting developments likely to come from these molecular biology studies is the ability to manipulate signalling pathways and hence stomatal responses using transformation strategies. For example, an inducible specific promoter of guard cells might allow

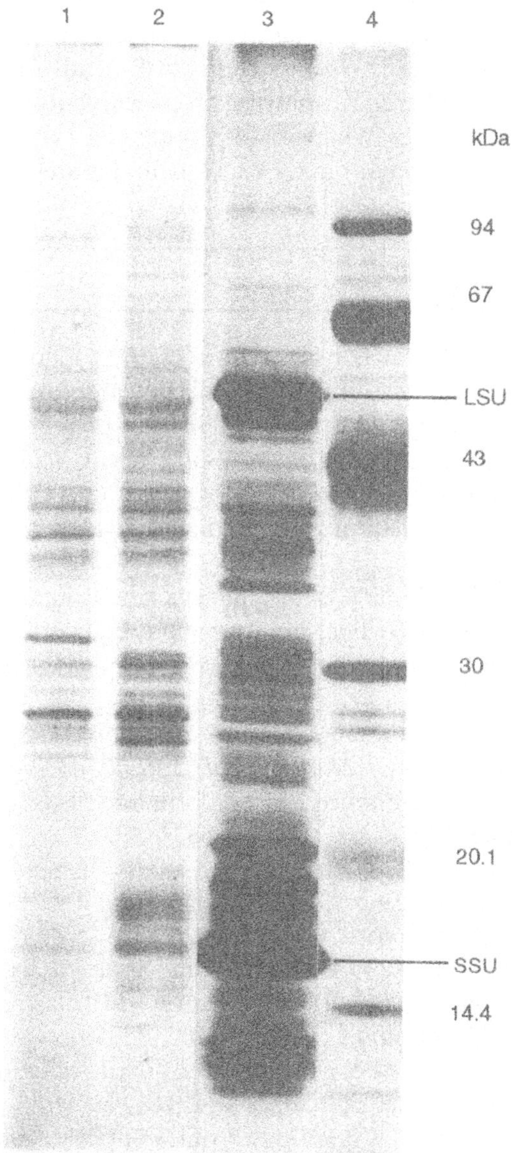

Figure 10.4 SDS polyacrylamide gradient gels showing polypeptides separated from guard cell protoplasts of an unknown *Commelina* species (lane 1) and *Commelina communis* (lane 2) and from mesophyll cell protoplasts of *C. communis* (lane 3). The polypeptides were stained with silver. Lane 4 is of molecular weight markers. LSU and SSU are the large and small subunits of RuBisCo, respectively. The data were provided by courtesy of V. Cotelle and A. Vavasseur.

experimental pertubation of key signalling proteins or enzymes Such developments might enable stomatal movements to be artificially manipulated which may be a more effective means of stomatal regulation in the field compared to spraying crops with anti-transpirant or pro-transpirant compounds Whatever the outcome, however, the next 10 years are set to see a major burst of activity in this area

References

Becker, D , Zeilinger, C , Lohse, G *et al* (1993) Identification and biochemical characterisation of the plasma-membrane H^+-ATPase in guard cells of *Vicia faba* L *Planta*, **190**, 44–50

Colasanti, J , Cho, S -O , Wick, S and Sundaresan, V (1993) Localization of the functional p34 ' homolog of maize in root tip and stomatal complex cells association with predicted division sites *Plant Cell*, **5**, 1101–1111

Dylenok, L A , Khotyleva, L V and Yatsevich, A P (1981) Use of monosomic lines of spring wheat for studying the genetic control of stoma frequency and size in the flag leaf *Dokl Akad Nauk BSSR*, **25**, 753–755

Gradmann, D , Blatt, M R and Thiel, G (1993) Electrophysiology of ion transporters in plants *J Membr Biol* , **136**, 327–332

Heichel, G H (1971) Genetic control of epidermal cell and stomatal frequency in maize *Crop Sci* , **11**, 830–832

Jones, H G (1987) Breeding for stomatal characters, in *Stomatal Function*, (eds E Zeiger, G Farquhar and I R Cowan), Stanford University Press, Stanford, CA, pp 431–443

Key, G and Weiler, E W (1988) Monoclonal antibodies identify common and differentiation-specific antigens on the plasma membrane of guard cells of *Vicia faba* L *Planta*, **176**, 462–481

Neill, S J and Hey, S J (1993) Analysis of guard cell gene expression in *Pisum sativum J Exp Bot Suppl* , **44**, 3

Ohya, T and Shimazaki, K -I (1989) Profiles of proteins in guard-cell and mesophyll protoplasts from *Vicia faba* L Fractionation by sodium dodecylsulfate–polyacrylamide gel electrophoresis *Plant Cell Physiol* , **30**, 783–787

Pfeffer, S R (1992) GTP-binding proteins in intracellular transport *Trends Cell Biol* , **2**, 41–45

Samac, D A and Shah, D M (1991) Developmental and pathogen-induced activation of the *Arabidopsis* acidic chitinase promoter *Plant Cell*, **3**, 1063–1072

Suh, H W, Dayton, A D , Casady, A J and Liang, G H (1976) Diallel cross analysis of stomatal density and leaf-blade area in grain sorghum, *Sorghum bicolor*. *Can. J. Genet. Cytol.*, **18**, 679–686.

Taylor, J.E., Renwick, K.F., Webb, A.R. *et al.* (1995) ABA regulated promoter activity in stomatal guard cells. *Plant J.*, **7**, 129–134.

Terryn, N., Arias, M.B., Engler, G. *et al.* (1993) *rhal*, a gene encoding a small GTP-binding protein from *Arabidopsis*, is expressed primarily in developing guard cells. *Plant Cell*, **5**, 1761–1769.

Webb, A.A.R., Renwick, K.F., Shirras, A.D. *et al.* (1993) The presence of LEA (late embryo abundant) protein homologues in guard cells. *J. Exp. Bot. Suppl.*, **44**, 3.

Yang, M. and Sack, F. (1993) An *Arabidopsis* mutant with multiple stomata. *Plant Physiol. Suppl.*, **102**, 122.

Index

Page references in bold refer to figures and page references in *italics* refer to tables

The manufacturer's authorised representative in the EU is Springer Nature Customer Service Centre GmbH, Europaplatz 3, 69115 Heidelberg, Germany. If you have any concerns regarding our products, please contact ProductSafety@springernature.com

Printed and bound by CPI Group (UK) Ltd, Croydon, CR0 4YY
30/09/2026
02225494-0018